An Alternative [illegible] the Distant Past

Charles Giuliani

Introduction

Examining remote history--whether it be the history of the Earth or that of mankind--is a most fascinating and awe-instilling pursuit. There are so many mysteries surrounding the distant past that cry out to us for close inspection, haunting us with such questions as these: How were fossils really formed? Why is it that fossils and fossil formations are so abundant on the continents, but virtually non-existent on the ocean floors? Why do the continents look as though they once all fit together like pieces of a puzzle? Why do most of the ancient world's great civilizations--those in the Middle East, the Andes, Mesoamerica, the Indus Valley, etc.--appear to have risen up suddenly, out of nowhere? Why were the ancients so preoccupied with studying the stars and the planets? How did early man build colossal structures that, in some cases, were comprised of stones that weighed as much as 2,000 tons?

All of these questions, and so many others like them, remain unanswered today, in a satisfactory manner, in spite of all the "experts" in the world who specialize in areas that are supposed to deal with questions like these, and in spite of all the volumes that have been written on related matters. One of the reasons for this is that key pieces of evidence have been completely ignored by these "experts." These ignored pieces of evidence exist in two main forms: 1) Physical phenomena that don't match up with prevailing paradigms and 2) Ancient oral and written traditions that contain important elements of truth, but are neglected because they often employ mythical and symbolic language to convey their messages.

When conducting an investigation, any good detective will gather up all available pieces of evidence--no matter if some of them are tainted--recognizing that they all contain, or at least have the potential to contain, important clues for solving the mystery that is being looked into. Any scrap of evidence that gets ignored, regardless how trivial it may appear to be, could jeopardize the entire investigation, resulting in the wrong conclusions being drawn.

If you have an open enough mind, I'm sure you will agree that the conclusions set forth in this study are the correct ones, because they are based on ALL available evidence, and because they fit more consistently with all this evidence than the conclusions offered to us by mainstream academia. The picture of early Earth and human history presented herein is revolutionary, but only because the one put on display in our great halls of learning has held our minds captive for far too long, and has diverted our attention away from important clues about the past.

Section I: From heaven to hell
Part 1: A history of Earth and its surroundings

The environment of the early Earth--a primeval paradise

When we look to ancient legends from around the world, we find that there are often far too many common threads that run through them to be written off as mere coincidences. Also, many ancient legends contain enough verifiable facts to warrant us affording them more than casual attention. In fact, ignoring them would severely handicap our ability to formulate a well-rounded picture of the past.

A good example of a commonly-repeated theme throughout many ancient legends, which stands up to close scientific scrutiny, is the concept that the ancient Earth, dating back to the earliest ages of man's presence upon it, was a virtual paradise compared to its current condition. A couple good samples of such legends are cited in the following quote from Boston University Professor Robert Schoch, in his book, *Voyages of the Pyramid Builders*: "The ancient Egyptians told of the First Times (Zep Tepi), an ancient epoch when order and harmony prevailed and then, sadly, broke down. As elaborated upon by the Greek poet and scholar Hesiod (circa eighth century B.C.) in his *Theogony*, the Greeks saw the world long before their own time as a succession of ages in which perfection was replaced step by step by deepening chaos and disorder"(1). The ancient writings of the Chinese, Hindus, and many other peoples from around the world also spoke of how their earliest ancestors inhabited a paradisal world in distant, bygone ages (some further examples of which we will look at shortly). General living conditions in this primeval dream world, according to these writings, were much more favorable than they are today.

One particularly unusual feature of the early utopian Earth to make note of, as described in the ancient Hebrew book of Genesis, is the arrangement of the atmosphere. Notice what we are told in the proceeding passage: "And God said, Let there be a firmament [expanse, i.e. the sky] in the midst of the waters, and let it divide the waters from the waters. And God made the firmament [sky], and divided the waters which *were* under the firmament from the waters which *were* above the firmament: and it was so. And God called the firmament heaven" (again, the sky). Genesis 1:6-8.

What we are being told here is that there was an expanse (or canopy) of water (or perhaps water vapor, or ice crystals) suspended above the atmosphere [Endnote 1]. Speaking of this canopy, two researchers of early Earth history, John Whitcomb and Henry Morris, wrote: "There is no question that a vapor blanket of indefinitely great extent could be supported by the lower atmosphere, since water vapor weighs only 0.622 times as much as dry air....Furthermore, the amount of vapor that could be maintained in any given volume of space in the vapor blanket would not be significantly affected by the presence or absence of air or other gases in the region....It may also be possible that the vapor blanket could have been in the upper troposphere, below the stratosphere"(3).

The book of Genesis, by the way, is not the only ancient source that talks of waters above the heavens. The ancient Egyptians, for example, believed that at one time the world, along with the heavens that enveloped it, were surrounded on all sides by water(4). But now the question arises: Did this aquatic canopy have anything to do with the Earth being a paradise in ancient times?

The water (or vapor, or ice) canopy suspended by Earth's ancient atmosphere, if we assume that such a thing did in fact exist [Endnote 2], would have created the most ideal, uniform living conditions for all life on the planet. First of all, it would have filtered out harmful gamma, ultraviolet, and x-ray radiation from the Sun and other, more distant stars. It also would have increased atmospheric pressure, and thus the amount of oxygen present in the atmosphere and, consequently, in living tissues. Bacteria could not have thrived in this type of oxygen-rich environment, like they do today. Therefore, many ailments afflicting modern man would have been unheard of in the very ancient world. It would also appear that these canopy-created conditions would have resulted in longer lifespans for all living organisms in the world.

Elaborating on these unusual conditions of the early Earth, along with introducing a new one (the idea that the Earth may once have orbited closer to the Sun, which we shall elaborate on shortly), one researcher, Richard Mooney, wrote: "[E]arth was at this time nearer to the sun, with a consequent greater evaporation of surface water, resulting in a high-altitude water vapor 'screen' surrounding earth. This would have had the effect of a deeper, denser atmosphere. The surface pressure may have been slightly higher, with a consequent higher atmospheric pressure....Another effect of this screen would have been to filter out a great deal of solar radiation, and certainly cosmic and hard x-radiation from extrasolar regions of the galaxy. Cosmic rays, it has been thought, may have a harmful effect on the human organism, which may be a factor both in the aging process and in certain diseases, particularly cancers....If earth was shielded by a denser atmosphere from the harmful effects of radiation, [the human race may have] conquered disease and [had] extended...life spans"(5).

Is there any mention of longer lifespans in ancient records? Indeed, there is, and from many archaic cultures around the world.

For instance, the famed Hyperborea of the ancient Greeks, which they described as a land that existed in their distant past, was said to have been a paradise of sunshine and fruits, whose inhabitants were a race of "gods" that lived for over a thousand years without growing old.

The ancient Hebrew scriptures tell a similar story. The book of Genesis relates how the early Earth was inhabited by people who lived to be many hundreds of years old: "And all the days that Adam lived were nine hundred and thirty years: and he died....And all the days of Seth were nine hundred and twelve years: and he died....And all the days of Enos were nine hundred and five years: and he died....And all the days of Cainan were nine hundred and ten years: and he died....And all the days of Jared were nine hundred sixty and two years: and he died....And all the days of Methuselah were nine hundred sixty and nine years: and he died....And all the days of Noah were nine hundred and fifty years: and he died." Genesis 5:5, 8, 11, 14, 20, 27; 9:29.

Emperor Ho-ang-ti of China, who is said to have lived in the fourth millennium B.C., inquired in his medical book as to "whence it happened that the lives of our forefathers were so long compared with the lives of the present generation."

The Sumerians had a legend that told of an ancient land (Dilmun) that was pure, clean, and bright, where there was no old age. This same legend also stated that there was no sickness in this ancient paradise(5), which was probably due, as mentioned a moment ago, to the abundance of bacteria-killing oxygen in the Earth's early atmosphere, resulting from the canopy-created increase in atmospheric pressure.

Finally, Hesiod, who we referred to earlier, wrote about a distant "Golden Age," or a "time of the gods," when the Earth was a near perfect world. He further stated that the inhabitants of this world "lived like gods with carefree heart, remote from toil and misery. Wretched old age did not affect them either, but with hands and feet ever unchanged they enjoyed themselves in feasting, beyond all ills, and they died as if overcome by sleep. All good things were theirs, and the grain-giving soil bore its fruits of its own accord in unstinted plenty, while they at leisure harvested their fields in contentment amid abundance"(141).

There were additional interesting benefits that would have resulted from the water canopy. Another one would have been the creation of a greenhouse effect, causing a globally moderate climate, even at the poles. Venus has such a greenhouse effect (although its climate is extremely hot, being much closer to the Sun, and its canopy is comprised of sulfuric acid). Many ancient legends confirm this--that the Earth was universally warm in the remote past.

For instance, the Irish have an old legend about a lovely realm where there was once continuous summer weather(2).

Eskimo legends also tell us of an archaic, blissful world where lakes never froze, where tropical animals roamed in herds, and where birds of many colors clouded the sky. This world was also described as a land of perpetual youth.

Plato gave a similar description of the environment of Earth long ago: "[T]he earth gave them [our early ancestors] fruits in abundance, which grew on trees and shrubs unbidden, and were not planted by the hand of man....[T]hey dwelt...mostly in the open air, for the temperature...was mild"(141).

A globally moderate climate would have meant the absence of strong winds, storms, and sand and ice deserts. And, as the above-cited Richard Mooney quote indicated, this "global warming" effect may have also been assisted somewhat by the possibility of the Earth having once resided slightly closer to the Sun--a subject to which we shall return momentarily.

The notion of a globally warm climate in prehistoric times is substantiated by the fossil record. For instance, fossils of tropical plants have been found in Antarctica, and tropical plant materials have been seen preserved in the mouths and stomachs of mammoths in Siberia, which indicate, of course, that these regions were not always frozen wastelands, but were in fact quite warm and hospitable to tropical life, just like the rest of the planet was.

As for the claim that the early Earth's atmosphere contained a higher concentration of oxygen, which was alluded to above, the fossil record confirms this as well--specifically fossilized tree sap (known as amber). Under specially-controlled conditions in the laboratory, government scientists were able to obtain uncontaminated samples of ancient air from bubbles inside specimens of amber, which were found at various locations around the world. It was discovered that there was roughly a 35% oxygen content in the early atmosphere, as compared to today's average of only about 21%(6, 155).

According to what we discussed a few moments ago, it seems that the increased oxygen and pressure of the early Earth's atmosphere, along with the filtering out of cosmic radiation, may very well have been the cause of the longevity of early man that we find mentioned in so many ancient legends. A sizable number of experiments have been performed to try and reproduce these ancient atmospheric conditions, in order to see if they do indeed have an effect on lifespans. One such experiment was conducted by Dr. Kei Mori of Keo University in Tokyo, Japan, which produced some truly amazing results. Inside a large hyperbaric chamber (with pressurized oxygen and carbon dioxide, as well as lead-lined walls to filter out harmful radiation from outer space), Dr. Mori grew a tomato plant that lived for over 8 years. But the effects on this tomato plant inside the chamber extended beyond mere longevity. To his amazement, Dr.

Mori discovered that this plant grew to 30 feet tall, and it had yielded over 13,000 tomatoes within the first 6 months of its life(8).

Similar experiments have been performed by Dr. Robert Dudley at the University of California, at Berkeley. The *New York Times*, commenting on these experiments, reported: "Dr. Dudley and others have conducted experiments raising fruit flies and other insects in oxygen-rich environments. Some have shown size increases; others have not.

"Dr. Dudley has focused on pressure because, in addition to having a higher concentration of oxygen, the [ancient] atmosphere would have had much more of the gas. 'Plants were pumping oxygen into the atmosphere,' he said. The amount of nitrogen would have been undiminished, so overall pressure would have risen.

"Though the results have yet to be published, his experiments with fruit flies raised under elevated pressures show a 20 percent increase in body mass over five generations"(155).

While ancient records tell us that unusual longevity was the norm in the very distant past, do they also tell us about large growth size being a common occurrence at this time? Indeed, they do--particularly regarding humans.

The ancient Chinese, for example, had written that long before them there were "giants...men twice as tall as us"(9), and that they inhabited the "realm of delight"(132).

The Genesis record tells us, speaking of man's remote past, that "There were giants in the earth in those days..." Genesis 6:4.

Claudius Aelianus, writing in the second century A.D., stated that the paradisal land of Atlantis was inhabited by "men twice as tall as those common to our climate, and they lived twice as long."

In the early second century A.D., a pseudepigraphical work known as *The Apocalypse of Baruch*, which was preserved in the sixth-century Syriac Vulgate, made the claim that "Men began as giants. These first giants were very highly developed, intellectually, artistically, and physically..."

The Japanese have a tale about a race of giants, which they call the Ainu, who their distant ancestors had fought against and finally beaten in battle.

There are also many North American Native tribes that preserve tales of a race of giants that long preceded them(132).

The ancient people of Iran believed that the Earth in its early history was universally warm, and that the men who inhabited it lived to a very ripe old age, and that they grew to an unusually large size(20).

Finally, the Aztecs of Mexico claimed that way before their time their land was inhabited by the Tzocullixeco--a race of giants(132).

As it turns out, we need not rely solely on ancient records for confirmation of the large growth size of life-forms (human or otherwise) in the distant past--the fossil record confirms this for us as well. Let us look at a brief sampling of fossil finds of this sort, to illustrate this point.

Today the lycopsis plant grows to about 16 inches in height. Yet, according to the fossil record, it once grew to be as tall as 120 feet.

Some ferns, which today grow in small clumps near to the ground, grew long ago to be as large as a tree.

Horsetails today reach to about 3 meters in height. Yet the fossil record contains the remains of horsetails that achieved heights of about 15 meters.

Certain mosses had reached statures of over 90 feet, and their trunks were often over 3 feet wide.

Leaving the flora world, let us now move on to some examples from the realm of fauna.

Today's dragonflies have a wingspan of about 4 inches. In the fossil record, however, one species, *Meganeuropsis permiana*, had a 40-inch wingspan.

A millipede-like creature, *Arthropleura*, grew up to an astonishing five feet long in archaic times.

Scorpions, as found in various fossil formations around the world, were up to 4 times larger than the largest examples living today.

The modern rhinoceros grows to about 5 ½ feet tall. But one rhino fossil, discovered in Germany, was over 17 feet in height.

Modern crocodiles can grow to over 20 feet long. But in the Sahara Desert, fossil remains of what has been dubbed "SuperCroc" have been unearthed. This massive reptile measured about 40 feet long and weighed roughly 10 tons.

Remains of an 8-foot-long beaver (North America) and 5-foot-tall penguins (New Zealand and Peru) have also been uncovered.

There have even been fossil human footprints found, in several locations around the globe, that are of an unusually large size.

As a case in point, many such footprints have been dug up in Glen Rose, Texas, that are up to 15 inches in length [Endnote 3].

Numerous human skeletal remains have also been discovered around the planet that are giant in size(8).

For instance, there was a large race of people that lived in Australia in prehistoric times, known as *meganthropus*, which ranged between 7 and 12 feet in height.

In China, remains of another race were found, bearing the name of *gigantopithecus blacki*, which stood between 10 and 12 feet tall.

Large stone tools have also been found in association with remains from both of these races, which rule out the possibility that these fossils represent human-like apes or some other type of non-human species.

"It is quite clear...that we live in a zoologically impoverished world, from which all the hugest and fiercest...forms have disappeared." - Alfred Russell Wallace, *The Geographical Distribution of Animals*, Vol. 1. New York: Harper and Brothers, 1876.

"'[G]igantism' and massiveness may have been a general or at least a widespread character of early mankind." - *Time*, July 3, 1944.

"[T]he discovery of Homo floresiensis, the miniature-size humans from the remote island of Flores, is genuinely unique. Never before has a fundamentally different kind of human being been found who lived at the same time as our own species, Homo sapiens. That raises an intriguing question: Is there truth after all in the many stories from many lands of other humans, extra large or extra small, living in the mountains or the forests, which have been dismissed as myths and fantasies?" - Christopher Chippendale, curator, University of Cambridge Museum of Archaeology and Anthropology, *Discover*, March 2005.

As indicated above, another cause of the universally-warm climate on Earth in the distant past, aside from the aquatic canopy, may have been the fact that the Earth was apparently positioned a bit closer to the Sun than it is today. Evidence for this comes from a variety of sources. But perhaps the most intriguing piece of evidence is this: The ancient Egyptians, Chinese, Indians (India), Babylonians, Incas, Hebrews, Assyrians, Greeks, Romans, Aztecs, and the Mayas all had a calendar year of 360 days [Endnote 4]. By no means was this due to the fact that these cultures were ignorant of the true length of the year. On the contrary, they measured the year's length with astonishing accuracy. Just take the Mayas, who calculated the Earth's year to be 365.242 days, which is remarkably close to its actual 365.2422 days, as we know it today.

Though these civilizations had 360-day calendar years, at the end of each year they attached an additional five days on, before the new calendar year kicked in. So we see, once again, that they were definitely aware of the true length of the year now being 365 days (or a little more than that). But with that being the case, why, you may ask, did they observe a 360-day calendar year in the first place? Could it be that they were honoring an age-old tradition of their forefathers, who once had a year-length that was 5 days shorter than their own, because the Earth at that time was slightly closer to the Sun?(11, 12)

Incidentally, it is because of this ancient 360-day reckoning for a year that we still measure a circle in 360 degrees.

A great disruption

If it is indeed true that the Earth was a bit closer to the Sun in ancient times, it would not require too much common sense to conclude that something major must have happened to cause our home planet to move to its current, slightly more distant position from its parent star. It would also not require too much common sense to ascertain that the Earth most likely changed its axial orientation at this time, and with great force at that [Endnote 5]. In other words, if enough energy existed to move the Earth into a larger orbit, such energy would also, by natural consequence, have knocked the Earth off its rotational axis, causing its current 23.5-degree tilt. Evidence for such a tilt having occurred in the distant past is quite overwhelming, and is supported by both ancient records and modern research conducted within many branches of the physical sciences.

We have already discussed how tropical fossils have been found in Antarctica and Siberia, and have attributed this phenomenon to the global greenhouse effect caused by the ancient aquatic canopy, coupled with the then-closer distance between the Earth and the Sun [Endnote 6]. But another cause of this universally warm climate in ancient times--perhaps an even more significant cause--would have been the absence of the Earth's current 23.5 degree axial tilt. As author and scientist W.B. Wright wrote: "The simplest and most obvious explanation of great...changes in climate, and of former prevalence of higher temperatures in northern circumpolar regions, would be found in the assumption that the earth's axis of rotation has not always had the same position..."(13).

How exactly would the absence of an Earth axis tilt have made a difference in the global climate, amounting to higher global temperatures, even at the poles? If the Earth's axis of rotation was once indeed perpendicular to the plane of its orbit, as all the evidence clearly suggests, its surface would have received a constant, even distribution of the Sun's light and warmth, aside from the greenhouse effect caused by the aquatic canopy.

Richard Mooney supplies us with further reason to believe that the Earth's axis could not have been disoriented in its earliest ages, like it is today: "Remnants of palm, fig, and magnolia trees have been found in arctic lands, and coral reefs once flourished in Spitzbergen. The presence of coal in Antarctica shows that the region was once covered in forests. It is not enough to hypothesize a warmer climate in this region. Vegetation of this kind could hardly have survived under the present conditions of six months day and six months night"(5). That is to say, in order for this type of vegetation to have grown in arctic regions in prehistoric times, these areas must have received a steady daily dosage of sunlight throughout the entire year. There could not have been a situation, as there is today, where the polar regions alternated between 6 months of being tilted toward the Sun (and thus experiencing six months of daylight), and then six months of being tilted away from the Sun (and thus experiencing six months of darkness). These plants simply could not have gone six months with no sunlight.

Let us proceed to examine some fascinating pieces of legendary evidence that testify to the Earth's rotational axis having tilted at some point in the remote past:

- Herodotus said that the priests of Memphis had told him that in very ancient times "the sun had moved from his proper course...and had risen where he now sets, and set where he now rises"(15).

Is this an indication of the Earth's axis tipping over? Could it be that today's south pole was actually oriented at one time to the north, and vice versa, so that it could indeed have been said that the Sun "rose where it now sets, and set where it now rises?"

- The Egyptian Ipuwer Papyrus describes an early cataclysm that "turned the earth upside down."

- Another Egyptian papyrus, speaking of this same colossal disaster, states that "the south became north and the world turned over." [Endnote 7]

- Caius Julius Solinus, a Roman writer from the third century, wrote thusly about the people of southern Egypt: "The inhabitants of this country say that they have it from their ancestors that the sun now sets where it formerly rose."

- The *Tractate Sanhedrin* of the Hebrew *Talmud* says that "Seven days before the deluge [a catastrophic, global flood], the Holy One [God] changed the primeval order and the sun rose in the west and set in the east"(12).

- The Tarahumara people of northern Mexico preserve world destruction legends that are based on a change in the Sun's path(20). This, of course, is what we would expect from the Earth having changed its axial orientation long ago.

- Plato wrote in *Timaeus* about a catastrophe that made the Earth move irrationally "forwards and backwards, and again to right and left, and upwards and downwards, wandering every way in all six directions"(17).

- Plato also made reference in *The Statesman* (*Politicus*) to a "change in the rising and the setting of the sun and the other heavenly bodies, how in those times they used to set in the quarter where they now rise, and they used to rise where they now set."

- In a short fragment of a historical drama (*Atreus*), Sophocles wrote: "Zeus...changed the course of the sun, causing it to rise in the east and not in the west"(12).

- The Greenland Eskimos told early missionaries that long ages ago the Earth had "turned over"(17).

- The Pawnee Indians say that there was a time long ago when the north and south polar stars "changed places"(12).

- The Native American Utes tribe of Utah has a very old legend of the Sun god, which states: "The sun-god was...conquered, and he appeared before a council of the gods to await sentence. In that long council were established the days and the nights, the seasons and the years, with the length thereof, and the sun was condemned to travel across the firmament by the same trail day after day till the end of time"(4).

- Another Native American Indian tribe, the Kutenai of Idaho and Montana, appears to have a lingering memory of an ancient fear that the Earth might again tilt from its current alignment, resulting in a repeated massive global catastrophe. As the author of *Ethnology of the Kutenai* put it: "The Kutenai look for Polaris [the North Star, which is aligned almost perfectly with the Earth's north celestial pole] every night. Should it not be in place, the end of the world is imminent"(21).

- The Kutenai were not unique in their fear of the Earth being knocked once again off its axis. In fact, a great many ancient cultures had such a fear. Among them, as researcher David Childress wrote, were the "Egyptians, Chinese, Mayas and many other ancient cultures [who] were obsessed with...planetary-solar phenomena. It is believed that they associated catastrophes...with planetary movements....Perhaps the ancient Egyptians, Mayas and other civilizations thought they could predict the next cataclysm by monitoring...the positions of the planets in relation to the earth"(15).

- Yet another group that was plagued by such a fear was the ancient priests of

Machu Picchu, who practiced a ritual in which a mystical cord was attached to a stone pillar in order to guide the Sun in its journey across the sky, preventing it from going off course(22).

- The last example of this sort of fear that we will look at comes to us from *Timaeus*, in which Plato stated that the stars and planets "periodically hide each other from us, disappear and then reappear, causing fear and anxious conjecture about the future of those not able to calculate their movements"(141).

Could we so easily write off all of this legendary evidence--from all corners of the globe, all saying the same thing--as pure fantasy? Yet this is exactly what mainstream academia does, even though common sense tells us that there has got to be some truth to all of this--and all the more so, when we consider the scientific evidence for an Earth axial tilt in bygone ages.

As we saw in one of the examples cited above, some ancient legends that talk of the Earth tilting associate this disaster with a universal flood. Let us now look at several more examples of ancient legends that make this same connection:

- In the Era-Epos tablets, we find this account given by the Babylonian god Marduk: "When I stood up from my seat and let the flood break in, then the judgment of Earth and Heaven went out of joint....The gods...trembled, the stars of heaven--their position changed, and I did not bring them back"(19).

- An ancient Chinese text warned that "if the five planets err on their routes," the land will be engulfed by "a great flood." Obviously the Chinese knew this could happen in the future because it had already happened in the past.

- We made reference above to an ancient Egyptian papyrus that tells of how, in the remote past, "the south became north" and "the world turned over." This document then goes on to state that, during this tremendous "celestial upheaval," the Earth was nearly destroyed by fire and water(17). [Endnote 8]

- Another such account is given by the Shasta Indians of northern California, who talk of how the Sun fell from its normal course at the time of a great deluge in the ancient past(16).

- The Hopi Indians describe how this same Earth-tilting, flood-producing disaster was followed by a great freeze: "[T]he world lost balance, spun around, and rolled over twice....Mountains plunged into seas with a great splash, seas and lakes sloshed over the land; and as the world spun through lifeless space it froze into solid ice. This was the end of the Second World....[A]ll the elements that had comprised the Second World were...frozen into ice..."(18).

What's interesting about the above-cited legends is that they describe exactly what we would expect if the Earth did indeed abruptly change its axial orientation in the past. We would find tremendous crustal upheavals, which would cause the Earth's waters to be violently tossed about, creating a global deluge. Could this be why we have so many legends--over 230 of them, in fact, which span the entire globe--that tell of how the Earth was inundated in the very distant past by a tremendous catastrophic flood?

We would also expect the abrupt tilting of the Earth's axis to bring about a sudden onset of brutally-cold temperatures in the polar regions, causing the flood waters in these areas to freeze over in a very short time, just as one of the above-cited legends suggested. Could this have been the cause (or at least part of the cause) of what scientists refer to as the Ice Age? As we shall see in a later subsection, the onset of the Ice Age was so precipitous that only a catastrophe of monumental proportions can account for it.

Yet another consequence that we would expect from the tilting of the Earth's axis, a most obvious one, would be the yearly seasonal changes of winter, spring, summer, and fall, which was suggested in the legends cited above. The biblical account of the ancient global deluge also suggests this. It indicates that the seasonal extremes of temperature that the Earth now undergoes each year were not known before this great flood catastrophe. The Genesis record has God making the following declaration just after this watery disaster: "While the earth remaineth, seedtime and harvest, and cold and heat, and summer and winter...shall not cease." Genesis 8:22.

The mention in this archaic Hebrew passage of "seedtime and harvest," as being a consequence of the changes brought about by the flood, would seem to further indicate that there had indeed existed before this great deluge a global, year-round warmth that made it possible to grow crops uninterrupted by the ravages of bitterly cold winters. Given this new, post-flood development of extreme seasonal changes, man's life would have become dependent upon a keen understanding and measurement of the new length of Earth's year. This could only have been accomplished by a careful study of the positions of the Sun and other astronomical bodies, to know when to plant and reap, and when to begin storing up for the coming winter season.

Can we not see in this scenario a sound explanation for why the ancients were so obsessed with studying the heavens? Is this not why the whole globe is covered with ancient astronomically-aligned structures (such as Stonehenge) for carefully monitoring the movements of heavenly bodies? And can't we see why these constructions were built with such enormous stones--so that they would remain in place for centuries, if not millennia, and thus be reliable calendars for many generations? Can we not also see how the building of stone calendars, which had originally begun as a vitally-important and practical science, had later mutated into an obsessive religion? What better explanation can there be for the development of Sun worship and astrology in ancient times? Since man was so dependent upon the movements of the Sun and stars (along with their constellations), it is easy to see how he would begin to look to these celestial entities, over time, as his "gods" who were the sources and sustainers of life.

We could fathom that a sudden tilting of the Earth's axis had so completely disrupted its former utopian environment that its atmosphere, just like its crust and its waters, would have been adversely affected. In fact, it would be difficult to imagine how the atmosphere could NOT have been affected. Thus we can speculate that the Earth may have lost its greenhouse-producing aquatic canopy at this time (another topic that we shall cover in more detail at a later point). The loss of the greenhouse effect would obviously have made a further significant contribution to the rapid onset of the Ice Age.

There is one final consequence of the Earth's sudden axial tilt that I wish to draw attention to at this time--a shift in the alignment of the Earth's magnetic field. With the Earth being knocked to one side, or tipped completely over, the magnetic field alignment would obviously have been affected, but not necessarily in tandem with the offsetting of the rotational axis alignment. That is, because the Earth's magnetic field is believed to be generated by an iron core "floating" inside a molten rock fluid (magma)[Endnote 9], when the Earth's rotational axis tilted, the core of the planet may not have tilted in unison with its solid outer surface, but may rather have sloshed about chaotically from within. The obvious effect of this action of the core would be an incongruity between the alignments of the Earth's magnetic and rotational axis poles. And indeed, this is just what we find. Today the Earth's magnetic axis is offset to its rotational axis by about 15 degrees. And there is ample evidence that the magnetic axis had undergone even more radical inclinations in the past--that it even reversed polarity at one time. Mainstream scientific journals, by the way, do not argue that there is a relationship between a wobble or tilt of the Earth and a change (or a complete reversal) in its magnetic pole orientation. *Nature*, for example, had this to say: "[A]n earth wobble [or tilt] may be of a magnitude sufficient to cause reversal of the magnetic field"(23).

A magnetometer enables scientists to measure very weak magnetic fields. It is through the use of this instrument that past changes in the orientation of the

Earth's magnetic pole have been measured. In a nutshell, the process works like this: Most ancient lava deposits around the world contain microscopic iron filaments. When the lava was in a liquid state, the iron filaments, floating freely in the liquid rock, were attracted by the Earth's magnetic pole, which caused them to align themselves all in one direction--toward the north, much like a compass needle. Later, when the lava hardened, the position of the filaments became frozen in time, enabling scientists today to track the shifting of the Earth's magnetic pole in ages past. Magnetometer tests of various lava deposits in England, for instance, have revealed that England's magnetic inclination was once 30 degrees, whereas today it is 65 degrees. In India's Deccan plateau, similar tests were conducted which showed that this area had once been inclined 64 degrees south, then 60, then 26, and finally, a major shift occurred, flip-flopping this area's inclination 17 degrees north(24).

The explanation often given for these shifts in orientation involves large time-frames. It is believed that they occurred over many millions of years, as the Earth's core was occasionally tugged by a passing asteroid, or by some other rare, gradualistic event. But it could just as easily be argued that these shifts happened in relatively rapid succession, during and immediately following the great flood catastrophe.

Of course, these magnetometer readings might not be exclusively indicative of a reorientation of the Earth's iron core. Some of them may be simply attributed to changes in the positions of the continents, which also appear to have been consequences of the flood catastrophe (more about this later)[Endnote 10].

Is it not becoming obvious that there is an abundance of legendary and physical evidence--evidence which cannot be ignored--that collectively paints an entirely different picture of prehistoric Earth from what the mainstream scientific establishment has painted, or has allowed to be painted? While it is true that catastrophism (the notion that the Earth suffered one or more major catastrophes in the remote past) has received a more favorable rating by academia in recent decades, certain staunchly-held biases have forced a distorted and limited picture upon us of the scope and timing of the Earth's catastrophic past, as we shall now see.

"The earth is utterly broken down, the earth is clean dissolved, the earth is moved exceedingly. The earth...reel[s] to and fro like a drunkard..." - Isaiah 24:19, 20.

The scope of the catastrophe

If the Earth was indeed kicked into a larger orbit that placed it slightly further from the Sun, if it was knocked off its rotational axis, and if it had the speed of its rotation altered (all of which resulted in a global flood disaster), a tremendous amount of energy must have been involved to accomplish these things--energy that originated from outside the Earth, out in space. As one catastrophist book, *The Day the Earth Nearly Died*, put it: "Orbital and rotational changes...could be expected if the tilt of the earth's spin were altered--although that could probably only occur as a response to some powerful influence outside earth itself"(17).

So now we are left with the question: What was this powerful influence, or source of energy, outside the Earth that brought on its orbital and rotational axis disturbances, resulting in a global flood catastrophe? Whatever it was, it must have indeed been of a colossal nature, which affected more bodies in our solar system than just the Earth. Do we find evidence elsewhere in the solar system that a tremendous calamity occurred in the past? Indeed, we do--and a great abundance of it.

Let's take a look at some of the key pieces of this evidence:

- Like the Earth, all planets in our solar system--even their satellites--have rotational and magnetic axis tilts (for those that have magnetic fields), as well as elliptical and inclined orbits (that is, the shapes of their orbits are oblong instead of circular, and none of them lie in the same plane).

Mercury has the most extreme orbital eccentricity (.206) and orbital inclination (7 degrees), except for Pluto, which technically isn't a planet (it's probably an escaped satellite of Neptune, having later picked up a satellite of its own, Charon).

The planet Uranus has a most extreme axial tilt (97 degrees), literally rotating on its side, and its magnetic pole is severely inclined as well (by 58.6 degrees). Not only that, but Uranus' magnetic pole isn't even located in or near the center of the planet (it's offset from the core by 30% of the planet's radius).

The same is true for Neptune's magnetic pole, which is tilted 47 degrees from its rotational axis, and it's offset from the planet's center by 55% of the planet's radius.

And then there's Venus, which has a day (equal to 243 Earth days) that is longer than its year (equal to 224.7 Earth days), and it rotates in the opposite direction from pretty much all the other planets (Pluto and Uranus being the other two exceptions). Is it possible that Venus' rotational axis was knocked completely over, so that its current north pole was once its south pole, and vice versa? This would certainly account for it rotating in a clockwise direction.

Other examples of aberrant movements of bodies in our solar system include both of Neptune's large moons, Triton and Nereid, which have retrograde orbits (they orbit in the wrong direction--clockwise). These are 2 out of only 8 moons that do this (there are a total of 72 moons in our solar system)[Endnote 11]. Also, Nereid, the smaller of Neptune's two largest satellites, has an extremely elliptical orbit, taking 359 Earth days to complete a full circuit around its parent planet--the longest orbital period of any moon in the entire solar system.

- The outer planets, known as the "gas giants" (Jupiter, Saturn, Uranus, and Neptune), all have ring systems (Saturn's being the most notorious) that are indicative of a satellite (or satellites) that broke up in the past, existing now as mere particles of dust and small chunks of ice and rock.

- Many of the satellites of the outer planets are marked by enormous rifts from various types of crustal deformations and breachings, indicating that these celestial bodies underwent tremendous tectonic and/or tidal stresses in their past, which nearly pulled them apart. Most notable are Saturn's moons Tethys and Enceladus; Uranus' Miranda, Titania, and Ariel; Jupiter's Europa and Ganymede; and, finally, Neptune's Triton.

And then there are the two moons of Saturn--Dione and Rhea--which have unique, wispy crustal crack systems that are spread across an entire hemisphere on each of them. Saturn's Iapetus has an unusual feature on it as well. Nearly one half of its near-equatorial circumference is hugged by a buckled ridge that runs in almost a perfect straight line throughout its whole length. The cause of all these features must have been of a very broad-scale and disruptive nature.

- Mars, like Earth, is covered with global-scale wounds which testify to a devastating flood disaster in its remote past. But we will put off elaborating on this matter until a later subsection.

- Venus, though not exhibiting signs of once having suffered a global flood (it probably never had oceans of any kind, except oceans of lava), has a planet-wide tortured surface that gives testimony to a cataclysmic past. Most outstanding is its vast array of globally-distributed volcanoes. Venus, in fact, has the largest number of volcanoes of all the planets (and moons) in the solar system (it has over 1,600 major ones, and over 100,000 minor ones)[Endnote 12]. Venus also has rift valleys as large as the East African Rift (the largest on Earth). Obviously this planet has had a very tumultuous history.

- All planets (except the outer ones which have no solid surface, and Earth which has a high erosion rate), moons, and asteroids are literally plastered with craters.

- At least 4 moons of the outer planets were once hit by objects so large that the resulting craters cover almost an entire hemisphere of their surfaces, and nearly smashed them to pieces upon impact. These moons are Jupiter's Callisto and Saturn's Mimas, Tethys, and Iapetus. The planet Mercury also has such an enormous hemispherical crater, known as Caloris. Our own Moon, as well, has some of the largest craters in the solar system. One of them is Mare Orientale, approximately 600 miles in diameter. The largest impact crater in the solar system, in fact, is on our Moon--the south pole's Aitken Basin, on the far side, which is just over 1,300 miles across (more than half the Moon's diameter!).

So pervasive is the distribution of cratering on celestial bodies throughout the solar system, that mainstream scientists have come to the conclusion that there was a time in the distant past when impacts had occurred on a much grander scale than they do at present. In fact, they refer to this heightened state of crater-producing impacts as the Great Bombardment period. This period, where an enormous amount of debris was being violently flung throughout the solar system, was also surely the cause of all the orbital, tidal, tectonic, and magnetic/rotational axis alignment disturbances in the solar system that were described above, and thus the cause of the flood disaster here on Earth (as well as an apparent global flood on Mars). But the question is, What could have been the source of all this debris that reeked so much havoc throughout our entire planetary neighborhood?

Between the orbits of Mars and Jupiter there is a huge gap. Technically there should be a planet there, but instead we find in this region a bunch of orbiting debris known as the asteroid belt. Could the asteroids in this region be the remains of a planet that once existed there, which somehow exploded, perhaps due to some type of tidal force or a devastating collision with a chunk of debris that wandered in from outside our solar system [Endnote 13], possibly from a supernova explosion? [Endnote 14] And could it be that the Great Bombardment period represents this time when this planet exploded, sending its broken fragments in every direction, striking every planet and moon that got in the way? [Endnote 15] Of course, mainstream science claims that all of these pieces of flying Great Bombardment debris were "leftovers" from the formation of the solar system. But it could just as easily (and legitimately) be argued that all these debris chunks were "leftovers" from a planet that violently broke apart [Endnote

16].

This brings us to another fascinating discussion: It is believed that most meteorites that fall to Earth today had originated in the asteroid belt. They are either pieces broken off asteroids from ongoing collisions with other interplanetary debris, and/or they are remnants from the original catastrophe that destroyed their parent planet that once orbited between Mars and Jupiter. In any case, meteorites, according to our "exploded planet" hypothesis, would represent pieces of this extinct heavenly body. But do we find any evidence from meteorites themselves that they may indeed have originated from this destroyed planet? Let's take note of some curious facts about meteorites that appear to support this conclusion:

- All meteorites fall into one of three main categories: stony, iron, and stony-iron. Could the stony meteorites represent the crust of this destroyed planet, the irons represent the core, and the stony-irons represent the mantle--the transition layer between the outer stony crust and the innermost metallic core?

- Most stony meteorites fall into the classification of chondrites. Chondrites contain tiny, spherical inclusions called chondrules. Put simply, chondrules are hardened droplets of liquid rock (mostly silicates) that are now crystallized. They obviously formed under extreme catastrophic conditions, first melting from a tremendous amount of heat (produced by their exploded parent planet?), and then quickly resolidifying in the deep freeze of interplanetary space, thus maintaining their spherical shape from their liquid state.

Chondrites, for the most part, contain a high volume of tiny metal (nickel-iron) flakes that are evenly and closely distributed throughout their matrices. Could these flakes have been blown out from the exploding planet's core, impregnating the liquefied rocky crust material before it had a chance to harden?

- The achondrite stony meteorites, as their name implies, are those that do not contain chondrules. One specific subclass of this group is known collectively as Ureilites. These meteorites are particularly interesting because they are saturated with microscopic diamonds. Diamonds, of course, are formed when carbon is subjected to tremendous heat and pressure, like what you would expect to find deep in the interior of a large planet's crust. Thus this class of meteorite must have originated well beneath the surface of a large planetary body, i.e. the extinct trans-Martian planet.

- Brecciated, shock-veined, and impact-melted stony meteorites also testify to an ancient catastrophe. These meteorites show signs of having once been fragmented and/or melted, due to collisions in space, and then having recompacted together, forming a conglomerate of broken rock fragments. While it is true that some of these types of stony meteorites may have been formed by more recent, ongoing interplanetary collisions, a good many of them, particularly the impact-melt specimens, appear to have come about from collisions involving a tremendous amount of energy, and probably date back to the original trans-Martian planet demise and its resulting Great Bombardment period. Also, the fact that brecciated, shock-veined, and impact-melted meteorites are fairly common indicates that they must have formed at a time when asteroid-to-asteroid impacts were occurring more frequently than they do today.

- One sub-class of stony-iron meteorites, known as pallasites, exhibits an unusual blend of nickel-iron alloy and clear silicate globs (olivine), the combination of which produces a sort of pimento loaf appearance when cut into slices. Pallasites are most unusual because it is unknown how their mixtures of nickel-iron and olivine could have come about in this fashion, considering that both of these substances must have solidified together from a molten state. The problem here lies with the fact that the nickel-iron alloy, having a much higher melting point than the olivine crystals, should have caused the olivine to vaporize, and to become chemically bonded with the molten metallic matrix. Instead, however, we find that the olivine inclusions remained separate from, and unscathed by, the much hotter surrounding molten nickel-iron alloy. Obviously this arrangement came about under extremely unusual conditions. Could these conditions have resulted from the destroyed planet catastrophe? Was the molten nickel-iron quickly solidified before it had time to incinerate the olivine encased within it, just after its parent planet exploded and ejected it out into the deep freeze of space?

- Iron meteorites also tell a story of a catastrophe in the distant past. When cut and etched in an acid solution, most iron meteorites will exhibit what is called the Windmanstatten Pattern--a criss-crossed crystalline structure that could only have been formed under conditions of extraordinary heat and pressure, such as what exist in the core of a planet. So it would appear that iron-class meteorites (or most of them, anyway) may indeed have been a part of the core of the proposed destroyed planet that once orbited beyond Mars.

The further back in time we go toward the Great Bombardment period, the more frequently meteoroid impacts on Earth (and all other bodies in the solar system) had occurred, and the larger were the sizes of these meteoroids (some large enough to be called asteroids). Most scientists today now believe that a major asteroid impacted the Earth in prehistoric times, which, among other things, had caused the extinction of the dinosaurs. The location where many researchers believe this impact had taken place is just north of the Yucatan peninsula, in the Gulf of Mexico. But this is not likely the actual location where the flood-causing, dinosaur-extinguishing culprit had struck. For, as we shall see, the destruction that the flood-causing impact inflicted upon the Earth was so devastating and pervasive that all traces of any crater it made would have been completely obliterated. Nevertheless, this significant impact site in the Gulf of Mexico probably at least represents a strike upon the Earth from roughly the same period--probably the early post-flood era--involving a rather large chunk of debris from the destroyed trans-Martian planet [Endnote 17].

Let me point out that it was not necessarily required for an asteroid to have actually struck the Earth in order to produce the effects of an Earth axis tilt and a global flood (although I personally believe that this is what happened, for reasons that will be presented later on). All that may have been required is the gravitational tug of a large piece of the destroyed planet passing in close proximity to the Earth. As one scientist, J.D. Mulholland, wrote: "If a planet-sized object were to pass close to the earth, then giant tides would be raised; there would be global earthquakes; the north pole would change direction; the day, the month, the seasons, the year would all change. Faith is not involved here; these are unavoidable consequences of the laws of motion as we presently know them"(26).

Of course, knowing the nature of the Great Bombardment catastrophe, it may even have happened that the Earth was affected by both phenomena--that the Earth may have been disrupted by an asteroid (chunk of the destroyed planet) hitting it, and another chunk passing close by it and affecting it gravitationally. Perhaps there were many pieces that both struck and passed close by the Earth at about the same time. There's really no way of knowing the exact number of events or their sequences. However, we do know, as we shall later document, that the degree of destruction wrought upon the Earth at the time was monumental [Endnote 18].

Like comet Shoemaker-Levy 9 that hit Jupiter, a large asteroid impacting the Earth would have thrown a tremendous amount of debris into the upper atmosphere, darkening the skies around the world for an extended period of time. And, as it turns out, many ancient legends tell of this very thing having happened during the great flood catastrophe. One such example comes from the Quiche Maya of Guatemala, who discussed in their holy book, the *Popol Vuh*, a time in the distant past when "...it was cloudy and twilight all over the world...the faces of the sun and the moon were covered." Other Mayan sources, talking of "the time of the ancients," confirm that this darkening of the Sun was indeed connected with the great flood, saying that the "earth darkened....It happened that the sun was still bright and clear. Then, at midday, it got dark....Sunlight did not return till the twenty-sixth year after the flood"(20).

The Aztecs tell a similar story of the Sun disappearing for years, and that when it returned, they were surprised to see it rise in the east (an obvious reference to a complete polar reversal, which we discussed earlier): "There had been no sun in existence for many years....[The Chiefs] began to peer through the gloom in all directions for the expected sight, and to make bets as to what part of heaven [the Sun] should first appear [in]...but when the sun rose, they were all proved wrong, for not one of them had fixed upon the east"(12).

Likewise, the Cherokees have an ancient flood legend that makes this assertion: "[T]he world [at the time of the flood] was dark all the time, because the sun would not come out"(28)[Endnote 19].

While it is true that powerful volcanic eruptions also darken the sky, the darkness is often localized, and is thus usually of a short duration of time. The fact that the darkness mentioned in the above-cited legends was of such a long duration, was apparently universal, and that it was associated with the great flood disaster, shows that we may indeed have legendary confirmation here that the cause of the great flood disaster was of a celestial nature. But, at the same time, we cannot rule out the possibility that the darkening of the sky mentioned in the last few legendary references may have been due, at least in part, to volcanism. For as we shall see later on, heightened volcanic activity was one of the consequences of the flood catastrophe.

But there are other flood legends that more clearly reveal a direct connection between the darkening of the sky and large meteoritic impacts.

In 1643, for example, a bishop in Iceland discovered an old document containing ancient Germanic myths. One section read: "The sun turns black, earth sinks into the sea. The hot stars down from the heavens are whirled..."(27).

There are other, similar catastrophe legends which, though they fail to mention the darkening of the sky, nevertheless make it clear that the source of this flood calamity was from outside the Earth.

Greco-Roman myths tell of the flood of Deucalion, which was immediately preceded by awesome celestial events. These events are described in the story of Phaeton, child of the Sun: "[T]he glorious sun, instead of holding his stately, beneficent course across the sky, seemed to speed crookedly overhead and to rush down in wrath like a meteor....[T]he grass withered, the crops were scorched; the woods went up in fire and smoke; then beneath them the bare earth cracked and crumbled and the blackened rocks burst asunder under the heat"(20).

The Ute Indians have a tradition that declares: "[T]he sun was shivered into a thousand fragments, which fell to earth causing a general conflagration....[There was great destruction from] the burning earth...[but then] a flood which spread over the earth...extinguished the fire"(17)[Endnote 20].

Finally, the Selungs, a people of the Mergui Archipelago (off southern Burma), have an old legend that tells of how "the daughter of an evil spirit threw many rocks [large meteorites or asteroids?] into the sea. Thereupon the waters rose and swallowed up all the land. Everything alive perished..."(Ibid.).

In regards to earlier revelations about Great Bombardment period debris from the destroyed planet reeking tremendous havoc throughout the entire solar system, let us take note of a particularly relevant ancient Chinese legend, which says that "the planets altered their courses. The sun, moon, and stars changed their motions."

On a fairly similar note, Plato, in *Timaeus*, portrays Egyptian priests explaining to Solon that "There is a story...that once upon a time Phaeton, the son of Helios, having yoked the steeds in his father's chariot, because he was not able to drive them in the path of his father, burnt all there was upon the earth, and was himself destroyed by a thunderbolt. Now, this has the form of a myth, but really signifies a declination of the bodies moving around the earth and in the heavens, and a great conflagration of things upon the earth..."(29).

Before closing this subsection, there is one more fascinating point that needs attention, regarding the solar system-wide catastrophe: It would appear that the destroyed planet once had a sizable ocean, or oceans, just like Earth and Mars, according to ancient sources. Mesopotamian texts, for example, make frequent reference to this extinct planet (Tiamat) having had a watery nature to it, often applying to this planet an alternate name, *Tehom-Raba*, which means "watery monster." *Tehom-Raba*, incidentally, means "great Tiamat" in Hebrew(17).

Some catastrophists believe that comets represent the remains of this planet's ocean (or oceans)--that these waters instantly froze in the extreme cold of deep space when Tiamat blew up. Others believe that comets might represent an exploded icy moon (or moons) of this destroyed planet. However, such is not the position of this study, for two main reasons.

First of all, a massive explosion that would involve the destruction of a planet and its moon (or moons) would most likely have vaporized any liquid water thereon, and would have disbursed this vapor into space as diffuse gases instead of solid blocks of ice.

Secondly, and finally, comets can't be remnants of a former ocean or icy moon because they are not comprised of ice, as is commonly believed. Instead, comets appear to be made up of the same materials as asteroids--mostly silicates and nickel-iron. In fact, comets and asteroids may very well be the same exact thing.

When the Deep Space 1 spacecraft flew by the nucleus of Comet Borrelly, it detected absolutely no ice on the surface at all. Instead, it found its surface to be hot and dry. The only water found was in the coma and tail, which can easily be accounted for as being the result of a chemical reaction between negatively charged oxygen ions from the nucleus and positively charged hydrogen ions from the Sun.

Another mission to a comet that confirmed that comets are not balls of ice was NASA's Stardust probe, which collected samples of material that blew off Comet Wild-2 when it flew by it in 2004. The samples were later returned to Earth and tested, revealing that the material this comet is composed of--silicates embedded with metal and sulfides--is more akin to a typical asteroid.

If comets were in fact balls of ice, we would expect them to have fairly smooth surfaces from all the melting that supposedly results from their close passages by the Sun, which is said to be responsible for the formation of their tails. However, on the contrary, the surfaces of comets are very rough, rocky, cratered, jagged, and ridged.

With all of these facts in mind, it would seem that comets are, just like asteroids, simply broken pieces of the destroyed trans-Martian planet, except that they have more radical orbits (highly eccentric and elliptical) and produce tails. But why, the question arises, do they thusly differ from regular asteroids?

As far as the highly eccentric and elliptical orbits of comets go, this can be accounted for by the fact that comets probably represent pieces of the exploded planet that were flung so far out into space, and so far outside the plane of the ecliptic, that they developed unstable orbits.

But what about the tails of comets? If comets are made of the same substance as asteroids, why do they have tails, whereas regular asteroids do not? The answer to this question is simple: Because a comet comes much closer to the Sun than a regular asteroid, it enters an area of the Sun's plasmasphere that is more highly electrically charged, which results in the formation of a visible plasmasphere around the comet. The tail results from a distortion of the comet's plasmasphere, where streams of ions from the Sun (the solar wind) stretch the comet's plasma [Endnote 21] out and away from the direction of the Sun. This high-voltage interaction with the Sun's plasmasphere, by the way, is the reason why comets are jet black in color--they have acquired carbonized, charred surfaces from such interactions.

There is much evidence to support the idea that the tails of comets are the result of electrified plasma. Here are but a few pieces of such evidence:

- When NASA's Deep Impact spacecraft impacted Comet Temple 1 in 2005, it produced a bright flash before ever striking the comet. This was because the craft penetrated its plasma field(191).

- On March 27, 1996, the European Space Agency's ROSAT satellite observed x-rays being emitted from Comet Hyakutake. Now how do you suppose a comet--an alleged ball of ice and rock--could emit x-rays, unless this was the result of a plasmic discharge?

- During its close flyby of Comet Wild-2, NASA's Stardust probe was struck by two bursts of energy--obviously electrical discharges from its plasmasphere. Anthony Truzollino, a senior scientist at Chicago University's Enrico Fermi Institute, said of these bursts: "These things were like a thunderbolt"(190).

In the laboratory, electrified plasma experiments have been performed which have successfully simulated comet tail-like features, demonstrating beyond any doubt that comet tails can indeed be easily accounted for by solar plasmaspheric interactions.

"The planets ran against the sky and created confusion....The celestial sphere was in revolution....The planets...dashed against the celestial sphere, and mixed the constellations; and the whole creation was as disfigured as though fire disfigured every place and smoke arose over it." - *Bundahis*, an ancient Indo-Iranian text(12).

* * * * * * *

In this subsection we have dealt with the huge scope of the flood disaster, involving the entire solar system. But we still have many details to cover regarding the extensive destruction caused by this disaster right here on Earth, as well as a few more effects this calamity inflicted throughout the solar system. Before covering these details, however, we first need to address another important matter.

The timing of the catastrophe

We have discussed the Great Bombardment period and its connection with the asteroid impact that brought about the extinction of the dinosaurs. As you may know, mainstream scientists do not associate these events with each other, nor do they place them within the time-frame that they believe man had walked the Earth. The Great Bombardment period is said to have occurred around the time of the formation of the solar system, or just thereafter, at about 4.3 billion years ago, whereas the asteroid impact that wiped out the dinosaurs is believed to have taken place roughly 65 million years ago. But is there any truth to these time-setting assumptions?

It is important to understand that these assumptions have more to do with psychology than established scientific facts. The situation was quite similar with the belief in the flat Earth dogma in the Dark Ages--it was required belief, regardless what the facts demanded. Anyone who questioned this belief was vehemently attacked, and even threatened with death, by governmental, religious, and academic authorities. While there were some who knew the truth and did not approve of the censorship, they only acknowledged their "heretical" belief privately to trusted companions, and denied it if questioned publicly. Though the situation might not be as extreme today for those who question modern science's dating and sequencing of prehistoric events, the thought of being branded an academic heretic is usually considered serious enough of a threat to discourage almost everyone from even inquiring skeptically into the matter. Thus modern dating methods and their consequential time-frame assignments have gone virtually unchallenged by mainstream scientists, despite all of the major obvious problems and inconsistencies with them, many of which we will examine herein.

The focus in this subsection (and the one that proceeds) will not be concerned so much with setting a time for the asteroid impact/Great Bombardment catastrophe, although the reader should be aware that most catastrophists place it at somewhere between 5,000 and 11,500 years ago. The choice is left to each reader to decide which date seems most likely. All that we will concern ourselves with here is debunking the popularly-held chronology of the Great Bombardment period and the asteroid impact that brought the demise of the dinosaurs (and many other species of animals, as well as many plants). The goal of this task is to demonstrate that there is no sound scientific reason for placing these events in the extreme distant past, in the order of billions or millions of years. With this goal accomplished, it is hoped that the reader will be able to see that both of these events--the Great Bombardment period and the extinction-producing asteroid impact (and the resulting global flood)--were contemporaneous with each other and with man's existence upon the Earth.

Before we begin our discussion of the problems with modern dating methods and the resulting extreme ages assigned to various phenomena, there is one final point that needs to be made: Let there be no confusion that an attempt is here being made to establish the age of the Earth. That is entirely another matter that the reader will be left to decide for him/herself, since it lies totally outside the reach of this study.

So how do modern dating methods work? On what basis do today's scientists claim to be able to ascertain the age of a rock, fossil, artifact, or bone? Simply put, it is believed that it can be determined how old an object is by testing how much radiation is currently left in a given chemical element found within that object, as compared to how much radiation had originally been present in that element. Admittedly, this dating technique seems to make sense, at least on the surface, and should, theoretically, yield somewhat reliable results. However, this is not at all how it plays out in reality. In fact, many mainstream scientists, at the risk of their careers, have actually publicly admitted this on occasion. One such scientist was Dr. William D. Stansfield, animal breeding instructor of biology at the California Polytechnic State University, who stated: "It is obvious that radiometric techniques may not be the absolute dating methods that they are claimed to be. Age estimates on a given geological stratum by different radiometric methods are often quite different (sometimes by hundreds of millions of years). There is no absolutely reliable long-term radiological 'clock.' The uncertainties inherent in radiometric dating are disturbing to geologists and evolutionists"(31). Here's another similar admission by a mainstream geologist, Curt Teichert: "No coherent picture of the history of the earth could be built on the basis of radioactive datings"(32).

It is important to understand that different chemical elements are utilized for the dating of organic and inorganic materials. Carbon-14 (C-14) is used for dating organic materials, whereas other radioisotopes are employed for dating inorganic materials. But the principle behind both is the same--measurement of radiation levels to determine a particular object's age.

Let us first look at C-14 dating, the method used to date organic materials.

Because of the half-life of C-14 being only 5,730 years (a half-life is the time required for half of an unstable element, in this case Carbon-14, to break down into its "daughter element"), it is known to be totally unreliable for dating anything beyond 50,000 years old. Some don't even trust it for dates beyond 5,000 years(7). Thus C-14 is only used for dating organic objects, like bones from an archaeological dig, which date to the fairly recent past. But there is a significant number of scientists who even doubt its competency here, arguing that this method of dating has far too many serious (some say fatal) flaws, rendering it entirely unreliable and thus altogether useless. Notice the following admission that one mainstream scientist, Robert Lee of Canada, made: "The troubles of the radiocarbon dating method are undeniably deep and serious....Continuing use of the method depends on a 'fix-it-as-we-go' approach, allowing for contamination here, fractionation there, and calibration wherever possible. No matter how 'useful' it is, though, the radiocarbon method is still not capable of yielding accurate and reliable results. There are gross discrepancies, the chronology is uneven and relative, and the accepted dates are actually selected dates"(33).

What is it about C-14 dating that drives scientists like this to make such critical claims about its reliability? To begin with, it is based upon several rather lame assumptions, namely:

- The formation of C-14 in the atmosphere has been constant for at least 70,000 years.

- C-14 formation is the same all over the world.

- The content ratio is the same in all kinds of specimens worldwide.

- Ancient specimens are not contaminated with solutions containing modern amounts of C-14.

- There is no loss of C-14 except by radioactive decay(7).

As it turns out, every single one of these assumptions is entirely false. Yet the academic world, as a whole, still continues to sell these assumptions as facts, and every new textbook edition carries on the tradition of towing the party line.

Quite surprisingly, however, every so often a mainstream scientific journal will carry an article which admits that these assumptions, or at least some of them, are not indisputable, and that there are thus serious problems with the validity of the C-14 dating method. One such example of this was the September 2000 issue of *Scientific American*, which actually conceded that carbon dating cannot be relied upon because the amount of C-14 in the atmosphere has not been constant over time. An interesting admission indeed!

Let's examine some of the specific factors that alter the C-14 content in the atmosphere (and thus in specimens that are dated using C-14), thereby overturning all of the five assumptions listed above, and thus invalidating C-14 dating as a dependable scientific tool (most, if not all, of these factors, incidentally, were particularly prominent at the time of the great flood disaster):

- Decay of Earth's magnetic field. This is an ongoing process which, for unknown reasons, is occurring at an alarming rate. It is estimated that, at the current speed of decay, the Earth will have nothing left of its magnetic field in another thousand years or so. And as it continues to weaken every year, more and more cosmic rays are able to enter the upper atmosphere, resulting in an increase in the decay rate of C-14, which consequently messes up the C-14 time-clock.

It's been suggested by some researchers that this particular factor (the weakening of Earth's magnetic field) may indeed have resulted from the flood catastrophe. However, this cannot be said with any degree of certainty.

- Air pollution from volcanic activity (and even industrial burning). Both of these factors can shield solar input and alter gas ratios in the air, thereby knocking the C-14 time-clock out of sync. And, as we shall see later on, volcanic eruptions definitely did accompany the flood disaster, on a massive scale.

- Meteors or asteroids falling to Earth. Just like volcanic eruptions, falling meteors and asteroids can create a blanket of dust high in the atmosphere that shields solar input and alters gas ratios in the air, thus affecting the C-14 clock. And it doesn't even need repeating that falling asteroids/meteors had accompanied (indeed, instigated) the flood disaster.

- Changes in solar activity, solar flares, and sunspots. Like cosmic rays, these factors can cause an increase in the decay rate of C-14. And as we have already seen, an increase in solar radiation most likely occurred at the time of the flood, not necessarily due to an increase in actual solar activity, but to the breakdown of the protective aquatic canopy, which would have amounted to a major increase in solar radiation reaching the Earth's surface.

- Changes in cosmic radiation levels reaching the upper atmosphere from extraordinary events in our galaxy, like a supernova (an explosion of a star). This is another C-14 clock-resetting factor that may have accompanied the great flood disaster, if the exploded trans-Martian planet was destroyed by an impacting piece of debris from a supernova explosion, as suggested earlier(7).

Seeing that most of the above factors persist today, and that most (or all) of them occurred at the time of the flood, and on a significantly heightened level at that, can you not see how completely useless C-14 dating actually is? How can this "clock" possibly be accurate when it is constantly being reset, and was obviously reset on an even grander scale as a result of the flood?

But let us move on to radioisotopic dating of inorganic materials.

Is inorganic dating any more reliable than organic dating? First of all, it is important to point out that the data from radioisotopic dating of inorganic materials is wide open to interpretation. And when it is discovered that this interpretation is biased by a desire to conform to a preconceived time-scale, the resulting dates arrived at can be called anything but accurate. As one catastrophist scientist, Harold Coffin, put it: "[C]hronologic interpretation of radioisotope data is not a simple, straightforward venture. It requires considerable interpretation and selection to develop a radioisotope-calibrated time scale. Scientists naturally use only that data that will fit into the generally accepted paleontological and geological theories"(34).

Though it might seem strange, scientists use radioisotopic dating of inorganic materials to date fossils (which, of course, are the remains of ancient organic materials). One of the reasons for this was already mentioned--that C-14 has a half-life that is far too short for dating materials that are believed to be multiplied millions of years old. But the other reason, which is even more important, is that most fossils do not contain organic materials, since such materials have long since decomposed and have been replaced by minerals, in a process known as petrification. Thus scientists are left with no choice but to rely upon the dating of inorganic materials in order to determine a fossil's age. And this is where their biggest problem lies.

By the way, the reason we are focusing specifically on fossils, in our discussion of the fallacy of radioisotopic dating of inorganic materials, is because, as we shall clearly later see, it is the position of this study that nearly all fossils were formed as a consequence of the relatively-recent flood catastrophe. Thus, if we can demonstrate that the conventional dating of fossils is incorrect, then we can more easily place the fossil record in (what most catastrophists believe to be) its proper historical context, in the relatively recent past (on the order of thousands of years, rather than millions). This is a task that can be accomplished, fortunately, with the use of simple logic.

Put in the simplest terms, the radioisotopic method used for dating fossils is not valid because dating the minerals that make up a fossil, or even the strata that contain the fossil, tell us absolutely nothing about when the fossil was formed, or when the strata were deposited that the fossil was found in. All that such a dating method can tell us, at best, is how old the minerals are. Understand that the minerals that make up a given fossil and its encasing strata existed long before the formation of the fossil, and long before the strata were laid down. Do you see the point? The "logic" behind radioisotopic dating of fossils and fossil strata is ridiculous! Catastrophist Harold Coffin put it this way, adding a few more details: "Science usually presumes the age of a fossil to be at least as great as the radioisotope age of the mineral that has replaced its organic material, of the mineral in the geologic formation that surrounds the fossil, or of the mineral in a formation that overlies or penetrates the one the fossil is in. It also begins with the premise that the end-products of those spontaneous radioactive transformations that have occurred since the mineral became a part of its present surroundings are readily identifiable. Another way to state it is to say that radioactive 'clocks' were 'set to zero' (that is, the accumulated results of all previous radioactive transformations were removed) when geological activity either formed or deposited the material in its present situation or matrix. Because such interpretation yields results that fit into the expectations of long ages, scientists have not analyzed this premise as critically as they should have.

"It is just not reasonable to presume that a natural transport mechanism such as molten lava or water will deposit chemically pure substances. Instead it is much more likely that even after erosion and igneous processes have moved the mineral to a new location, it will still retain some of the elements created by radioactive decay that it previously possessed. In general, we should expect the radioisotope age of a mineral associated with a fossil to reflect, at least partially, its physical history before its association with the fossil and thus not give a valid indication of the actual time that really elapsed since it came into contact with the fossil. A graveyard provides a handy illustration. The radioisotope age of the minerals in graveyard soil give no indication of the time when someone actually buried the bodies there..."(34).

Hopefully the point of the unreliability of modern dating methods has been driven home sufficiently enough so that we can begin to think a bit more "outside the box" in terms of dating the formation of the fossil record and the occurrence of the asteroid impact/flood disaster as having been contemporaneous, and having been in the far more recent past than we have been led (or misled) to believe. But this discussion of the fallacy of modern dating methods would remain incomplete without addressing the "geologic column" and its relevancy (or lack thereof). Therefore, before going on to a more detailed discussion of the flood catastrophe, let us first turn our attention to this important subject.

The geologic column

According to "establishment" scientists, the designated time-frames for Earth's prehistoric ages have been so "firmly established," with "pinpoint accuracy," that sedimentary rock formations can be confidently dated, on sight, simply by looking at the fossils contained within them. And how is it that they know how old the fossils are? Well, by looking at the rock formations that contain them. Yes, that's right--scientists utilize circular reasoning, dating rocks by the fossils in them, and then turning around and dating the fossils by the rocks that they are found in. Can you not see how illogical and downright ridiculous this line of reasoning is? Are we to understand this to be science? Speaking in reference to this matter, R. H. Rastall, formerly a lecturer on Economic Geology at Cambridge University, once made the following admission: "It cannot be denied that from a strictly philosophical standpoint geologists are...arguing in a circle. The succession of organisms has been determined by a study of their remains embedded in the rocks, and the relative ages of the rocks are determined by the remains of organisms that they contain"(35).

To justify such groundless "logic," scientists from this school of thought are quick to point out that the initial dates that were arrived at had been obtained through the "infallible" method of radioisotopic dating. But, as we saw in the last subsection, there is absolutely nothing infallible about this dating method.

The "circular reasoning" methodology described above is built around the concept of what is called the "geologic column." The geologic column is an imaginary mix-and-match jigsaw puzzle time-chart that scientists have assembled, which they claim outlines the earliest ages of Earth's history and the development of life upon it during those ages. This chart divides the Earth's distant past into various eras, which are further divided into periods, a couple of which are further divided into epochs. Furthermore, it places different life-forms within what is claimed to be the corresponding time-frames during which those life-forms had first appeared on Earth. This chart has become the accepted standard whereby fossils and fossil-bearing formations are dated. We only need, says modern science, to match up a fossil, or a rock that contains a fossil, with its corresponding location on the chart in order to determine the age thereof.

It is claimed and/or assumed that in sedimentary formations the world over, the sequencing of strata, along with the fossils found within them, smoothly correspond with the geologic column chart--that all strata and their accompanying fossils are found in the same order as shown in the chart, or that they can all be fitted neatly and smoothly somewhere into this chart. But is there any truth to this?

First of all, do understand that there is positively no place on Earth where the geologic column can be observed, in its entirety, in any rock formation. In fact, there is no place where even a significant portion of this column can be found in the geologic record. The whole column has literally been assembled according to the way that mainstream scientists want it to be, to support their preconceived notions. In their 1952 book *Geology*, two mainstream scientists, K. E. Caster and O. D. Von Engeln, made this confession: "If a pile were to be made by using the greatest thickness of sedimentary beds of each geologic age, it would be at least 100 miles high....it is, of course, impossible to have even a considerable fraction of this at one place. The Grand Canyon of Colorado, for example, is only one mile deep....By application of the principle of superposition, lithologic identification, recognition of unconformities, and reference to fossil successions, both the thick and the thin masses are correlated with other beds at other sites. Thus there is established, in detail, the stratigraphic succession for all the geologic ages"(36). In other words, scientists have forced a square peg to fit into a round hole. It's the old trick of starting with a premise and then making the evidence conform to that premise--plain and simple.

Caster and von Engeln, cited above, went on to identify exactly what this premise is: "The geologist utilizes knowledge of organic evolution, as preserved in the fossil record, to identify and correlate the lithic records of ancient time"(36). The admission made here is that the geologic column has been constructed to make it fit in with mainstream science's belief in the theory of evolution--the idea that life-forms followed a pattern of development from simpler to more complex varieties, and having required many millions of years to do so. Now please understand that this study is not an attempt to refute the evolution theory. This is another issue that the reader must decide upon independently. We are here simply exposing faulty reasoning for date-setting that arose from a deeply-embedded bias.

Let's look at an additional, even more bold and commendable admission made by another mainstream scientist, Edmund M. Spieker, regarding the illegitimacy of the geologic column and the claims built around it: "Does our time scale [geologic column]...partake of natural law? No....I wonder how many of us realize that the time scale was frozen in essentially its present form by 1840....How much world geology was known in 1840? A bit of western Europe, none too well, and a lesser fringe of eastern North America. All of Asia, Africa, South America, and most of North America were virtually unknown. How dared the pioneers assume that their scale would fit the rocks in these vast areas, by far most of the world? Only in dogmatic assumption--a mere extension of the kind of reasoning developed by [Abraham Gottlob] Werner from the facts in his little district of Saxony. And in many parts of the world, notably India and South America, it does not fit. But even there it is applied! The followers of the founding fathers went forth across the earth and in Procrustean fashion made it fit the sections they found, even in places where the actual evidence literally proclaimed denial. So flexible and accommodating are the 'facts' of geology....And what essentially is this actual time-scale--on what criteria does it rest? When all is winnowed out, and the grain reclaimed from the chaff, it is certain that the grain in the product is mainly the paleontologic record and highly likely that the physical evidence is the chaff"(37).

Far from the world's sedimentary rock formations (and the fossils within them) coinciding with the geologic column chart, or bearing testimony to great passages of time for their accumulations, the truth is, they instead reveal, all too often, a total disconnect with the sequential order of the chart, and they strongly testify, instead, to very rapid deposition and fossilization rates--just what would be expected from a global catastrophic flood. We will spend the duration of this subsection elaborating on several of the major discrepancies that exist between the predicted and observed arrangements of the strata and fossils of the world, as well as other discrepancies that present unsurmountable problems for the geologic column.

Occasionally it happens, in sedimentary formations around the globe, that supposedly "older" strata are found superimposed over "younger" strata. In other words, sedimentary layers are sometimes observed to be in the wrong order--the very opposite order--from what they are supposed to be, if we go by the geologic column. The explanation offered for this particular discrepancy by geologic column advocates is that these reverse-order strata have been overturned--literally flipped over--in the distant past by a folding (or "upthrusting") of the Earth's crust. This, of course, is a possibility, perhaps even a probability (at least in some cases), considering the grand-scale tectonic disruptions that occurred during and after the flood catastrophe. However, it doesn't seem likely that crust folding can account for all of the many occurrences of reverse-sequence strata throughout the world. But regardless, the geologic column faces another reverse-sequence problem--a far more serious one--which lies this time with fossils inside the strata, rather than with the strata themselves. It is a serious problem because it cannot be accounted for by a folding of the Earth's crust.

The problem I am referring to involves a somewhat frequently-occurring, out-of-place arrangement of fossil remains within sedimentary formations the world over--out-of-place, that is, according to the reckoning of the geologic column.

For instance, certain types of pollen that are claimed to have first arisen, or "evolved," in the Tertiary period have been found in (what is claimed to be) the more archaic Cretaceous of the East Netherlands. Some types of spores, which are said to have first appeared in the Jurassic, have been discovered in the "older" Precambrian of the Ukraine(106). How can this be? Is the sequencing of fossils in the world's strata telling us a different story from what academia has told us?

The two reverse-sequence examples cited in the last paragraph represent a small sampling of such findings in the fossil record. But notice that they both have one thing in common: They deal with species appearing in strata that purportedly long predate the time that such species were supposed to have "evolved." Yet the very opposite of this scenario has also been observed. That is, we occasionally find species in the fossil record that appear in layers which, mainstream science insists, long post-date the time they are said to have become extinct. Examples of this include certain types of foraminifers that are believed to have perished during the Cretaceous, but have been found in the "younger" Tertiary of West Israel, Australia, Sweden, and the Ionian Sea. Some brachiopods that we've been told had disappeared in the Pennsylvanian have been unearthed in the "more recent" Permian of the Yukon in Canada(106). Again, how can this be, if we accept the "established" chronology of the geologic column?

But we haven't even scratched the surface yet. For the problem of out-of-sequence fossils involves much more than mere pollen, spores, foraminifers, and brachiopods. As catastrophist Walt Brown wrote: "[I]n Uzbekistan, 86 consecutive hoofprints of horses were found in rocks dating back to the dinosaurs....Sometimes, land animals, flying animals, and marine animals are fossilized side-by-side in the same rock. Dinosaur, whale, elephant, horse, and many other fossils, plus crude human tools,...[have been] found in phosphate beds in South Carolina. Coal beds contain round, black lumps called coal balls, some of which contain flowering plants that allegedly evolved 100 million years after the coal bed was formed. In the Grand Canyon, in Venezuela, and in Guyana, spores of ferns and pollen from flowering plants are found in Cambrian and Precambrian rocks--rocks deposited before life supposedly evolved. A leading authority on the Grand Canyon even published photographs of horselike footprints visible in rocks that...[supposedly] predate hoofed animals by more than a hundred million years. Other hoofprints are alongside 1,000 dinosaur footprints in Virginia.

"Petrified trees in Arizona's petrified forest contain fossilized nests of bees and cocoons of wasps. The petrified forests are supposedly 220 million years old, while bees (and flowering plants which bees require) supposedly evolved almost a hundred million years later. Pollinating insects and fossil flies, with long, well-developed tubes for sucking nectar from flowers, are dated 25 million years before flowers supposedly evolved. Most evolutionists and textbooks systemically ignore discoveries which conflict with the evolutionary time scale"(38).

And then there's the problem of the so-called "living fossils." These are species of plants and animals that were originally reported to have died off millions of years ago, but then were later found to still be alive in modern times.

The coelacanth fish is one glaring example. This species was initially thought to have become extinct during the Cretaceous, since no remains of this fish were ever found in higher (and thus "younger") levels of strata anywhere in the fossil record. However, in December of 1938, a living specimen was caught by a fisherman off the eastern coast of South Africa, which shocked the mainstream scientific community.

Another example we will cite comes from the plant kingdom--the dawn redwood tree (*metasequoia glyptostroboides*)--which was said to have become extinct in the Tertiary, but was later discovered, in 1941, to still be in existence in China.

These discoveries are the equivalent of finding a living dinosaur in the Congo or the Amazon--they amount to major assaults on one of the chief "pillars" of modern scientific thought, which have never been resolved.

To this list we could add fossilized human footprints (which we mentioned earlier) found in strata that supposedly date back millions of years. On this matter, Walt Brown wrote: "In 1968, 43 miles northwest of Delta, Utah, William J. Meister found...[an] apparent human shoe print inside a 2-inch-thick slab of rock. Also in that slab were...trilobite fossils, one of which was squashed under the 'heel'"(Ibid.). Subsequent searches in the Delta, Utah, area have turned up similar shoe prints, although none of them showed a trilobite having been stepped on. Keep in mind that trilobites are said to have pre-existed humans on this planet by some 240 million years, so finds like this present a great problem for mainstream scientists. But they are no problem for catastrophists, who are not bound by the blind biases of contemporary thought.

Most of the human footprint fossils that have been found, of course, were made by individuals who walked in their bare feet, which have been discovered in Kentucky [Endnote 22], Missouri, Texas, New Mexico, Pennsylvania, Nevada, the Gobi Desert, Africa (Laetolil), Australia (Mt. Victoria), Turkey (near Demirkopru), and various other places around the world. Sometimes such human tracks have been uncovered right alongside dinosaur tracks. A find of this very nature was reported back in 1983, in the *Moscow News*, No. 24, p. 10, in an article called "Tracking Dinosaurs." It stated: "This spring, an expedition from the Institute of Geology of the Turkmen SSR Academy of Sciences...found over

1,500 tracks left by dinosaurs in the mountains in the south-east of the Republic. Impressions resembling in shape a human footprint were discovered next to the tracks of the prehistoric animals."

Such finds--human prints alongside those of dinosaurs--have also been uncovered in Australia(38). But perhaps the most notorious examples of human tracks alongside those of dinosaurs are the ones that have been excavated in Glen Rose, Texas. Here we find long stretches of human and dinosaur trails that cross right over each other, in the same exact layer of Cretaceous limestone. These finds, of course, are impossible, according to the geologic column. Yet there they are.

Fossilized human skeletal remains have also been unearthed from various strata around the world that are supposed to long predate the time of human existence.

From the late Pliocene of Foxhall, England, for example, a human jaw was found by workers digging for coprolites (fossil dung), back in 1855.

In addition, various human bones were found in the middle Pliocene of Castenedolo, Italy, in 1860, 1880, and 1889.

Moving on to another example, a human vertebra was discovered in the early Pliocene of Monte Hermoso, Argentina, in 1908.

Going back further in time (or so we're told), to the Miocene, Professor Johannes Hurzeler of the Museum of Natural History in Basil, Switzerland, found a lump of coal in Tuscany, Italy, in 1958, which contained a completely flattened jawbone of a human child(150).

Journeying back even further into the past, according to the geologic column, a complete human skeleton from the late Eocene was discovered at Delemont in Switzerland, around 1883.

Looking back further still, to the "age of the dinosaurs," a complete adult male human skeleton was recovered from 90 feet below the surface, in a Carboniferous coal bed that was overlaid by a 2-foot-thick layer of slate in Macoupin County, Illinois, back in 1862(113).

Then we have the famous Malachite Man finds, initially discovered in 1971 in Lower Cretaceous Dakota sandstone, near Dinosaur National Monument in Utah. A total of 10 skeletons were found, partially replaced by a green mineral known as malachite, from whence these finds got their name. According to the geologic column, this sandstone is roughly 140 million years old(7).

A similar find is described in the following quote from the book *Secrets of the Lost Races*: "An interesting item appeared in many of the nation's newspapers on April 10, 1967, reporting the discovery of an artifact and human remains at the Rocky Point Mine in Gulman, Colorado. At a depth of 400 feet below the surface, according to an account in the *Saturday Herald* of Iowa City, the excavators found human bone embedded in a silver vein. By geological standards, the find was estimated to be several million years old. But in addition to the bones, they uncovered a well-tempered copper arrowhead four inches long. Neither bone nor arrowhead belonged there, according to our way of thinking, yet there they were--unexplainable and certainly unexpected. The historians and geologists are unable to fit these remains into the theoretical framework of evolution; partly because of this, the find has been conveniently forgotten"(9).

Speaking of manmade artifacts being found in what is alleged to be multi-million-year-old strata, here is another troubling example of this, as described by the authors of *Forbidden Archaeology*: "In 1877 Mr. J. H. Neale was superintendent of the Montezuma Tunnel Company, and ran the Montezuma tunnel into the gravel underlying the lava of Table Mountain, Tuolumne County [in California]....At a distance of between 1400 and 1500 feet from the mouth of the tunnel, or of between 200 and 300 feet beyond the edge of the solid lava, Mr. Neale saw several spear-heads [made] of some dark rock and nearly one foot in length. On exploring further, he himself found a small mortar three or four inches in diameter and of irregular shape. This was discovered within a foot or two of the spear-heads. He then found a large well-formed pestle and near by a large and very regular mortar. All of these relics were found the same afternoon, and were all within a few feet of one another and close to the bed-rock, perhaps within a foot of it. Mr. Neale declares that it is utterly impossible that these relics can have reached the position in which they were found excepting at the time the gravel was deposited, and before the lava cap formed. The position of the artifacts in gravel close to the bed-rock at Tuolumne Table Mountain indicates they were 33-55 million years old"(113).

In a later subsection of this study, we will be looking at several more examples of manmade artifacts that have been found in layers of sedimentary rock that are supposed to be many millions of years in age. But for now, let's take a look at some finds of a slightly different order, which reveal human traces among fossils that present yet further problems for geologic column fans. The type of finds I am referring to are fossil bones and shells of creatures that are claimed to have lived millions of years ago, and yet they have carvings and other incisions made in them that are obviously of human origin. And since such specimens have been found buried in undisturbed archaic layers of sediment, the claim cannot be made that humans who lived in more recent millennia made these etchings in fossilized bones and shells of animals that lived long before them.

Some good examples of this were found around 1848 in St. Prest, France, in a gravel pit that is said to be from the Pliocene-Pleistocene boundary, and a bit later. Bones were uncovered there of such extinct species as *Elephas meridionalis*, *Rhinoceros leptorhinus*, *Rhinoceros etruscus*, *Hippopotamus major*, and a giant beaver known as *Trogontherium cuvieri*. Many of these unearthed bones had nicks and scrapes on them that were unquestionably produced by manmade implements--by tools of hunters in the process of killing and removing meat from their catches.

Other examples of this nature, and from the same time-period, were discovered in Val d'Arno, Italy, around 1883. Found here specifically were nicked-up bones of *Elephas meridionalis* and *Rhinoceros etruscus*.

Also, in the late Pliocene of the Red Crag Formation in England, there have been found numerous shark teeth with perfectly-round holes bored through their centers, clearly for the purpose of making weapons or necklaces.

Dating back even further (or so we're told), a *Rhinoceros pleuroceros* jaw, nicked and scraped by manmade tools, was found in a middle Miocene layer of sandstone in Billy, France, sometime around 1868(Ibid.).

We could go on to cite many more examples like these, but quite possibly the most fascinating of all are those that contain actual artistic etchings, which simply cannot be denied as having been the work of human hands.

The first of two examples we will look at is a carving on a fragment of the joint of a bone of either a *Deinotherium* or a mastodon, found in Miocene strata from the Dardanelles in Turkey, sometime around 1874. It features the etched design of a horned quadruped with an arched neck, rounded chest, long body, straight fore-legs, and broad feet. Also seen on the piece are traces of several other figures that were nearly obliterated by the ravages of time. Found aside this bone were further signs of a human presence--a flint flake and several different kinds of animal bones that were fractured longitudinally for the purpose of extracting the marrow.

The second and final example that is worth mentioning here, discovered about 1881, is a shell found in the late Pliocene stratified deposits of the Red Crag Formation in England. It contained a crude but unmistakably human face carved on its outer surface(Ibid.).

And then we have this strange tale: Sometime between 1912 and 1914, Carlos Ameghino and his associates, working on behalf of the natural history museums of Buenos Aires and La Plata, made a most fascinating discovery in the Pliocene Chapadmalalan Formation at the base of a cliff extending along the seaside at Miramar, Argentina. *Forbidden Archaeology* picks up the story from here: "From the top of the Late Pliocene Chapadmalalan layers, Ameghino extracted the femur of a toxodon, an extinct South American hoofed mammal, resembling a furry, short-legged, hornless rhinoceros. Ameghino discovered embedded in the toxodon femur a stone arrowhead or lance point, giving evidence for culturally advanced humans 23 million years ago in Argentina [according to conventional wisdom]"(Ibid.).

Finds such as those described in the last several paragraphs force us to conclude that either man lived much longer ago than we have been led to believe, or dinosaurs and other extinct creatures have lived much more recently. Considering all the flaws we have been discovering with dating methods and the resulting extreme ages that scientists have assigned to the world's strata and their entombed fossils, and considering all the pieces of evidence we have been looking at (and will continue looking at) that testify to a relatively recent global flood catastrophe that created these strata and their fossils, which conclusion do you suppose is correct?

==========================

Further evidence of man and dinosaurs coexisting

As an interesting side note, let us look at some further pieces of evidence of the coexistence of man and dinosaurs, which do not involve the geologic column. These pieces of evidence fall into two main categories: ancient records and archaeological works of art. We will begin with ancient records.

Though the flood was most likely responsible for the demise of the dinosaurs and other large reptiles, it is possible--even probable--that some of these prehistoric creatures had survived this catastrophe, at least for a time. For we find that there are many ancient legends, from nearly every culture the world over, most of which date back to the earliest ages of the post-flood era, that talk of fearsome dragons that roamed the Earth, having scales, fangs, and claws, which breathed fire in some cases, and in other cases had wings. The frequency and similarity of these ancient, globally-distributed legends actually force us to ask if they are accounts of real creatures--dinosaurs, or related animals--that existed alongside the people who initially recorded these accounts.

Probably the most tantalizing of ancient accounts of this nature are found in the biblical book of Job, which is believed to be the oldest book in the Bible, written soon after the flood. This ancient text gives descriptions of two very unusual beasts that lived in Job's day, both of which seem impossible to affiliate with any other known creatures but dinosaurs (well, one appears to have been a dinosaur, while the other one was probably an aquatic relative thereof).

The first of these beasts is referred to as "behemoth" in Job 40:15-18, 23, where God says: "Behold now behemoth, which I made...he eateth grass as an ox. Lo now, his strength is in his loins, and his force is in the navel of his belly. He moveth his tail like a cedar: the sinews of his stones [thighs] are wrapped together [tightly knit]. His bones are as strong pieces of brass; his bones are like bars of iron....Behold, he drinketh up a river, and hasteth not: he trusteth that he can draw up Jordan into his mouth."

Does this not sound like an accurate depiction of a sauropod? Some species of this gargantuan dinosaur grew up to 60 feet tall and 130 feet long (perhaps even longer), and some weighed as much as 110 tons. They were known to be vegetarians, just as Job was told: "...he eateth grass as an ox."

The strength and force of these beasts truly were in their "loins" and "belly," where most of their body mass and muscular tissues were located.

Their massive tails could also be said to have moved, or swayed, "like a cedar"--sauropod tails were very long and quite wide at the base, as is the case with a tree trunk.

Additionally, their muscular thighs were very firm ("tightly knit"), and their huge bones were indeed like "strong pieces of brass," or "bars of iron."

Finally, the description of behemoth drinking up a river and drawing up the Jordan into his mouth seems quite applicable to a sauropod as well, which must have consumed tremendous amounts of water every day.

The second beast described in the book of Job is called "leviathan," in chapter 41. We will be quoting from verses 1, 2, 10, 12, 14-17, and 19-21, where God again speaks, saying: "Canst thou draw out leviathan with an hook? or his tongue with a cord which thou lettest down? Canst thou put an hook into his nose? or bore his jaw through with a thorn?...None is so fierce that dare stir him up....I will not conceal his parts, nor his power, nor his comely proportion....Who can open the doors of his face [his mouth]? his teeth are terrible round about. His scales are his pride, shut up together as with a close seal. One is so near to another, that no air can come between them. They are joined one to another, they stick together, that they cannot be sundered....Out of his mouth go burning lamps, and sparks of fire leap out. Out of his nostrils goeth smoke, as out of a seething pot or caldron. His breath kindleth coals, and a flame goeth out of his mouth."

One could scarcely deny that we have here a precise portrayal of a species of the mosasaur family. Although technically not dinosaurs, mosasaurs were giant aquatic reptiles ("sea monsters") that reached up to about 60 feet in length, and ate, amongst other things, sharks.

The book of Job spoke fittingly of this type of creature when it stated that it could not be caught with a fishing hook, or drawn in by the tongue. Truly, none was "so fierce that dare stir him up."

As further indicated by Job, mosasaur jaws were very powerful. They were used to crush both the shells and bones of prey, and the teeth of mosasaurs, as also stated by Job, were indeed very "terrible round about." For this reason, this same beast, "leviathan," was referred to in the book of Isaiah (chapter 27, verse 1) as a "piercing serpent." It is also called a "sea dragon" in this same verse.

Additionally, Mosasaurs had scales "shut up together as with a close seal." Their skin, in fact, was much like that of snakes.

The final point brought out by Job about leviathan is that it breathed fire from its mouth, and smoke from its nostrils. Here is where you are probably thinking that this narrative departs from reality. However, before you settle yourself on this opinion, take a look at a few important facts:

- The rest of the description of "leviathan" in this passage gives us no reason to doubt that what is being relayed is a legitimate delineation of a well-known living creature at that time. Except for the mention of a mouth breathing fire and nostrils breathing smoke, every other point that is made describes what we know to have been true about mosasaurs.

- As mentioned above, the concept of "fire-breathing dragons" was a common theme in the literary works of pretty much all ancient cultures. The sheer fact that there are so many widespread and consistent mentions of the fire-breathing aspect of dragons in ancient works should itself cause us to ask, Is it possible that mosasaurs, or some other similar, extinct creatures possessed the ability to exhale fire from their mouths and smoke from their nostrils, and that this marvel was seen and recorded by ancient eyewitnesses around the world, which we have wrongly dismissed as myths?

- If we are to take literal the fire-breathing and smoke-snorting attributes of leviathan (or "dragons," in general), we must look for some parallels in the animal kingdom today that can serve to illustrate how such phenomena could have been possible. But do such parallels exist?

Consider fireflies, for example. If these critters were extinct today, we might scoff at ancient tales of bugs that flashed lights in the darkness of night, writing them off as pure fairy tales. Yet we know that such bugs do indeed exist. With these thoughts in mind, notice another statement made about leviathan in Job 41:18: "By his neesings [snorting] a light doth shine..." Can you honestly say that this sounds mythical to you--completely outside the realm of possibility?

How about bombardier beetles? When these creatures sense danger, they create an actual explosion outside the posterior end of their bodies, to kill or frighten off predators. The online encyclopedia, "Wikipedia," explains the process this way: "Secretory cells produce hydroquinones and hydrogen peroxide (and perhaps other chemicals, depending on the species), which collect in a reservoir. The reservoir opens through a muscle-controlled valve onto a thick-walled reaction chamber. This chamber is lined with cells that secrete catalases and peroxidases. When the contents of the reservoir are forced into the reaction chamber, the catalases and peroxidases rapidly break down the hydrogen peroxide and catalyze the oxidation of the hydroquinones into p-quinones. These reactions release free oxygen and generate enough heat to bring the mixture to the boiling point and vaporize about a fifth of it. Under pressure of the released gasses, the valve is forced closed, and the chemicals are expelled explosively through openings at the tip of the abdomen. Each time it does this it shoots about 70 times very rapidly. The damage caused is fatal to attacking insects and painful to human skin"(114).

If these beetles were also extinct today, we would likewise view with skepticism any ancient legends that talk of bugs that gave off painful fiery emissions if touched. Yet we know that they, too, do exist.

The last example we will consider are the electric eels. If these creatures were also extinct today, just imagine how the modern academic world would laugh at ancient tales of fish that would stun (and in some cases kill) anyone who even lightly caressed them. Yet we know that these creatures exist as well.

So, considering fireflies, bombardier beetles, and electric eels, why couldn't there have been creatures in the distant past--perhaps mosasaurs--that breathed fire from their mouths, or smoke from their nostrils, which we hear so much about in the world's archaic accounts of dragons?

We should now be able to see that ancient references to land-dwelling "dragons" or "dreadful beasts" were most likely talking about dinosaurs of one type or another, whereas the mention of "sea monsters" seem to have been indicative of mosasaurs. But what about the allusions to flying dragons (or else winged or feathered serpents)? Well, the only candidates that appear to match up with these particular descriptions are pteranodons--winged reptiles.

Herodotus is one person who mentioned such creatures. In his *Historiae*, he wrote: "There is a place in Arabia, situated very near the city of Buto, to which I went, on hearing of some winged serpents; and when I arrived there, I saw bones and spines of serpents, in such quantities as it would be impossible to describe. The form of the serpent is like that of the water-snake; but he has wings without feathers, and as like as possible to the wings of a bat."

Based on this description, Herodotus was most likely talking about a

Rhamphorhynchus--a pterosaur that did indeed look like a water-snake with featherless wings.

Herodotus went on to mention how he was told that "at the beginning of spring, winged serpents fly from Arabia towards Egypt; but that ibises, a sort of bird, meet them at the pass, and do not allow the serpents to go by, but kill them." Then, when describing Arabia, he wrote that "winged serpents, small in size, and various in form, guard the trees that bear frankincense, a great number around each tree. These are the same serpents that invade Egypt."

This narrative is far too detailed and natural-sounding to be written off as myth. Herodotus was quite clearly talking about real creatures and real events involving them.

Another ancient source that mentions winged (or flying) serpents is the Bible. In fact, the two references that mention these creatures (Isaiah 14:29; 30:6) talk of a "fiery flying serpent." Can we infer from this that certain pterosaurs breathed fire, like leviathan?

Sometimes ancient legends make reference to very large beasts, but fail to provide specific details about what these creatures looked like. Yet it is quite apparent that they must indeed be allusions to dinosaurs and other contemporary animals of great size that are now extinct. Examples of this abound in Native American folklore.

For instance, the Ojibwa tribe talk of "giant animals" that caused trouble to their ancient ancestors.

The Brule tribe discusses how "water monsters" once lived near lakes and rivers.

Pawnee Indians tell stories of nonhuman "giants" that lived in the previous world.

As far as Native American folklore goes, the last example of this sort we will cite here comes to us from the Yuma tribesmen, who say that their remote ancestors used to struggle with "large, dangerous animals"(129).

On a similar note, an ancient Babylonian legend tells of the hero Gilgamesh, who decided to make a great name for himself by journeying to a far-away land to cut cedar trees that were in big demand in his city. While in the cedar forest with his fifty volunteers, he came across a huge reptile-like animal that ate trees and reeds. This story ends by stating that he killed the beast and cut off its head, to save it as a trophy(137).

Another legend of this same caliber comes from ancient Scandinavia. It describes a reptilian creature that had a body the size of a large cow. Its two hind legs were long and powerful, while its forelegs were remarkably short. It was also portrayed as having huge jaws(138). This sounds like some species of allosaurus.

A medieval Scottish legend mentions a certain "Fraser of Glenvackie," who is said to have killed the last of the "dragons" in Scotland in the year 1520.

A chronicle from 1405 mentions a giant reptile, or "dragon," that was seen at Bures in Suffolk, England. It declares: "Close to the town of Bures, near Sudbury, there has lately appeared, to the great hurt of the countryside, a dragon, vast in body, with a crested head, teeth like a saw, and a tail extending to an enormous length. Having slaughtered the shepherd of a flock, it devoured many sheep..." After discussing an unsuccessful attempt by local archers to kill this terrible beast, because of its impenetrable hide, the chronicle goes on to say: "[I]n order to destroy him, all the country people around were summoned. But when the dragon saw that he was again to be assailed with arrows, he fled into a marsh...and there hid himself among the long reeds, and was never more seen."

The details of this chronicle force us to conclude that a dinosaur is being referred to here. Yet the exact species cannot be determined, since there are several dinosaur classifications that could be matched with this description.

The last of such legends that we will consider is from ancient China. Actually, of all ancient lands, perhaps none had more tales of this sort than did this one. But the specific legend we will here consider is most fitting for our discussion, seeing that it pertains to a time just after the flood catastrophe. It talks about a man named Yu, who, "after the flood," had "built channels to drain the water off to the sea." He also helped to make the land livable by driving out "dragons" from the marshlands so that new farmlands could be created(139).

As already stated, dragons turn up, not only in ancient written records, but also in the artwork of the ancient world--on pots, cups, cave and cliff walls, temples, etc. Is it possible that these paintings and carvings might be representations (albeit exaggerated or stylized ones, in some cases) of dinosaurs and other large land, sea, and air reptiles that survived the flood, but have since become extinct? This question is difficult to answer in the negative when we examine these ancient artworks in detail. What follows is such an examination of some of the best examples of artwork of this nature.

- One of the most impressive ancient artworks of this type is a Mesopotamian cylinder seal, from about 3300 B.C., which contains the carved image of an animal that has a close affinity to a sauropod. The body, legs, and characteristic long tail and neck are attributable to no other classification, living or extinct. (Actually, this seal depicts 2 such beasts, with intertwined necks.) How could the artist who made this seal have invented, by chance, a mythical beast that, in every detail, looks exactly like what we now know sauropods to have looked like?

- Portraying long-necked “dragons” with their necks intertwined was apparently a common practice in the ancient world. Another example of this comes from Egypt, and dates to about 3200 B.C. Known as the Narmer Palette, it portrays two obvious sauropods that look much like those pictured on the Mesopotamian seal described above, and from almost the same period of time.

- An unusual handle is found on an ancient Chinese ornamental box. Dating to the Eastern Zhou Dynasty of the third century B.C., it is fashioned in the shape of a sauropod, complete with clawed toes.

- A further example of intertwined, long-necked “dragons” in artwork from the ancient world is found on a Roman mosaic that dates to around 200 A D. It depicts what appear to be representations of an extinct creature known as a Tanystropheus--an aquatic reptile with an extremely elongated neck that measured about 10 feet longer than its body and tail combined.

- There is a stunning petroglyph found at the Natural Bridges National Monument in Utah that bears an uncanny resemblance to a sauropod (specifically a Brontosaurus), complete with long neck and tail, small head, 4 large legs, wide body, and arched back. It is claimed to be the work of Anasazi Indians who occupied the area between the fifth and fourteenth centuries A.D. However, it may very well be much older, long predating the Anasazi.

- An equally-impressive work of art is found in Arizona’s Hava Supai Canyon, looking remarkably like an Edmontosaurus (a relative of the Tyrannosaurus). The age of this glyph is not known, but it is obviously quite ancient.

- In northwestern New Mexico there exists an ancient rock drawing of 2 hunters stabbing their prey. The victim’s features are exactly like those of a duck-billed dinosaur. Though its age is indefinite, it is evidently very old as well.

- An urn from Caria, in modern-day Turkey, which dates to about 530 B.C., contains the image of a serpentine aquatic beast that has the unmistakable appearance of a mosasaur. It has an elongated body with front and rear paddle arms, as well as thick jaws, long teeth, and large eyes. On its back, near the head, is a fin that mosasaurs are usually not pictured with in modern textbooks. However, some mosasaur fossils have been found with a narrow cranial crest behind the eyes that may have had a fin attached to it. Also pictured on the urn are dolphins above and below the "mosasaur," as well as an octopus underneath it and a seal behind it, which further indicate that this was indeed a sea creature.

- A figure of what clearly looks like a spiny-back anklyosaur was apparently created by the Jalisco culture, which lived along the Pacific coast of Mexico from around 300 B.C. to 300 A.D. Could anklyosaurs have survived until this late date? This shouldn't surprise us, in light of the so-called "living fossils" we have discussed.

- At Angkor Wat, an ancient site in Cambodia that dates to the twelfth century A.D., near the exit of a temple complex on the premises, is a carving of what can only be identified as a Stegosaurus. Everything from the head, body, tail, legs, and the plates on its back amount to a very accurate depiction of this extinct animal, looking as though it was copied right out of a modern prehistory textbook. Should you find it difficult to believe that the Stegosaurus survived in Cambodia up until the twelfth century, ask yourself why--why does this seem so strange? Are you allowing your thinking to be limited by a groundless, modern academic bias?

- An even later relic, from China, looking strikingly like a beaked Oviraptor (a rather small dinosaur only 6 feet long), dates to the fourteenth century of our current era. It comes complete with 3-toed feet, long tail, and a crest on the top of its head, just like its archaic counterpart that we know from the fossil record.

And then there are those ancient artworks that, though they are distinctly "dinosauresque" in appearance, do not seem to represent any specific dinosaur species. Instead, they look like composite beasts that possess characteristics of two or more species, probably the result of fading memories of creatures that had gone extinct generations earlier. We will here cite two such examples:

- On a rock wall near Middle Mesa at Wupatki National Park, Arizona, lies a Native American work of art that has the appearance of a combination between a triceratops (because of the crest on its head) and a stegosaurus (due to the "fins" on its back). Interestingly enough, this creature is also shown breathing out fire.

- An ancient Peruvian pot, found in a pre-Columbian grave, has a creature painted on it that looks like a cross between a sauropod (judging from its body shape) and a stegosaurus (considering the "fins" on its neck and back). This animal, like the one described above, is also pictured breathing out fire.

So, do you still have any doubt that man and dinosaurs (along with other reptilian giants) lived contemporaneously, both before and after the flood?

========================

Let's come back to our discussion of the many problems that exist with the geologic column.

There are many formations around the world that are missing whole sections of strata that should be represented there, according to the geologic column. Stated more plainly, instead of seeing sections A, B, and C in sequence, all we find is A and C, without section B, the middle section, being present at all. We observe this problem in the Grand Canyon, for instance, where more than 150 million years of the supposed geologic column are absent (the Mesozoic, Cenozoic, Pennsylvanian, Devonian, Silurian, and Ordovician are all missing from the Grand Canyon's strata)(Ibid.). How could this have happened?

Geologic column advocates claim to be able to explain such huge gaps in sedimentary formations across the globe. They say that the missing strata were eroded away in the distant past. But this is an entirely unacceptable proposition. For we find that there is almost never any sign of erosion having taken place in such instances (between sections A and C). On the contrary, what we find is a smooth transition of flat sedimentary layers from section A to section C. Erosion does not happen this way. Instead, wind and water, which are mostly responsible for erosion, scour the Earth's surface, digging through underlying layers of rock and sediment at varying depths, creating a rugged, uneven surface.

It is interesting to note, by the way, that almost all sedimentary formations all over the world (even the ones that aren't missing a "section B") are deposited in this fashion--one flat layer upon another, with almost no sign of natural, jagged-surfaced erosion between any of the layers. Considering that these sedimentary layers are supposed to have accumulated over many millions of years of time, there should be at least an occasional sign of erosion between them. In fact, you would expect signs of erosion to be quite frequent in the world's strata. However, once again, you almost never find this anywhere. We have here, then, yet another testament to a global flood catastrophe, where countless layers of sediment were flatly deposited, one upon another, in rapid succession.

Here's another problem with the notion of sedimentary rock formations having required multiplied millions of years to accumulate: In the bulk of the world's sedimentary deposits, the mineralogical composition is the same throughout each individual formation, in layer after layer, through scores of feet in depth and often hundreds of miles in length. If these deposits were formed gradually, over millions of years, how could the composition of the deposited silt or sand within them have remained consistent over such a long duration of time, and across such a large area of space? Compositional consistency on this kind of scale, seen throughout most of the world's strata, only makes sense in the context of a global catastrophe, where enormous waves gathered up colossal amounts of earth from location A and deposited it all at location B, repeating this pattern universally.

In Nova Scotia, and in several other places around the world, huge petrified tree trunks have been found embedded in cliffs of sedimentary rock, known as polystrate fossils (i.e. fossils that cross through multiple layers of the sedimentary formations that contain them). What is so unusual about these petrified trees is that they are positioned vertically, and stand, on average, about 60 feet tall. Why is this so unusual? The problem lies, again, with the geologic column, which imposes the idea that it took millions of years for sedimentary deposits to form. Clearly, such could not have been the case with formations that have 60-foot-tall petrified tree trunks standing straight up in them. These tree trunks could not possibly have stood around for millions of years, waiting to be buried by slowly-accumulating sand or silt. They were instead buried rapidly, during a catastrophe.

Some polystrate tree trunk fossils have even been observed in an inverted position in sedimentary formations, while others have been found passing vertically straight through intermittent seams of coal. For example, near Cookville, Tennessee, in the Kettles coal mines, there is a petrified tree that begins in one coal seam, protrudes upward through numerous layers of shale, and finally penetrates into another layer of coal. This is highly unusual, since, according to mainstream science, coal seams require a very long time to form (anywhere from thousands to millions of years, depending on the thickness of the seam). But at the time of the great flood catastrophe, coal formed in a very short time, as these coal-penetrating polystrate tree fossils clearly indicate. What happened is that tremendous masses of vegetation were swept up by turbulent wave actions and dumped in great heaps, later to be pressurized and "coalified" by the heat and pressure of further overlaid masses of rock and silt(34)[Endnote 23].

Petrified tree trunks standing vertically in sedimentary formations are not the only examples of polystrate fossils. Back in 1976, a mining company in Lompoc, California, discovered an 80-foot-long fossilized whale buried in a thick deposit of diatomite. Diatomite is a fossilferous material made up of tiny, microscopic sea creatures, similar to chalk. Since this diatomite deposit was so thick, it was believed that it took many millions of years for this material to accumulate on the sea floor. But the whale fossil that was discovered in this deposit was found in a vertical position, with its bones fully articulated (in the same position that they assumed when the whale was alive). Thus there is no way for this material encasing it to have taken millions of years to accumulate. Without a doubt, this diatomite deposit must have been laid down rapidly, under catastrophic conditions, and diatoms must have been much more abundant at the time of the flood (probably due to the utopian conditions that prevailed in the antediluvian world, as was discussed earlier)(43).

If the world's fossil formations were laid down rapidly, during a catastrophic global flood, we would expect to see, in at least some formations, the remains of organic materials arranged in a chaotic, jumbled, and broken fashion, reflecting areas where the actions of waves were more turbulent. And indeed, this is just what we find in the fossil record. In fact, there are, around the world, probably just as many jumbled arrangements as there are ones that were formed under less violent conditions.

But there is something else we would expect to find in the fossil record if there truly was a global flood: mass "graveyards" that contain both aquatic and non-aquatic remains mixed together in the same deposit. Do we find this in the fossil record as well? Indeed we do, and all over the world, as the following quote from catastrophist Harold Coffin reveals: "Beds containing both marine and terrestrial fossils mixed together are not rare....A flood model would expect changing source areas for sediments, especially during the waning stages of flooding. Currents would carry sediments, and retreating water would allow runoff from terrestrial sources. Tidal action could repeat the sequence many times....Marine fossils often accompany coal. They may be fish or sea creatures mixed directly with Carboniferous material or as marine fossils in thin shale partings (layers) in the midst of the coal seams. Frequently marine organisms lie in the shales or sandstones directly above or below the coal"(34).

Yet another sign that we would expect to see, if there truly was a global flood, would be similar directional orientations of fossil remains in strata, resulting from the flow of water currents that carried and deposited them thusly. As Coffin wrote: "If a flood of global extent actually occurred, we should be able to see evidences of water currents orienting the fossils. Such orientation can serve as a tool to determine whether plant and animal fossils are in position of growth, had a natural death and burial, or resulted from sorting, positioning, and burial by water. Petrified trees especially give opportunity for compass measurements and often show a prevailing direction or position. In one location 58 specimens of petrified trees (Cordaites) in the coal measures of Nova Scotia have a preferred orientation. *Stigmaria* (creeping stems or roots) from the same general region also indicate certain specific directions. Studies of the petrified trees in the...Chinle formation of northeastern Arizona and southeastern Utah have shown that seven out of eight sites (involving a total of 739 trees) have clear orientation. Plots of the directions of the petrified logs resemble those made for trees ripped out by the Birch Creek flood in Montana in 1964. The 1964 flood uprooted thousands of trees and spread them over the prairies east of the Rocky Mountains. Examples of orientation resulting from the activity of water are not difficult to find in the paleontological literature"(Ibid.).

The world is literally teeming with fossil formations that contain the remains of well-preserved, delicate plant and animal life. The fact that such remains were preserved at all, let alone in exquisite detail, bears strong testimony to the reality that these organic materials, of necessity, were buried suddenly, while the animals/plants were still alive, or else very quickly after they had died. They must have been completely enshrouded in a thick blanket of mud, deep within the Earth, before bacteria, predators, and weathering conditions had a chance to break them down.

We will now take a look, in the next several paragraphs, at some noteworthy fossil finds of this nature.

In the Western United States there is an enormous fossil-bearing deposit known as the Green River Formation, which runs for hundreds of miles, stretching across parts of Wyoming, Colorado, and Utah. In some places it is more than 60 feet thick. This formation is most famous for the countless millions of fish fossils that can be found all throughout its layers. To preserve fish in the perfect condition in which these specimens are found, with their bones fully articulated, there must have been a tremendous disaster that buried them suddenly, sealing them off from all factors that would have otherwise caused them to rapidly decay.

Not only that, but the fact that these fish are all flattened to the thickness of a sheet of paper, sandwiched between layers of silt that are themselves of the same thickness, demands a rapid burial process that involved an enormous amount of pressure. As Walt Brown put it: "Many fossilized fish are flattened between extremely thin sedimentary layers. This requires squeezing the fish to the thickness of a sheet of paper without damaging the thin sedimentary layers immediately above and below. How could this happen?"(38). These fish had to be crushed flat while they were still in a soft, malleable state, just like flowers pressed between the pages of a book. Fossils like these could not have formed gradually, through intermittent periods of silt deposition, as geologic column advocates claim. For it is a known fact that the soft bones of fish (cartilage) begin to disarticulate within a few hours after death. If these fish died in a continuous series of droughts, they would not have been preserved at all, let alone in such pristine condition, and in such great numbers. Where on Earth could we go today and find a dried-up riverbed with dead fish lying around, holding decomposition, weathering, and predators at bay while they wait to be buried by sand or silt? It just doesn't happen this way! A gradual deposition rate for fossil-bearing strata such as these is simply not an option [Endnote 24].

Not all fish fossils are preserved in a flattened, two-dimensional fashion, like the ones discussed above. In Brazil there are found, in large numbers, three-dimensional fish fossils, encased within mudstone concretions, which are so well preserved that you can still see their silvery skin and scales. In some cases, you can even tell, by sawing them in half, what they had for their last meal. Obviously such fossils were not formed under conditions of extreme pressure that most other fossils were formed under, as they are not flattened. But they were still clearly formed during a catastrophe, since they were buried rapidly enough to seal them off from all factors of decomposition.

In addition to delicate fish being perfectly preserved in the fossil record, we find across the globe many other extremely perishable types of organic materials, preserved down to the finest detail, such as insects, jellyfish, worms, coprolites (fossilized excrement), and leaves, often still retaining the green coloring from the chlorophyll. So pristine are the conditions in which such fossils are often found, in fact, that even some mainstream scientists have been forced to acknowledge that they must have been formed during very unusual and sudden circumstances. One such scientist, Wilfrid Francis, speaking about the renowned Geiseltal lignite fossil deposits in Germany, wrote: "Here...is a complete mixture of plants, insects, and animals from all the climatic zones of the earth that are capable of supporting life. In some cases leaves have been deposited and preserved in fresh condition, the chlorophyll being still green, so that the 'green layer' is used as a marker during excavations. Among the insects present are beautifully colored tropical beetles, with soft parts of the body, including the contents of the intestines, preserved intact. Normally such materials decay or change in color within a few hours of death, so that preservation by inclusion in an aseptic medium must have been sudden and complete"(41).

Fish fossils found in Indiana provide us with a further clear indication of a rapid sedimentary deposition rate, as the following quote from Harold Coffin will demonstrate: "In the early part of the 1960s scientists of the Chicago Natural History Museum did extensive digging and collecting in the black shales of Indiana. The shale beds, which contained coal seams, have horizons with abundant animal fossils--including many fish and sharks....Some of the fish remains from Indiana revealed little blemishes in their scaly covering. Careful examination showed what looked like little explosions that had caused the scales on the outside of the body to bend outward into the overlying shale. Something had forced its way through the body wall into the surrounding shale. Their conclusion, a reasonable one, was that gases, formed within the decomposing body, burst out after mud had covered the body. Where the gas penetrated through the body wall, it pushed the scales of the fish out into the sediment"(34). Clearly, in order for the bulging effect of the release of these gases to have been preserved, we must indeed be dealing with a rapid burial and fossilization process for these aquatic creatures.

Frequently found in the world's fossil record are the remains of fish and other creatures that, because of the unusual contorted positions they are in, exhibit clear signs of having died under very traumatic circumstances. For instance, Hugh Miller, a geologist, commenting on various fossil formations throughout the British Isles, stated that "at this period in our history [when these fossils were being formed], some terrible catastrophe [of]...sudden destruction [involved the death of fish over] an area at least a hundred miles from boundary to boundary, perhaps much more. The same platform in Orkney as at Cromarty is strewed thick with remains, which exhibit unequivocally the marks of violent death. The figures are contorted, contracted, curved; the tail in many instances bent around the head; the spines stick out; the fins are spread to the full as in fish that die in convulsions"(42). Can't you see that, no matter where we look on Earth, fossils tell us the story of a sudden, grand-scale catastrophe?

Bivalve shell fossils, distributed in great abundance worldwide, also show signs that the creatures that once inhabited them had died under traumatic conditions. They are almost always found complete, with both halves tightly closed together, looking as if the critter is still alive inside. This tells us that these shells must have been buried soon after death, or that they were buried alive, without time for the delicate muscles, which held both halves of these shells together, to relax or begin to decompose, resulting in the two halves opening, separating, and scattering(34). Actually, most bivalves usually spring open at the time of death, out of reflex. So why, then, is it the rule, and not the exception, to find bivalve fossils tightly closed? It is truly amazing how so many people blindly accept the geologic column and its premise of a gradual deposition rate for sedimentary formations, when the evidence is so hopelessly stacked against it all.

As you may know, oil, like coal, is a fossil, formed from enormous amounts of organic materials that have been chemically altered. Oil deposits, like coal formations, are believed to have taken millions of years to form, as plant and animal parts accumulated in an oxygen-free environment. The problem is, how do you get masses of organic materials to accumulate over very long time-periods, and remain oxygen-free? Once again we find that the mechanism of a global flood catastrophe, instantly burying masses of heaped-up organic materials deep inside the Earth, seems a much more likely explanation. And this option becomes all the more likely when one considers the fact that oil deposits are contained in underground pockets that are under tremendous amounts of pressure. If these deposits were millions of years old, how could such pressure have been maintained for so long?

Bacteria have been discovered in the bellies of bees preserved in amber, which were collected from strata that are claimed to be 25 to 40 million years in age. But, upon being cultured, the bacteria were found to grow, which means that they were still alive. Do you suppose they could have actually survived for 25 to 40 million years?(38)

As you may know, DNA is an extremely delicate cellular component that could not possibly survive for millions of years. This has been admitted by people like Brian Sykes of *Nature* magazine, who stated that the rate in which DNA breaks down in the laboratory demands that it couldn't possibly last more than 10,000 years, under the best of conditions(133). Yet, intact DNA has been found in magnolia leaf fossils, as well as fossilized oak, cypress, and tulip trees that are said to have lived 20 million years ago. Also, DNA sequences have been extracted from termites and stingless bees embedded in pieces of amber that are alleged to be 30 million years of age(134).

Several fascinating dinosaur fossil finds have been made in which the bones thereof were not petrified, or, in some cases, were only partially petrified. One find of this sort was uncovered in New Mexico--bones from Seismosaurus, the largest dinosaur ever found to date. Also, a rather large array of unpetrified and partially-petrified dinosaur bones have been collected in northern Alaska, on the banks of the Colville River(136). Now how do you suppose that bones which supposedly date back millions of years in the past could remain in their original state, or even partially so, without disintegrating?

Even stranger than these findings, the March 24, 2005 edition of *New Scientist* reported that, in South Dakota, "Paleontologists have extracted soft, flexible structures that appear to be blood vessels from the bone of a Tyrannosaurus rex that died 68 million years ago [or so they claim]. They also have found small red microstructures that resemble red blood cells." Similar finds were also made in Montana(135). There has also been found a protein called osteocalcin in some dinosaur bones. Does it seem possible to you that these delicate, highly-perishable biological substances could have been preserved for scores, or even hundreds of millions of years?

It is thought by most scientists that the process of fossilization, particularly petrification, generally requires a great length of time to occur. But this is not true. When the conditions are right, petrification can happen in a very short time. Petrification is simply the process of minerals replacing organic materials, as they break down and disintegrate. This process can be accelerated when there is a lot of water, especially pressurized water, saturating the mud encasing the organic materials. The cells in these organic materials act like sponges, absorbing the pressurized water which contains large amounts of dissolved minerals. Then, as the organic materials begin to break down, the dissolved minerals, at the same time, crystallize and replace them, with much of the original structure of the organic materials being retained, often down to the cellular level. During the flood catastrophe, of course, all of these conditions existed. These conditions, in fact, are what enabled the encasing sediments of these petrified organic materials to harden into stone themselves, much like the way cement is made. These petrification and cementation processes were further assisted and accelerated by the tremendous heat and pressure that these sediments and organic materials were subjected to, being buried under layer after layer of debris by the raging flood waters above. The more material that was deposited, the greater the heat and pressure became in the underlying layers, baking the sediments and their contents like clay in a potter's oven.

Looking at a recent example of what is called "rapid petrification" should suffice to demonstrate just how quickly organic materials can become mineralized. With this example, all that was involved was simple saturation with heavily-mineralized water, without the extremes of heat and pressure. So we can just imagine how much faster this process would have occurred during the flood, when these factors were both prevalent and universal.

The example being referred to here was discovered around 1980 in a dry creek bed near the town of Iraan, in western Texas, by Jerry Stone, an employee of the Corvette Oil Company. What he found was a rubber-soled boot with a petrified human foot and lower leg inside.

This boot was hand-crafted by the M. L. Leddy boot company of San Angelo, Texas, which began manufacturing boots in 1936. Gayland Leddy, the founder's nephew, grew up in the boot business and now manages Boot Town in Garland, Texas. He recognized the "number 10 stitch pattern" used by his uncle's company where he worked for several years. Mr. Leddy said he believes that the boot was made sometime in the early 1950's.

Only the human remains were petrified, and not the boot itself. This demonstrates that some substances quite plainly do not lend themselves to petrification so easily as others.

The bones of the foot and partial leg are replaced by limestone, having been impregnated with dissolved lime by the waters that once flowed in the dry creek where the boot was found. The internal marrow structure can clearly be seen at the point where the bones had broken off from the rest of the leg. Because of the limestone composition of this specimen, it has been dubbed the "Limestone Cowboy Boot."

Looking back over all that we have discussed in this subsection, it ought to be clear by now that the arrangement of strata in the world's sedimentary formations, along with the fossils contained therein, do not lend support to the popular notion that great passages of time were required for their deposition and petrification. It also ought to be clear that the world's strata, and the fossils encased therein, were deposited during a catastrophic global flood. Nevertheless, it must be acknowledged that fossils, in spite of the catastrophic conditions under which they were formed, are generally found in specific associations in formations around the planet, according to their species--they are commonly congregated together in recognizable patterned groupings, and these groupings tend to follow a somewhat predictable series of affiliations from one set of strata to the next.

Most scientists interpret this phenomenon to be indicative of evolutionary development from simpler life-forms in the lower layers of strata to more complex forms in the upper layers, which occurred, as they insist, over very long passages of time. However, as we have abundantly demonstrated, the world's sedimentary formations, along with the fossils they bear, do not represent great passages of time, but instead collectively denote a rapid deposition rate.

So how do we account for the grouped arrangements in the fossil record described above, in the context of a highly-destructive and chaotic flood catastrophe?

Part of the reason for these arrangements can obviously be attributed to certain habitat-specific integrations, where groupings of fossil remains are simply a reflection of how the plants and animals they represent had lived and died in the same habitat, and were thus all buried proximally together in the accumulating flood deposits.

Another reason for patterned arrangements in the fossil record could be the mobility capacities of the entombed life-forms. That is, creatures like mammals would be more likely found in upper layers of strata, since they possess a greater ability to have effectively struggled against rising flood waters and thus reach higher ground. Conversely, smaller creatures, especially aquatic ones, would be expected to appear in the lower layers of strata, since they are much more limited in mobility, and therefore could not have so easily escaped to higher elevations.

But not all patterned groupings of fossils in the world's sedimentary formations can be ascribed to habitat-specific integrations or mobility capacities. There was another mechanism at work responsible for many patterned arrangements, which can indeed be explained--and quite easily so--in the context of the flood catastrophe.

According to the extensive research of Dr. Walt Brown, a large percentage of the patterned arrangements in the fossil record appear to be the result of the densities/buoyancies of the original organic materials that are now preserved as fossils in the world's strata. Put more simply, different plants and animals, during the time of the flood catastrophe, would have settled at different depths in the mud deposits (which later hardened into sedimentary rock), depending on the individual densities/buoyancies of their bones, shells, trunks, stems, etc.

Understand that the "mud" we are talking about here acted more like a soupy, quicksand-like liquid than mere wet sand or silt--a phenomenon known as *liquefaction*. The tremendous pressure created by the titanic wave actions of the flood catastrophe literally impregnated sand and silt with water. This, coupled with the accumulating weight of layer after layer of mud being deposited by these waves, caused the organic materials trapped within these pressurized soupy mud layers to settle at different depths, floating upward or sinking downward, again, according to their individual densities/buoyancies.

Laboratory experiments conducted by Walt Brown and others have confirmed that various types of organic materials, embedded within liquefied mud, do indeed situate themselves in patterns that closely resemble those found in the fossil record. These same experiments, incidentally, have also shown that the liquefaction factor had played an additional role in stratifying sediments, apart from wave actions that deposited sand and silt in a layered fashion. What was observed in these experiments is that, as more and more liquefied mud was laid down, the overlying pressure would force the water within the mud below to rise higher and higher toward the surface, creating, in the process, a layering effect in the mud(38).

* * * * * * *

It is hoped that the information provided in these last two subsections has sufficed to convince the reader that the problems that exist with modern dating methods and their related chronologies are significant enough to warrant a rejection of their validity and reliability. With this task accomplished, we are now better prepared to take a closer look at the great flood catastrophe, and the asteroid impact that spawned it--events that nearly wiped out all life on Earth in the relatively recent past; that left telltale signs all over the planet (and indeed, all over the solar system); and that firmly etched ghastly memories of it all in the collective mind of mankind, bequeathed to us in the form of myths and legends.

Further legendary and physical evidence of an ancient, widespread catastrophe

We have seen, from both legendary and physical evidence, that the pre-flood Earth was a virtual paradise, where many life-forms grew larger and appear to have lived longer, seemingly because of the existence of an aquatic canopy in the upper atmosphere that caused a higher air pressure and a greater concentration of oxygen. This canopy apparently also filtered out harmful cosmic and solar radiation and contributed, in conjunction with an untilted Earth axis and a closer Earth-Sun proximity, to a year-round globally-warm climate.

In addition to this, we also talked about the possibility of rainfall being unknown before the great flood catastrophe--that an entirely different mechanism existed for the watering of plant life. From ancient records we saw how a great subterranean global water chamber existed that watered the Earth's flora from beneath, by a "mist" that came up "from the earth."

Next, we saw how this utopian environment was suddenly and terribly disrupted by the gravitational and/or impacting influence of an asteroid (or asteroids), which originated from a destroyed planet that once orbited between Mars and Jupiter.

Furthermore, we also discussed the fact that debris from this same destroyed planet had reeked havoc all throughout the entire solar system (the Great Bombardment period).

And finally, we examined modern dating methods, showing how there is absolutely no basis for dating fossils and fossil formations to multiplied millions of years in the past. In fact, to the contrary, we have discussed sound evidence that such formations and their accompanying fossils had formed rapidly, and in the relatively recent past, during the flood catastrophe (brought on, of course, by impacting and/or close-passing debris from the exploded trans-Martian planet) which has been placed somewhere between 5,000 and 11,500 years ago.

Having laid such a foundation, let us now turn our attention back to ancient records, to see what further details they can provide in helping us better understand the events that played out when this disaster struck the Earth.

Looking again to the ancient Hebrew scriptures, we are told that when the flood disaster broke out, "...all the fountains of the great deep were broken up, and the windows of heaven were opened, and rain fell on the earth for forty days and forty nights." Genesis 7:11. [Endnote 25]

Please notice how we are given two sources for the flood waters--the "fountains of the great deep" breaking up, and the opening of the "windows of heaven." Could the opening of the "windows of heaven" be a reference to the aquatic canopy being punctured from the incoming asteroid, causing this water (or condensed vapor, or melted ice crystals) to fall to Earth as rain? [Endnote 26] And could it be that the mention of the "fountains of the great deep" breaking up is a reference to the global subterranean water chamber being punctured by the asteroid as well, causing its pressurized watery contents to burst through the Earth's overlying upper crust layer, thrusting skyward and then falling back as rain, joining the falling waters of the collapsed aquatic canopy?

In such a scenario, the bulk of the flood waters would have actually originated from underground. On this matter, an ancient Chinese legend, which was conveyed to early Jesuit missionaries to China, makes this statement: "The Earth was shaken to its foundations. The sky sank lower toward the north [the collapsing aquatic canopy?]. The sun, moon, and stars changed their motions. The Earth fell to pieces ["fountains of the great deep" were broken up?] and the waters in its bosom rushed upwards with violence and overflowed the Earth..."(38).

An ancient text from India, known as the *Mahabharata*--a poem of vast length and complexity, which achieved its present form in the second century A.D., but dates back to a much earlier time--makes this statement: "Then...the earth split apart and sixty million people in great cities were drowned in one terrible night"(5).

Strabo, a Greek writer, tells us that Typhon (another name for the destroyed trans-Martian planet, or a piece thereof) descended underground, making fountains break forth(12).

The Hopi Indians of the American southwest have a legend which states that the angry serpent god "turned the world upside down [axial tilt?], and water spouted up through the *kivas* [sunken sacred dwellings--"fountains of the great deep"?]....The earth was rent in great chasms, and water covered everything..."(17).

Likewise, the Pawnee Indians, referring to this same time, speak of how "Rain began to fall, and water began to bubble up from deep in the Earth"(129).

The Brule Indians, members of the Lakota Nation, talk of how "with a great quake the Earth split open, sending great torrents of water surging across the entire world"(Ibid.).

Finally, the Mayan *Troano Manuscript* appears to tell the same story. An attempt to decipher this document yielded this translation: "The country of the hills of earth...[was] sacrificed. Twice upheaved, it disappeared during the night, having been constantly shaken. The land rose and sank several times in various places, and at last the surface gave way [as the underground waters spewed forth?] and the ten regions were torn asunder and scattered, the millions of inhabitants sank also [seemingly in a flood]"(5).

All of this makes total scientific sense. For if we accept the notion of an ancient global flood, the laws of physics absolutely demand that the bulk of its waters would have had to come from beneath the ground, since the atmosphere could not contain enough water to deeply inundate the entire planet (of course, already-existing bodies of water on the Earth's surface had also made their contributions to the sloshing flood waters, but their combined volume alone would surely not have been significant enough to account for a global-scale flood either).

As the waters shot up from underneath the Earth's upper crust layer, which by this time was fractured into pieces, these pieces (later to be known as the continents) began sinking beneath the waters, eventually meeting up with the lower crust layer. This resulted in all dry land on Earth being completely inundated by massive amounts of water. But if the entire planet was truly engulfed in a grand-scale flood, this raises several important questions that now need to be addressed.

The first and most obvious question is, If there was a global flood in the past, what happened to all the water? The answer to this is quite apparent: It now resides in the oceans of the world. But then the next logical question follows: How could the whole world have been inundated with the ocean waters, when the continents, along with the tall mountains on top of them, are currently miles above sea level? The answer to this question is also apparent: These high elevations, as mentioned earlier, were not always above sea level.

But now another question emerges: How is it that the continents and their mountains came to be above sea level? The answer to this question is apparent as well. But before presenting it, an elaboration on some important background information is in order.

Scientists today talk about how the continents look as though they had at one time fit together like pieces of a jigsaw puzzle, having once been part of a single super-continent known as Pangaea. Interestingly, the Genesis record, in the heart of the creation story, appears to lend support to this idea of the existence of a single continental land-mass, along with a singular oceanic body of water (of unknown size and depth), in the pre-flood world: "And God said, Let the waters under the heaven be gathered together unto one place, and let the dry land appear: and it was so." Genesis 1:9.

Most mainstream scientists today subscribe to a theory known as continental drift, which holds that the original continent, Pangaea, broke up into several pieces that moved across the world's ocean floors, which we now call the seven continents. Pretty much all catastrophists embrace this same basic hypothesis.

Not so surprisingly, the Genesis record seems to make reference to this spreading apart of the single Pangaea continent: "And unto Eber were born two sons: the name of one *was* Peleg; for in his days was the earth divided..." Genesis 10:25. The Hopi Indians appear to allude to this ancient, Earth-splitting calamity as well. As mentioned earlier, they claim that, at the time of the flood, or sometime thereafter, "...the Earth was rent in great chasms"(17).

Continental drift advocates point out that the mid-oceanic ridges appear to be the point of contact from whence the continents once separated. This seems rather logical, as the contours of the continents (and, better yet, the contours of the continental shelves) match up very closely with those of the mid-oceanic ridges [Endnote 27].

After separating, it is believed that the continents drifted across the ocean floors, eventually coming to rest where they are today. Some believe that this process occurred rapidly, in just a few hundred years or so, while others believe it took a very long time, and still continues today at a very slow pace. But is this actually how the seven continents came about? Did they really journey across the world's ocean floors to get where they currently are? While there is no question that the continents were once connected where the mid-oceanic ridges now lie, the question begs an answer: How could they have moved to their current positions, thousands of miles away?

There is another theory of continental origins that fits with the available evidence far better than continental drift--the expanded Earth theory. It agrees that there was indeed a single continent at one time, and that it did become segmented (as a result of the flood-producing asteroid impact). However, it does not agree with the notion that the continents moved across static ocean floors. Instead, it asserts that the entire planet expanded by about 40% its current size, much like a balloon inflating, causing its outer, less-flexible, granite-based crust layer (Pangaea) to spread apart at its weakest points (the cracks created by the asteroid impact). As the Earth expanded, its lower, more dense and flexible basaltic crust layer (what we now call the ocean floors) stretched out, remaining essentially intact but thinning out as the expansion progressed [Endnote 28].

The balloon analogy will serve us quite well, to illustrate further how this worked. Suppose we were to take a blue balloon, blow it up a little more than half-way, and then paint it brown. Next, we wait for the paint to dry, and then continue blowing the balloon up the rest of the way. As we do so, we will see that the brown paint will start to crack and separate into large patches. The more we blow up the balloon, the more blue color we will see, and the farther apart the brown patches will separate. And so it was when the Earth expanded--the blue color of the balloon represents the ocean floors stretching out from underneath the overlying continents, which are represented by the brown paint separating into patches that spread apart in the process of expansion.

There is actually an ancient document that makes reference to the Earth having expanded in size in the remote past. In the *Zend-Avesta*, an archaic Iranian (Persian) saga, we read: "Then Yima ["the first king of men"] stepped forward, in light, southwards, on the way of the sun, and [afterwards] he pressed the earth with the golden seal, and bored it with the poniard, speaking thus: 'O Spenta Armaiti, kindly open asunder and stretch thyself afar....And Yima made the earth grow larger by one-third than it was before..."(49)[Endnote 29].

If we could shrink the Earth down by about 40% (the above quote was off a bit on this) of its current diameter, the continents would fit together almost perfectly--far better, as it turns out, than if we could move the continents across the ocean floors and attempt to fit the puzzle back together. This is one of the many advantages that the expanded Earth theory has over continental drift.

Another advantage, of course, is that we don't have to explain how the continents could have moved across thousands of miles of ocean floors. With expansion theory there was no movement of the continents at all--they simply stayed pretty much in place (although they were slightly stretched out) as the Earth expanded beneath them, widening the distances between them in the process [Endnote 30].

A further advantage of the expansion theory is that it doesn't have to explain (like continental drift theory does, but never has been able to) how the continents could have maintained any discernable traces of their original shapes as they moved so far across the ocean floors. With all the resistance that they would have encountered in such a journey, their contours would have become massively distorted so as to make it difficult, if not impossible, to recognize that they had once fit together.

Still another advantage of the expansion model is that it explains the almost total lack of silt and sedimentary deposits on the ocean floors. How so? As stated earlier, nearly all sedimentary deposits were laid down during the flood. This was at a time before the Earth had expanded, which brought about the separation and spreading apart of the continents. Hence, up until the time of the flood and its aftermath when the Earth expanded, there were no sea floors yet in existence upon which silts and sediments could be deposited. And by the time the sea floors were created (or exposed) by the expanding Earth, the mechanism of massive, flood-induced sedimentary deposition had ceased.

The expansion theory also explains why we see virtually no craters on the ocean floors, but instead find them almost exclusively on the continental land masses. Once again, this is because the lower crust level (oceanic floors) was not yet exposed when the bulk of these impacts were occurring, at the time of the flood catastrophe.

So now it ought to be clear why the continents are no longer below sea level. As the Earth expanded, spreading the continents (broken pieces of the former upper crust layer) further and further apart, the ocean basins were created, into which the flood waters had drained, thereby forming today's oceans. The more the Earth expanded, the further the continents were separated, and the lower the sea level had dropped [Endnote 31].

Interestingly, there is actually a passage in the ancient Hebrew scriptures that bears witness to this very thing having happened (along with so many other things we've been looking at). This fascinating account was written by David the Psalmist, who, referring to the flood and the events that followed it, wrote: "Thou [God] coveredst it [the Earth] with the deep [the flood] as *with* a garment: the waters stood above the mountains. At thy rebuke they [the waters] fled; at the voice of thy thunder they hasted away. They go up by the mountains; they go down by the valleys unto the place which thou hast founded for them [ocean basins and trenches?]. Thou hast set a bound [the continental land masses?] that they [the flood waters] may not pass over; that they turn not again to cover the earth." Psalm 104:6-9 [Endnote 32].

Though the expanded Earth theory has its advantages, it nevertheless presents a problem of its own, or at least what appears to be one at first glance--the problem of how the Earth could have expanded in the first place. But the solution to this apparent dilemma is really quite simple, if we consider the possibility of there having been a massive increase in the Earth's internal temperature. After all, heat does, of course, cause things to expand. And we know that there was indeed a tremendous amount of heat generated within the Earth during and after the flood-producing asteroid impact. We can say this with confidence because of the massive amount of volcanic and tectonic activity that accompanied this disaster (which we have already touched upon, and will be addressing in more detail later on). In fact, it was so hot at this time, as we are told in ancient legends (which we shall shortly examine), that the oceans had actually boiled, obviously due to heat emanating from below.

In order for the Earth to have expanded in the manner that this theory proposes, it should be plain to the reader that no additional material was required to be added to the Earth. All that was needed is for the existing material in its interior to decrease in density, which is precisely what we would get from a massive increase in thermal energy.

But what precise type of mechanism (or mechanisms), you may rightly be asking, could have raised the Earth's internal temperature high enough to cause the entire planet to expand, and to such a large degree? Consider the fact that the Earth, as we have seen, was knocked off its axis at the time of the asteroid impact--a factor that alone would have powerfully offset the centrifugal force of our planet's rotation, resulting in an enormous increase in its internal heat and pressure. As the ancient Hebrew book of Isaiah put it, the Earth "moved exceedingly" and reeled "to and fro like a drunkard." Isaiah 24:19, 20.

Yet there were other factors at work besides an Earth-axis tilt.

Another one may have been a sudden alteration in the Earth's rotation rate that came about, again, from the flood-producing asteroid impact [Endnote 33]. Like the Earth being knocked off its axis, this factor alone could likewise have single-handedly caused the Earth's internal heat and pressure to dramatically increase.

There was also the obvious jolting effect that would have resulted if the Earth did indeed suffer a sudden orbital dislocation, being pushed slightly further out from the Sun (which we discussed earlier)[Endnote 34]. This, too, all on its own, would have substantially raised the heat and pressure of the Earth's interior, as it would have altered the Earth-Sun gravitational relationship, tugging with ferocity at it's inner liquid mantle and core.

For that matter, all three of the above-mentioned factors--the Earth's axial, orbital, and rotational speed readjustments--would have affected the Earth-Moon gravitational relationship, which would have also definitely created powerful tidal forces that acted upon the Earth's liquid rock interior, perhaps even more significantly than what had been the case from the disruption of Earth's solar gravitational relationship.

It's been suggested by some researchers that the Earth may contain a plasma core. This being the case, the expansion of the Earth would be even more easy to explain, since plasma, in a super-heated state, significantly expands. And as it turns out, the seismic wave readings that scientists have used to determine that the Earth's core is made primarily of iron may not indicate this at all. For, according to British scientist Hugh Owen, it happens that "the behaviour of waves passing through a plasma core would be similar to that in a solid iron-sulphur core"(183).

If the Earth does have a plasma core, it's feasible that it could have been significantly heated by the close passage of a large piece of the destroyed trans-Martian planet, sometime after the flood. Powerful electrical plasmic discharges may have occurred between the two bodies, boiling the Earth's plasmic core in the process, resulting in the expansion of our planet's size [Endnote 35]. There is actually much legendary evidence which strongly hints that ancient man had perhaps witnessed such powerful discharges in the heavens. These spectacles would have been prominent features in the ancient skies, not only because there was much more debris venturing close to the Earth at that time (from the exploded trans-Martian planet), but because the Earth's plasmasphere and plasma core (if there is one) were much stronger. We know this because, as we saw earlier, the Earth's magnetic field, which largely results from the Earth's electrical properties, was far stronger in the distant past. (More information on ancient eyewitness records of plasmic discharges in the heavens can be found in Appendix D.)

Given all of these ancient heat-inducing factors in the Earth's deep interior, it's difficult to imagine how the Earth could NOT have expanded in its past. But it must be borne in mind that this expansion surely could not have happened too quickly. If it did, the Earth would probably have exploded into pieces, which may very well have been how the trans-Martian planet was destroyed (more on this later). The Earth, instead, probably "reeled to and fro like a drunkard" over many years--perhaps several centuries--as it tried to settle itself from the initial impact that knocked it out of kilter. During this time, its heat would have built up more and more, perhaps egged on further by additional asteroidal impacts and/or close asteroidal passages that jolted and/or tugged on the Earth gravitationally, resulting in our planet expanding larger and larger (see Appendix B for a discussion of the prospect of multiple post-flood Earth tilts spawned by numerous asteroidal encounters).

The Earth is not the only body in the solar system that exhibits signs of having once expanded. In fact, there are a great many of them that exhibit such signs. Jupiter's moon Ganymede is one. It shows raised brown patches ("continents," or pieces of a former upper crust layer) that look as if they once fit together, but are now separated by a lighter, lower-elevation tan-colored material. Here we see a possible close parallel with what seems to have happened on Earth. Calculations based on computer simulations that "shrink" Ganymede down (so that all the former upper crust layer pieces fit tightly back together) reveal that Ganymede was once one-third smaller than its current size. We also find that the brown patches on the surface of this moon are more heavily cratered (just like Earth's continents), while the younger, tan, former lower crust layer is very sparsely cratered (just like Earth's ocean floors).

Another moon of Jupiter to take close notice of is Europa. There's no question that the cracks on this moon resulted from crustal spreading, since computer manipulations of close-up photos of it reveal that the edges of these splits did indeed once line up perfectly. As is always the case with the expansion of celestial bodies, Europa exhibits spreads of its older, upper crust that expose a smoother, younger, former lower crust layer. Upon closer inspection, this moon also displays, in some places, a chaotic assortment of shattered icy crust fragments that obviously once fit together like a puzzle, before the expansion of this world took place.

Venus, like Earth, also appears to have the same type of continent/ocean-basin geological arrangement. As researcher David Noel wrote: "Venus, similar in size to Earth, has a clear distribution of raised 'continents' [Aphrodite Terra and Ishtar Terra] and wide, flat 'seabeds,' although of course the latter do not contain any water. And the interesting thing is that the 'seabeds' cover about 70% of the planet--the same as on Earth"(145).

Additionally, in the Aphrodite Terra region there even appears to be a Venusian equivalent of mid-oceanic ridges, from whence the "continents" tore away when this planet expanded.

Venus also has what are called "fracture belts," which look like stretch marks, similar to "fracture zones" here on Earth. And, as mentioned earlier, Venus also has huge rift valleys, like Earth's. Could these features constitute further evidence that this planet expanded in ancient times, tearing or breaching its crust in the process?

Mars possesses many signs of having once expanded as well. But we will hold off from addressing this matter until later on.

Our own Moon may also have expanded in the past. The lunar lowlands (the dark blotches spread mostly across its Earth-facing side) appear to be an exposed underlying layer of the Moon's crust that stretched out from beneath the overlying lighter-colored highlands (upper crust layer--lunar "continents," if you will)(156). Here again, just like on Earth and Ganymede, we find that the older upper crust layer (highlands) is heavily cratered, while the exposed younger, lower crust layer (lowlands) is lightly cratered--just what we would expect if the Moon expanded after, and as a result of, the Great Bombardment period.

It makes sense that the exposed lower lunar crust layer (lowlands) would be located primarily on the Earth-facing side, since the Earth would have tugged most heavily on this hemisphere of the Moon, causing its underlying lowland crust layer in that specific hemisphere to expand and bulge up through the overlying highland layer, more so there than on the other side of the Moon.

What appears to be further proof of lunar expansion in the past are the various valleys (or "canyons") that wind along the Moon's surface, looking almost like they were carved out by an extinct river. What they seem to represent are instances of crustal breaching, where the upper layer was torn apart to expose an underlying crustal layer.

Even the Sun is acknowledged by most scientists to have expanded in the past. In fact, they claim that it is still in the process of expansion. Furthermore, it is believed that the Sun will eventually enlarge to such an extent that it will engulf the Earth and Mars, before eventually dying out.

Our current discussion of expansion theory would be incomplete without mentioning the enigma of the outer planets--the gas giants--Jupiter, Saturn, Uranus, and Neptune. These planets are truly massive, yet they have no solid surface to walk on--they are essentially comprised of nothing but gases. Saturn is so low in density, in fact, that it would literally float in water. Why are these planets like this? It might be that an expanding planet, if it doesn't expand too rapidly to the point of exploding, could decrease in density so much that it becomes gaseous. And the fact that only the outer planets expanded to this extent might be due to their greater distance from the Sun. The inner, denser planets were probably unable to expand too far because of the gravitational effects from the Sun's closer proximity.

Interestingly, the exploded trans-Martian planet was located right in the midst of the transition zone between the rocky inner planets and the gaseous outer planets. Could there be some connection here? Was this planet more susceptible to exploding because of its orbital location, which made it impossible for it to remain a rocky planet or become a gaseous one? In other words, was it too far from the Sun to remain a rocky planet but too close to become a gaseous one? Did it explode because it expanded too rapidly as a result of an object from outside the solar system either impacting it or passing too closely to it, exerting powerful tidal forces upon it, and significantly heating it up internally? We'll obviously never know the answers to these questions with certainty. But it's intriguing to speculate on such matters.

Let us now come back down to Earth.

With the expansion theory, the tectonic plates that comprise the world's ocean floors are obviously the result of this lower crust layer of the Earth having been thinly stretched out and breached in certain places. Because these breaches represent weak spots in this lower crust layer, earthquakes and volcanic eruptions have more commonly occurred in these regions.

When these breaches were first created, the new lava that oozed out provided fresh crustal material that quickly hardened in the undersea environment, and thus served as "glue" to hold these breaches together. At the same time, this "glue" helped to facilitate the ongoing expansion of the Earth's crust by continuously replenishing the crust in areas where it became too thin to stretch any further. This was real "sea floor spreading" in action.

As already mentioned, the expansion of the Earth created what are called "fracture zones" (stretch marks) on the ocean floors. In fact, Earth expansion theorists argue that all crustal breaches, such as faults and rifts, both on the ocean floors and the continents, are the result of crustal stretching and tearing as the Earth expanded. Some of the best examples of this are the Mariana Trench in the Pacific (roughly seven miles below sea level), the East African Rift, and the San Andreas Fault.

The expansion theory presents us with a new viewpoint about the origin and nature of the mid-oceanic ridges. Instead of them being the borders of tectonic plates, they are seen simply as scars that were formed when the continents first split apart. But how exactly did these raised scars come about?

When the bulk of the flood waters shot up from below, through the cracks in the Earth's upper crust layer, this created weak spots through which pressurized magma was able to push its way up from under the lower layer of the Earth's crust, creating the mid-oceanic ridge system (which followed the contours of the cracks in the upper crust layer).

Assuming that all of the above-described events did indeed occur as a result of the great asteroid impact/flood catastrophe--the massive fracturing of the Earth's upper crust layer, the knocking of the Earth off its axis, the expansion of the Earth's size, etc.--would it not seem logical to conclude that all the ongoing destructive geologic processes, like earthquakes and volcanic eruptions, are simply "aftershocks" from the initial great flood catastrophe?

We could also add to this list the slight wobbling of the Earth's rotational axis (known as precession), along with the Earth's equatorial bulge, as additional "leftovers" from the tilting of the Earth's axis that occurred at the time of the flood [Endnote 36]. Indeed, there are a great many abnormalities with the Earth that appear to be aftershocks from this catastrophe in its fairly recent past. In fact, I would even venture to say that many of these abnormalities cannot be explained except in the context of this calamity. As we proceed, this point will become all the more clear.

The expansion of the Earth, of course, would also have resulted in an expansion of its atmosphere, and thus a great loss of air pressure. With the pre-expansion higher pressure, we can see how the early atmosphere could have much more easily supported an aquatic canopy. This would also explain the previously-discussed greater concentration of oxygen found inside air bubbles contained in amber samples, and thus the longer lifespans and larger growth sizes that were prevalent before the flood. Furthermore, it would explain how such large winged creatures like pterodactyls (which had wingspans as big as 40 feet across) were capable of flight, whereas today they would not be able to get off the ground because the atmosphere is far too thin (today's largest wingspans, found on albatrosses, are only about 10 feet across). Also, large dinosaurs like the Brontosaurs could not have moved around very easily in our current relatively thin atmosphere. But in a much denser atmosphere, such huge animals like this would have been provided the necessary buoyancy they would have needed to keep from falling over so easily. The same, of course, can be said about many large plant species that were known to have lived at that same time.

Decreases in the Earth's atmospheric pressure and atmospheric oxygen content may not be the only reasons for the reduction in growth size of life-forms in the post-flood world. With the diminishing of the Earth's density that resulted from the planet's expansion, there would have been a consequent decrease in the Earth's gravitational strength as well. Thus the smaller growth size of flora and fauna after the flood may have also resulted from adaptation to this new environment of gravity depletion. As the Earth expanded and gravity decreased, there was no need for growing so large in order to counteract the immense strength of the force of gravity. Large bones and muscles (or trunks and branches) were no longer needed. In fact, they became quite cumbersome. Consequently, plants and animals adapted themselves to the new circumstances by not growing so large--a process that was assisted, once again, by the change in atmospheric pressure and composition (less oxygen content)[Endnote 37].

There were other adaptations taking place in the post-flood world. A further obvious example would be winter migrations of birds. Before the flood there was no winter, and thus no need for migration. Two other examples, which arose for the same reason as bird migrations, are winter hibernation and fall defoliation.

At the same time, it would seem that some species of plants and animals, perhaps due to their genetic makeup, were able to maintain their large growth size into the post-flood world--surviving even to our day. Take the California Redwood trees, for example. Giants such as these are rare in the plant kingdom of today, but were common in the antediluvian world. In the animal kingdom, the whale stands far above the rest as the most giant of creatures inhabiting our contemporary world, having survived and maintained its large size from the Earth's pre-flood era. It also appears that certain giant races of humans continued to live after the flood catastrophe, at least for a time, but eventually died off. For there exists a wealth of archaeological and legendary evidence of such people who continued to live for centuries, and even millennia, into the post-flood world (see Appendix C).

* * * * * * *

We will proceed in this subsection to examine more details of the flood disaster, along with their sustained side-effects upon the Earth and its environment. But before we do this, let us first take a closer look at a few more flood legends from around the world. Doing so will help us to further see that this disaster was of such colossal impact that it has remained firmly and permanently etched upon the memories of cultures from every corner of the globe. We will also be able to see that, though some of the details of these legends may vary or even be tainted by mythical flavorings, the basic theme remains essentially the same everywhere we look--that a devastating ancient calamity had nearly wiped out all life on Earth, and had drastically and irreversibly altered our planet's geography--which makes it difficult, if not impossible, to write off these accounts as mere fiction.

Speaking in this regard, one catastrophist author, Hugh Miller, wrote: "There is...one special tradition which seems to be more deeply impressed and more widely spread than any of the others. The destruction of well-nigh the whole human race, in an early age of the world's history, by the great deluge, appears to have impressed the minds of the few survivors, and seems to have been handed down to their children, in consequence, with such terror-struck impressiveness that their remote descendants of the present day have not even yet forgotten it. It appears in almost every mythology, and lives in the most distant countries and among the most barbarous tribes..."(46).

Let us now turn our attention to a few fascinating excerpts and points of interest from some of the world's great flood legends:

- The ancient Hebrew account, found in the book of Genesis, states: "And the waters prevailed exceedingly upon the earth; and all the high hills, that were under the whole heaven, were covered. Fifteen cubits upward did the waters prevail; and the mountains were covered. And all flesh died that moved upon the earth, both of fowl, and of cattle, and of beast, and of every creeping thing that creepeth upon the earth, and every man: All in whose nostrils *was* the breath of life, of all that *was* in the dry *land*, died. And every living substance was destroyed which was upon the face of the ground, both man, and cattle, and the creeping things, and the fowl of the heaven; and they were destroyed from the earth..." Genesis 7:19-23.

- A Peruvian legend tells us that "the sky made war on the earth"(17).

- A Brazilian legend of the Cashinaua people states: "The lightnings flashed and the thunders roared terribly and all were afraid. Then the heaven burst and the fragments fell down and killed everything and everybody. Heaven and earth changed places. Nothing that had life was left upon the earth"(Ibid.).

- Philo, writing from Alexandria back in 30 A.D., made the following comments: "The vast ocean being raised to a height which it had never before attained, rushed with a sudden inroad upon islands and continents. The springs, rivers and cataracts, confusedly mingling their streams, contributed to elevate the waters....For every part of the earth sank beneath the water, and the entire system of the world became...mutilated, and deformed..."(24).

- The Iroquois Indians have a tradition which states that the sea had infringed upon the land, so that all human life was destroyed(47).

- The Chickasaws assert that the whole world was once completely demolished by water(Ibid.).

- The Sioux say that there was a time when there was no dry land, and all men had disappeared from existence(Ibid.).

- The Native Cahto tribe of California preserves a very old flood legend, which informs us that "The sky fell. The land was not....The waters of the ocean came together. Animals of all kinds drowned"(48).

- The Navajo tell us that "Suddenly a fierce storm of black sand and ash blew in from the east [the direction of the asteroid impact?]....Not long after that, a storm of stinging ice and blinding snow that looked like white sand came down upon them [the start of the Ice Age?]....Then, without warning, a torrent of water poured out of the Earth and began to rise, swirling all around them higher and higher [another reference to the "fountains of the great deep" breaking up?]....Before long, nearly everything in the world had drowned"(129).

- Certain other Native legends of North America state: "The ocean...is contracted, and that which lately was sea is surface of parched sand, and the mountains which the deep sea has covered start up and increase the number of scattered Cyclades [islands], the fishes sink to the bottom, and the crooked dolphins do not care to raise themselves on the surface for air as usual. The bodies of the sea-calves float lifeless on their backs on the top of the water"(17).

- Another ancient flood legend, that of the Mexican Toltecs, has this to say: "[M]en were destroyed by tremendous rains and lightning from the sky, and even all the land, without the exception of anything, and the highest mountains, were covered up and submerged in water fifteen cubits (caxtolmolatli)....[M]en came to multiply from the few who escaped from this destruction..."(47).

- The Quiche Mayan *Popol Vuh* proclaims: "The waters were agitated by the will of the Heart of Heaven (Hurakan), and a great inundation came upon the heads of these creatures....They were engulfed, and a resinous thickness descended from heaven;...the face of the earth was obscured, and a heavy darkening rain commenced--rain by day and rain by night....There was heard a great noise above their heads, as if produced by fire. Then were men seen running, pushing each other, filled with despair; they wished to climb upon their houses, and the houses, tumbling down, fell to the ground; they wished to climb upon the trees, and the trees shook them off; they wished to enter into the grottoes (caves), and the grottoes closed themselves before them....Water and fire contributed to the universal ruin at the time of the last great cataclysm..."(Ibid.).

- Mexico's Mixtec Indians recall: "In a single day all was lost, even the mountains sank into the water...subsequently there came a great deluge in which many of the sons and daughters of the gods perished"(17).

- Finally, an ancient Chinese text declares: "[T]he pillars supporting the sky crumbled and the chains from which the earth was suspended shivered to pieces. Sun, moon and stars poured down into the northwest, where the sky became low; rivers, seas and oceans rushed down to the southeast, where the earth sank. A great conflagration burst out. Flood raged. Wild beasts and terrible birds made men their prey"(Ibid.).

We shall now look in detail at some of the major physical evidences of this great flood catastrophe. As we do so, we will begin to see the above-cited legends, and others like them, less as myths and more as recollections of an actual historical event that struck a bygone generation of humanity with a most justifiable, paralyzing fear.

Mountain formation

Although there may have been mountains in the pre-flood world, they were likely nowhere near as high as those existing today. In fact, they were probably rolling hills in comparison. The reason for making this claim is that it was not until the flood disaster occurred that a grand-scale mechanism existed to create the colossal, globally-distributed mountain chains we have today. But exactly how, you may ask, could the flood have created the Earth's great mountain ranges?

Adherents to the continental drift theory claim that the mountains of the world were pushed up as the continents moved across the ocean floors. But as we have seen, the continents could not have thusly moved. It was, instead, the process of Earth expansion that was responsible for orogenesis (the process of mountain formation).

The simple stretching of the continental land masses, which resulted from the expansion of the oceanic floors beneath them, would have caused a thinning out of the continents in their mid-sections, and a buckling at their outer edges. A good example of this can be seen in the United States, where we find the Great Plains bracketed by the Rockies on one side and the Appalachians on the other.

Here's another possible explanation of orogenesis under the expansion model, which could have occurred instead of, or in addition to, the above explanation: We could imagine that, after the Earth had finished expanding and had begun to cool, it may have shrunk slightly, causing the continental land masses above to buckle, especially at their coastal boundaries. This slight shrinking of the Earth may also have made a contribution to the raising of mid-oceanic ridges to their current elevations, through horizontal compression [Endnote 38].

Whether it be the continental drift or the Earth expansion theory, the one thing that most defenders of both of these concepts have in common is the time factor. As always, mainstream (and even a good many non-mainstream) scientists love to attach enormous lengths of time to anything geological. They claim that the world's largest mountain formations came about gradually, over many millions of years. However, as we shall see, a lengthy time-frame, having begun in distant past ages, is not only unnecessary for explaining orogenesis, but downright nonsensical.

If the world's mountain formations came about as a result of the buckling effect from the Earth's expansion (and perhaps some contraction as well), we would expect each formation around the world to be oriented parallel to other formations in the same region. We would also expect that these parallel sets of mountain chains would themselves be parallel with the coastal boundaries of their respective continents. And, as you may not be surprised to discover, these are the precise arrangements that the world's mountain ranges are found in. The proceeding quote from catastrophist author Donald Wesley Patten cites several important examples of such arrangements: "[M]ountain systems often occur in parallelism--that is to say that major systems are frequently parallel. The cascades and the Rockies are one example, the Sierra Madre Occidental and

Sierra Madre Oriental are another example. The Himalayas and the Kun Luns are a third example. Other examples are too numerous to mention"(24).

Some mountain ranges appear to have come about as a result of strips of land, which protrude from continental land masses, being pushed into their parent continents when the Earth expanded and/or contracted. One good example of this is the Alps, where the protruding landmass of Italy looks as though it was pushed northwards into the European mainland, thrusting up this group of mountains in the process. The same may have occurred with India pushing up the Himalayas and, through a rippling effect, the Kun Luns as well.

When we look specifically at the mountain ranges on the coastal edges of the continents, where we would expect the buckling effect to have been most prominent, we find that many of them do indeed exhibit a buckled, or folded, pattern. Such a pattern clearly indicates that these mountains were formed when the rocks that they are made of were in a soft and malleable state (as either lava or mud), and then later hardened into solid-rock, after being folded. We can be certain of this because the folding pattern is too smooth and consistent throughout. Solid, hardened rocks are too rigid to fold without cracking and splintering, which would thus distort the folded pattern, if not obliterate it altogether. Consequently, we have evidence here for our flood catastrophe, which brought about the folding of these mountains swiftly, rather than over an extended duration of time, as "conventional wisdom" dictates.

Of course, not all mountains were formed at a time when the materials that comprise them were in a soft, malleable state. A great many mountains are composed of massive chunks of rock that were broken and thrusted up from the underlying bedrock, under conditions of tremendous crustal stresses, which, once again, only our flood disaster can account for. So obvious is it that such mountains have formed rather quickly, under tumultuous circumstances, in fact, that even some mainstream scientists have been forced to rethink their support of the standard model that insists they formed slowly, over countless eons. For example, one geologist, the late Eduard Suess, author of *The Face of the Earth*, wrote: "The earthquakes of the present day are certainly but faint reminiscences of those telluric movements to which the structure of almost every mountain range bears witness. Numerous examples of great mountain chains suggest by their structure...episodal disturbances of such indescribable and overpowering violence, that the imagination refuses to follow the understanding"(50).

Not only is it most apparent that mountain chains formed quickly, under catastrophic conditions, but it is equally clear that they originated in the relatively recent past. The certainty of this claim is secured simply by looking at the rugged, jagged shapes of most mountains around the world, especially their peaks.

If these geologic features were created many millions of years ago, erosion would have smoothed out their rough textures by now. Yet we almost never find significant amounts of erosion atop any of the world's great mountain ranges.

In light of the evidence we have looked at, would it not be safe to say that the bulk of the Earth's mountain formations must indeed have come about as a result of the "Great Bombardment" and its aftermath, which brought about so many other disruptions on our planet's surface, as well as the surfaces of many other celestial bodies in our solar system? And would it also not be safe to say that the specific mechanism responsible for orogenesis, both on Earth and apparently all other bodies in our solar family, was crustal expansion and/or contraction from internal heating and/or cooling?

"Taken in their entirety, the orogenic belts are the result of worldwide stresses that have acted on the crust as a whole. Certainly the pattern of these belts is not what one would expect from wholly independent, purely local changes in the crust." - Dr. Walter Bucher, Columbia University(17).

Volcanoes and related phenomena

It's impossible to take in the scope of the carnage of the flood disaster without considering the obvious grand-scale lava eruptions that both accompanied and followed it. Though we have, in fact, already mentioned this issue, it has not been given the attention it needs in order to help us understand the important role it played in completely reshaping the surface of the Earth.

With the tilting of the Earth's axis and the breaching of certain areas of the lower layer of the Earth's crust, which occurred during as well as after the flood disaster, we would expect to find enormous lava flows in various places around the world. And indeed, this is just what we find. Quoting again from Donald Wesley Patten: "In the intermontane plateau west of the Rockies and east of the Cascades, there was a great outpouring of lava during the flood catastrophe. In some places, the lava deposits exceed 8,000 feet in depth upon original bedrock. There are many successive layers of lava, anywhere from a few inches to hundreds of feet thick. They are often separated by minute layers of shale. This lava plateau [the Columbia Plateau] covers approximately 150,000 square miles, and covers parts of five states--almost half of Washington, nearly two thirds of Oregon, and lesser parts of Idaho, Nevada, and California.

"We contend that at this time of cataclysmic upheaval, lava flowed or bled out upon the earth's surface in several parts of the world, including the Abyssinian plateau of Africa, the Deccan plateau of India [with lava flows that are

10,000 feet thick], and also lesser lava plateaus in Arabia and Brazil"(24).

Another formation similar to these is the Canadian Shield, which is comprised of 200,000 square miles of lava flows. Now what type of mechanism do you suppose could have been responsible for such colossal-scale arrangements like these?

The above-described ancient lava flows did not necessarily result from outright volcanic eruptions. In fact, there are often no volcanoes found anywhere at all near such huge lava deposits. Flows like these appear, instead, to have oozed out of deep fissures in the Earth's lower crust layer, which then tore through the upper crust layer--actions brought on, of course, both by tidal forces and Earth expansion. And, as indicated in the last quote, these types of flows frequently contain intermittent layers of shale. What this reveals is a series of tidal waves that laid down beds of sediment in between massive layers of lava that seeped up from below.

But let us now move on to a discussion of actual volcanic eruptions.

There can be no doubt that volcanoes were being formed at the time of the flood, and were erupting violently the world over, compounding the horrors of this disaster. To assist us in comprehending the swift and widespread devastation that these geologic outbursts are capable of inflicting, and most certainly did inflict on a grand scale during the flood, an examination of some documented, relatively-recent volcanic explosions will prove most useful:

- The massive eruption of Thera (Santorini), in about 1500 or 1600 B.C., seems to have dominated the writings of the Mediterranean region during that period, as this next quote from catastrophist author Richard Mooney reveals: "The story of the Argonauts of Jason contains passages which appear to refer to some great eruption in the Mediterranean. Apollonius says: 'The next night caught them well out in the wide Cretan Sea, and they were frightened, for they had run into that sort of night that people call the pall of doom. Black chaos had descended on them from the sky, or had this darkness risen from the nethermost depths? They could not tell whether they were drifting through Hades or still on the waters.'

"A papyrus written by Ipuwer, who must have been an eyewitness, contains a description of the effects of an earthquake in the wake of the eruption: 'The towns are destroyed. Upper Egypt has become a waste. All is ruin....The residence is overturned in a minute.'

"A stele at El-Arish bears this hieroglyphic inscription: 'The land was in great affliction. Evil fell on this earth. There was a great upheaval in the Residence....Nobody could leave the palace during nine days, and during these nine days of upheaval there was such a tempest that neither man nor gods could see the faces of those beside them'"(5).

This eruption, which occurred in the Aegean Sea between Greece and Turkey, involved a 4,900-foot mountain that shook with the force of hundreds of hydrogen bombs. Debris shot miles into the air and rained 100 feet of ash and dust on nearby islands. The rest of this island sank into the sea, causing tidal waves hundreds of feet high, rushing outward at 200 miles per hour. Such waves smashed again and again into Crete and other nearby shores, killing hundreds of thousands. The Minoan civilization was wiped out.

Because of this calamity, Santorini is no longer round in shape, but is rather shaped like a crescent, having a large portion missing from the blast that has also left a gaping hole in the sea floor that is hundreds of feet deep(29).

Notice how this volcanic eruption spawned both earthquakes and tidal waves. Now imagine this process being repeated all over the Earth, all at one time. Are you beginning to comprehend the immense scope of the flood catastrophe?

- Iceland was met with tremendous convulsive forces in 1783. About 30 miles offshore, a submarine volcano erupted which released so much pumice that the sea was covered with it out to a distance of roughly 150 miles. A new island was created from this event, comprised of high cliffs. But less than a year later it sank back into the sea, leaving in its wake a reef of rocks 30 fathoms below sea level(47).

- Rising and sinking islands, brought on by erupting submarine volcanoes, are not so uncommon. In some areas, such as the Azores, this phenomenon has occurred rather frequently. Researcher Charles Berlitz brought this point out, stating that "A number of islands have appeared, disappeared, and sometimes reappeared from the restless depths of the Atlantic. In 1808 a volcano on Sao Jorge in the Azores crested several thousand additional feet and in 1811 a large volcanic island appeared in the Azores which, after being given a name--Sambrina--and charted on maps, suddenly returned to the sea"(29).

Here's another, more recent example: In 1963, a new island, Surtsey, was formed off the coast of Iceland, in just a few days time, by a volcanic eruption. It started from the ocean floor, 400 feet below sea level, and rose 600 feet above the sea. This particular island still remains to this day(51).

- On October 8, 1822, a powerful volcanic eruption befell the island of Java. A loud explosion was heard as the ground shook, and huge columns of hot water and boiling mud, mixed with burning sulfur and ashes, were shot high in the air like a water-spout. Debris was flung as far away as the Tandoi River--40 miles distant--and even further. But then, just 4 days later, a second eruption occurred, even more devastating than the first, in which hot water and mud were again expunged, and great blocks of basalt were tossed up to 7 miles from the volcano. As a result of this second eruption, over 4,000 people were killed, and 114 villages were destroyed. There was also a violent earthquake that accompanied this eruption, which breached one side of the volcano(47).

- The island of Krakatoa was destroyed by a volcanic eruption in 1883, disappearing in just 12 hours. Deposits of ash fell over thousands of square miles, and 3 massive tidal waves caused havoc in Java and Sumatra, killing thousands of people. A thick blanket of dust hovered in the atmosphere, causing many days of darkness over thousands of square miles. The sound of the explosion was heard as far away as Australia, a distance of over a thousand miles(5).

- On May 18, 1980, Mount St. Helens in the state of Washington erupted. Within hours, thousands of square miles of forest were completely obliterated--leveled flat and buried by a thick blanket of ash and dust(51).

As devastating as all the above-described events were, they had all occurred on a rather limited, localized scale. Remember that the eruptions, earthquakes, and tidal waves of the flood disaster were happening all at once, all over the Earth. These actions gave rise to the great sedimentary and igneous rock formations around the globe. But it is also most likely that a great many of the metamorphic rock formations in the world were created around this time as well, since the flood and its after-effects caused tremendous compression of the Earth's continental land masses, both horizontally (as the continents were stretched and compressed by the expanding and slight contracting of the Earth) and vertically (as billions upon billions of tons of accumulating debris were heaped up by tidal waves and volcanic eruptions). These compressions would have subjected sedimentary and igneous rock deposits to enormous heat and pressure, creating chemical changes therein (i.e. creating metamorphic rocks).

Let us conclude this discussion of volcanoes and related phenomena by citing some ancient flood legends which record some very intriguing details that parallel what we have been looking at:

- An old Persian narrative (the *Zend-Avesta*) asserts that "The sea boiled, and all the shores of the ocean boiled, and all of the middle of it boiled." The cause of this heat was attributed to the "star" Tistrya, "the leader of the stars against the planets"(147).

Could this "star" be anything but a reference to the destroyed trans-Martian planet, pieces (or "stars") of which struck the Earth and the other planets, causing, amongst other things, tremendous volcanic outbursts?

- Hesiod wrote in his *Theogony*, speaking in reference to an upheaval caused by a celestial collision: "The huge earth groaned....A great part of the huge earth was scorched by the terrible vapor and melted as tin melts when heated by man's art...or as iron, which is hardest of all things, is softened by glowing fire in mountain glens."

- In his recollections of the widespread devastation wrought upon the Earth by Phaeton (the extinct trans-Martian planet), Ovid penned these words: "The earth bursts into flame, the highest parts first [volcanoes?], and splits into deep cracks [which separated and formed the continents?], and its moisture is all dried up. The meadows are burned to white ashes; the trees are consumed, green leaves and all, and the ripe grain furnishes fuel for its own destruction....Great cities perish with their walls, and the vast conflagration reduces whole nations to ashes.

"The woods are ablaze with the mountains....Aetna is blazing boundlessly...and twin-peaked Parmassus....Nor does its chilling clime save Scythia; Caucasus burns...and the heaven-piercing Alps and cloud-covered Apennines....

"Then also Libya became a desert, for the heat dried up her moisture....The waters steam; Babylonian Euphrates burns; the Ganges, Phasis, Danube, Alpheus boil; Spercheos' banks are aflame. The golden sands of Tagus melt in the intense heat, and the swans...are scorched....The Nile fled in terror to the ends of the earth....The same mischance dries up the Thracian rivers, Hebrus and Strymon; also the rivers of the west, the Rhine, Rhone, Po and the Tiber....Great cracks yawn everywhere....Even the sea shrinks up, and what was but now a great watery expanse is a dry plain of sand [the Sahara?]. The mountains, which the deep sea had covered before, spring forth, and increased the numbers of the scattered Cyclades"(12).

- The Mayan *Popol Vuh* states that a great universal calamity took place in the distant past, at which time a certain god "rolled mountains" and "removed mountains," and that "great and small mountains moved and shaked." It also says that "mountains swelled with lava"(Ibid.).

- The Zuni Indian flood legend declares: "Earthquakes shook the world and rent it....The heights staggered and the mountains reeled, the plains boomed and crackled under the floods and fires, and the high hollow places [caves], hugged of men and the creatures, were black and awful, so that these grew crazed with panic and strove alike to escape or to hide more deeply....

"Waters washed the face of the world, cutting deep trains from the heights downward, and scattered abroad the wrecks and corpses of stricken things and beings, or burying them deeply....[I]n the trails of those fierce waters, cool rivers now run, and where monsters perished lime of their bones we find....Gigantic they were...shriveled or contorted into stone. Seen are these all along the depths of the world....

"There were vast plains of dust, ashes and cinders....There were great banks of clay and soil burned to hardness--as clay is when baked in the kiln-mound--blackened, bleached or stained yellow, gray, red or white, streaked and banded, bended or twisted. Worn and broken by the heavings of the underworld and by waters and breaths of the ages, they are the mountain-terraces of the Earth-Mother, dividing country from country"(17).

A deep freeze

Imagine the scene: The skies darken over the entire Earth as debris is sent high into the atmosphere from the asteroid impact and resulting volcanic eruptions. And, all the while, the subterranean water chamber spews up massive amounts of water from beneath the broken upper layer of the Earth's crust. With the Sun blocked out, with the sudden tilting of the Earth's axis, and with the destruction of the greenhouse effect from the loss of the aquatic canopy, the air temperature was quickly and dramatically lowered over a good portion of the planet, causing much of the precipitating waters to fall back to Earth in the form of snow and hail. Thus all life-forms in the polar and circumpolar regions instantly froze, having been enveloped in a blanket of muddy ice and snow. And so began what scientists call the "Ice Age."

As the centuries passed after the flood and the Earth continued to expand, the breaching of the ocean floors that resulted must have given rise to further volcanic eruptions that spewed Sun-blocking clouds of dust high into the air, maintaining a low temperature over a large area of the Earth during these centuries. This brought about the formation of giant glacial ice sheets across broad sections of the continents, as the remaining flood waters thereon, unable to drain out to sea because they were trapped in low-lying basins (or because the Earth hadn't yet expanded enough for the sea level to drop substantially), had frozen solid. Because of the prevailing low temperatures and the immense size of the glacial ice sheets, numerous centuries went by before these frozen flood waters were finally able to melt and drain off.

It's possible that there were several "mini-ice ages" that occurred over the centuries immediately following the flood, resulting from what appears to have been multiple tilts of the Earth's axis, probably brought on by impacts from other large asteroidal bodies, or else close passages by such bodies that tugged on the Earth gravitationally (the likelihood of multiple post-flood Earth tilts is discussed, as stated before, in Appendix B).

One of the reasons for believing in multiple ice ages (or at least in a second one that took place centuries after the initial Ice Age that began at the time of the flood) is the fact that Ice Age fossil remains are commonly found on top of petrified strata that were obviously laid down at the time of the flood, indicating that there was a secondary mass extinction post-dating the deluge. Further proof of this lies in the fact that Ice Age fossils are rarely ever mineralized, revealing that they weren't buried deep in the Earth like the plants and animals that died during the flood. Also, the sediments that Ice Age fossils are buried in generally aren't petrified either, but are rather loosely compacted and/or in a frozen state.

But regardless if there was one, two, or several ice ages, one thing is indisputably clear, which will be our main focus here--that all Ice Age fossils show tell-tale signs that the animals and plants they represent had died very suddenly, under extreme catastrophic conditions.

Most fossils found in the polar and sub-polar regions are preserved as frozen carcases encased within layer upon layer of frozen "muck"--a mixture of ice, dirt, and volcanic ash. Studying these fossils tells us much about the very unusual conditions under which these frozen animals and plants had died. Speaking on this matter, Walt Brown wrote: "Mammoth carcases are almost exclusively encased in frozen muck. Also buried in muck are huge deposits of trees and other animal and vegetable matter....

"At least three mammoths and two rhinoceroses suffocated....

"[One mammoth's] digestive and respiratory tract contained silt, clay, and small particles of gravel. Apparently, Dima [the name given this mammoth] breathed air and ate food containing such matter...[just] minutes or hours before his death"(38).

Brown then went on to describe another unusual condition in which some fossil mammoths have been found, making it even more clear that Ice Age animals died under swift and tumultuous circumstances: "Several frozen mammoths, and even mammoth skeletons, were found upright. Despite this posture, [the Berezovka mammoth] had a broken pelvis and shoulder blade, and a crushed leg. Surprisingly, he was not lying on his side in a position of agony"(Ibid.).

Similar points were brought out in a 1960 *Saturday Evening Post* article, which noted: "About a seventh of the entire land surface of our earth, stretching in a great swath around the Arctic Circle, is permanently frozen....[T]he greater part of it is covered with a layer, varying in thickness from a few feet to more than a thousand feet, composed of different substances. It includes a high proportion of earth of loam, and often also masses of bones and even whole animals in various stages of preservation or decomposition.

"The list of animals thawed out of this mess would cover several pages....[T]he greatest riddle, however, is when, why and how did all these creatures, and in such absolutely countless numbers, just killed, mashed up and frozen into this horrific indecency [get this way]?

"These animal remains were not in deltas, swamps or estuaries, but were scattered all over the country. Many of these animals were perfectly fresh, whole and undamaged, and still either standing, or at least kneeling upright.

"Vast herds of enormous, well-fed beasts, beasts not specifically designed for extreme cold, [were apparently] placidly feeding in sunny pastures at a temperature in which we would probably not even have needed a coat. Suddenly they were all killed without any visible sign of violence and before they could so much as swallow a last mouthful of food [buttercups and bean pods], and then were quick-frozen so rapidly that every cell in their bodies is perfectly preserved"(5).

These descriptions of frozen arctic fossils, as you can see, contain details that are exclusively indicative of a sudden catastrophe. There is positively no room for debate here. Many of these animals, as mentioned above, were frozen so quickly that there was no time for their remains to decay, even on the cellular level. In fact, there wasn't even time for some of these frozen beasts to fall to the ground after dying, as some, once again, were frozen in a standing position. And, as was also mentioned above, many of these animals didn't even have time to swallow the last bite of their last meal before being encased in frozen muck. This is so reminiscent of our earlier discussion about certain fish fossils in the Green River Formation that were also fossilized while in the process of eating their last meal, showing smaller fish sticking out of their mouths.

To further show just how quickly the freezing of mammoths found in the arctic regions must have come about, consider this: Refrigeration experts who have studied the remains of mammoths that show no sign of decay indicate that the temperature of their environment at the time that they died must have dropped from at least 70 degrees Fahrenheit (a subtropical environment) to around -150 degrees (an arctic wasteland) within no more than a two-hour time period(38). It goes without saying that this simply cannot happen under normal circumstances.

As we have seen, the frozen muck encasing these fossils contains a large amount of volcanic ash. Commenting on this issue, the book *When the Earth Nearly Died*, referring specifically to frozen fossil remains in Alaska and Siberia, states: "Interspersed in the muck depths, and sometimes through the very piles of bones and tusks themselves,...[are] layers of volcanic ash....[T]here were volcanic eruptions of tremendous proportions....Toxic clouds of gas from [these] volcanic upheavals cause[d] death on a gigantic scale"(17).

The poor animals in this region didn't stand a chance of survival. On account of the noxious volcanic gases, the drastic drop in temperature, the rushing muddy water currents, the falling ash, snow, dust, etc., death came swiftly and with a vengeance, sparing none.

It's important to point out that bones, tusks, and/or horns of mammoths, mastodons, bison, and a whole host of other Ice Age animals have been found with tiny, round, iron micro-meteorites embedded in them. Judging by the "craters" that these iron balls made in these fossil remains, it's quite clear that they were flung at a tremendous velocity when they impacted their encasing organic materials, and that they were very hot at the time of impact, since they produced burn marks around the areas of impact. Such fossils have been found all over the northern hemisphere, from Canada to Siberia, and down into the United States and Europe. This indicates that the object that struck the Earth, in this particular instance (either during the flood or sometime thereafter), had impacted somewhere in the northern latitudes.

Though no specific crater exists large enough to account for such a widespread distribution of these micro-meteorites, there need not be one. For we find that large, devastating meteoritic encounters with Earth don't always produce craters. In 1908, for example, a rather large meteor exploded about six miles above Evenkia, Siberia, which scientists refer to as the Tunguska Event. Though it leveled 60 million trees over a 2,150 square km area, it left no crater behind. But what it did leave behind, besides the fallen trees and huge brush fires, were tiny micro-meteorites embedded in trees all over the entire region, looking exactly like the tiny meteorites found in the fossils described above(129).

Earlier on we talked about how, in the fossil record preserved in the world's sedimentary formations, we sometimes find organic materials preserved in jumbled heaps, indicating transportation and deposition by violent water currents. We also discussed, on the other hand, how we sometimes find the most delicate organic materials perfectly preserved, without any sign of disturbance, likewise indicating a rapid burial (but obviously under less violent water currents). And, not so surprisingly, when we look at the "Ice Age" fossils, we find the same thing--some are piled up in frozen, jumbled heaps (indicating that powerful currents had deposited them), whereas other remains are found fully intact, with few signs of having been disturbed (indicating that these remains were frozen and preserved in a location where the water currents weren't quite so powerful).

We have already seen examples of essentially undisturbed frozen animals, found in a still-standing position in the muck. So let us now look at some examples of chaotic, jumbled masses of frozen organic materials, which provide, in the words of Charles Hapgood in his book *The Path of the Pole*, "...evidence [of]...disturbances of unparalleled violence. Mammoth and bison alike were [found in Siberia that were] torn and twisted as though by a cosmic hand....In one place we can find the foreleg and shoulder of a mammoth with portions of the flesh and toenails and hair still clinging to the blackened bones. Close by is the neck and skull of a bison with vertebrae clinging together with tendons and ligaments and the chitinous covering of the horns intact. There is no mark of knife or cutting instrument [as there would be if human hunters had been involved]. The animals were simply torn apart and scattered over the landscape like things of straw and string, even though some of them weighed several tons. Mixed with piles of bones are trees, also twisted and torn and piled in tangled groups; and the whole is covered with a fine sifting muck, then frozen solid"(52).

Speaking of trees being found in the frozen muck, here's a thought-provoking description of such a find: "Though the ground is frozen for 1,900 feet down from the surface at Prudhoe Bay [Alaska], everywhere...oil companies drilled around the area they discovered an ancient tropical forest. It was in frozen state, not petrified state. It is between 1,100 and 1,700 feet down. There are palm trees, pine trees, and tropical foliage in great profusion. In fact, they found them lapped all over each other, just as though they had fallen in that position"(53). Many of these frozen trees were found to have frozen fruit (plums) still on their branches, and preserved so well that they were still edible!

When considering snow- and ice-covered arctic regions like Greenland and Antarctica, and the virtual non-existence of snowfall there today (these areas are literally snow and ice deserts), one would have to wonder from whence all this snow and ice came, except it all be a remnant from the Ice Age (or ice ages) that accompanied and followed the flood catastrophe. Consider the fact that the ice sheet that covers Antarctica is over 2 ½ miles thick at its deepest point, which makes Antarctica the highest continent in the world above sea level. Aside from the flood catastrophe and other subsequent related calamities, how else could all of this accumulated ice have come about? All of it is obviously frozen flood waters that were never able to melt and drain out to sea.

Throughout this study we have cited numerous fascinating examples of flood legends that faintly recount tantalizing memories of this ancient cataclysmic event. Should it be surprising, then, to discover that there are also numerous ancient recollections of the onset of the Ice Age (or ice ages), and that these recollections connect this same calamity (or calamities) with the global flood that spawned it (or them)? Let us take a look at a few of the best examples of these intriguing ancient accounts:

- We'll begin by taking a second look at an excerpt from an ancient Hopi Indian legend cited earlier, which states: "[T]he world lost balance, spun around, and rolled over twice....Mountains plunged into seas with a great splash, seas and lakes sloshed over the land; and as the world spun through lifeless space it froze into solid ice. This was the end of the Second World....[A]ll the elements that had comprised the Second World were...frozen into ice..."(18).

- The so-called Teutonic tribes of Germany and Scandinavia, a culture best remembered through the songs of the Norse scalds and sages, have transmitted a legend to us that declares: "[T]he world was enveloped in hideous winter. Snow-storms descended from all points of the horizon....the world trembled....Mountains crumbled or split from top to bottom....The earth itself was beginning to lose its shape....Flames spurted from fissures in the rocks; everywhere there was the hissing of steam. All living things, all plant life, were blotted out....[T]he earth was no more than cracks and crevasses. And now all the rivers, all the seas, rose and overflowed. From every side waves lashed against waves. They swelled and boiled slowly over all things. The earth sank beneath the sea....[F]rom the wreckage of the ancient world a new world was born. Slowly the earth emerged from the waves. Mountains rose again and from them streamed cataracts of singing waters"(20).

- In the Gran Chaco region that sprawls across the northern borders of Paraguay, Argentina, and Chile, the Toba Indians have a myth about the "great cold," which states: "When the great cold set in,...[t]he people were freezing, and they cried the whole night. At midnight they were all dead, young and old, men and women....[T]his period of ice and sleet lasted for a long time....Frost was as thick as leather"(Ibid.).

- Contained in the biblical book of Job is this fascinating statement, which can only be interpreted as a description of the post-flood Ice Age conditions that prevailed in Job's day: "Out of whose womb came the ice? and the hoary frost of heaven, who hath gendered it? The waters are [hard as stone]...and the face of the deep [the ocean] is frozen." Job 38:29, 30.

- The Mayan *Popol Vuh* talked of a time when there was "white hail and blackening storms. The cold was incalculable"(146).

- In the *Zend-Avesta* (cited earlier) we find these remarks: "And Ahura Mazda [a deity] spake unto Yima ["the first king of men"] saying: 'Yima the fair.....Upon the material world a fatal winter is about to descend, that shall bring a vehement, destroying frost. Upon the corporeal world will the evil of winter come, wherefore snow will fall in great abundance....

"'And all three sorts of beasts shall perish, those that live in the wilderness, and those that live on the tops of the mountains, and those that live in the depths of the valleys under the shelter of stables'"(54).

- Another ancient Iranian document, the *Bundahish*, talks about a "vehement destroying frost" that "assaulted and deranged the sky" and overspread it with darkness. This happened, we are told, as the great ice sheets tightened their grip all across the land(20).

- Rene Noorbergen, author of *Secrets of the Lost Races*, relates that "Egerton Stykes, in his *Dictionary of Nonclassical Mythology*, page 20, states his belief that the Norse legend of Fimbelvetr, the 'Terrible Winter' that launched the epic disasters of Ragnarok and the destruction of the gods of Valhalla, may reflect a historical fact: the obliteration of a prehistoric civilization in the boreal regions by the Ice Age catastrophe"(9).

The early post-flood world was obviously not a pleasant place to live, scarcely being an improvement over the flood disaster itself. Much damage was inflicted upon the Earth as the glaciers from the Ice Age (or ice ages) began to melt and shift. On this matter, let us contemplate what catastrophist Robert Schoch had to say: "As ice melted in the sudden warming, the land beneath the vanished glaciers rebounded at the loss of their weight, setting off tremors, earthquakes, uplift, and subsidence--something of a seismic chamber of horrors. The sudden collapse of ice sheets into the oceans could have set off superwaves, towering walls of water that spread for thousands of miles and swept suddenly onto low-lying coastlines. And on those same coastlines the sea would have edged up year by year, consuming the land and pushing its inhabitants toward higher ground"(1).

This melting of the glaciers, by the way, may explain why we find so many sunken ruins off the coasts of nearly every continent around the world [Endnote 39].

Canyon formation

We have discussed the Earth's rotational axis tilt, the massive fracturing of its outer crust layer (which resulted in the release of massive amounts of pressurized water contained thereunder), and the breaching of its inner crust layer (which resulted in the release of enormous quantities of pressurized lava from underneath), bringing about the inundation of the entire planet with gushing waters and colossal, oozing lava flows. We have also discussed how these factors made conditions ripe for global-scale earthquakes and tidal waves that inflicted even more devastation. We further talked, just moments ago, about a great freeze that began in certain regions at this time--the Ice Age (or ice ages). But these destructive forces, as bad as they were, do not account for all the damage wrought, directly or indirectly, by the initial asteroid impact that spawned the flood catastrophe. For we find that over a lengthy period after this calamity, perhaps spanning several centuries, some of the flood waters had continued to reek havoc on dry land, carving up the surface of the Earth with enormous, tell-tale scars that today are some of the greatest natural wonders of the world.

Just after the flood, remember, as the continents were separating and the waters began draining into the ocean basins, some of these waters were inevitably left behind on the continents, being trapped and retained by great depressions therein. It was at this time that many of the world's large lakes were formed. For example, Lake Titicaca, situated high up in the Andes mountains, was created by flood waters that were unable to escape out to sea as the Andes were being pushed up. Other large bodies of water, like the Great Lakes, located in more northern latitudes, probably formed as glaciers (frozen flood waters) melted and flowed into the basins that now contain them. And how did these basins form? Obviously as a result of the spreading and tearing of the Earth's upper crust layer, when the Earth expanded.

Some of the great bodies of water left behind by the flood no longer remain with us today. They lasted for a time--perhaps a few centuries--until one of their natural retaining walls became weakened and burst under the pressure of the waters pushing against them. This would have caused a sudden, violent release of billions upon billions of tons of water, carving up the Earth's surface as it drained, eventually finding its way out to sea. It was probably in this manner that Arizona's Grand Canyon and the Scablands of Washington state were formed

[Endnote 40]. Both of these water-cut formations are carved through enormous sedimentary rock deposits that were undoubtedly laid down at the time of the flood, and were later "canyonized" in the same way as just mentioned, over a relatively short period of time.

But an appropriate question that now arises, which demands an answer, is this: Could such enormous features as the Grand Canyon and the Scablands really have been carved out in a fairly brief time-span, rather than over a period of millions of years, as we have been told?

First of all, it is important to understand that rushing water, under tremendous pressure, is one of the most destructive forces known to man. Whether it's a flooding river, a bursting dam, or a tsunami, water can cut through and smooth over solid rock, level mountains, transport enormous boulders, reshape an entire coastline--there's really nothing that can withstand its destructive power. And all of this can be done in a surprisingly short duration of time.

To see just how destructive water can be, we only need look at the damage caused by modern, localized river floods that don't even involve very large quantities of water, as described in this next quote from scientists John Whitcomb and Henry Morris: "Even the relatively small amounts of water involved in river floods have caused damage that staggers imagination. Bridges, houses, immense boulders, and trees are torn up and swept along as mere pebbles and matchsticks. Such floods seldom obtain a depth of more than a few dozen feet..."(3).

So then, if relatively small-scale yet powerful water currents, created by modern river floods, are capable of inflicting tremendous damage in a short time, what do you suppose a massive current of billions upon billions of tons of water, rushing over the same given area for weeks, months, and even years on end, could have been capable of accomplishing, following the flood disaster?

Looking at a couple specific examples of gorges created by rushing waters in rather recent times can help us to better understand how formations like the Grand Canyon and the Scablands could have come about rather quickly.

The Burlingame Canyon near Walla Walla, Washington, is one good case in point. This canyon was carved out during a localized flood in 1926, where raging waters created a cavity in solid rock that is 1,500 feet long, 120 feet deep, and 120 feet wide--a process that took only 6 days. Now it is obvious that this canyon, as big as it is, is no way near the scale of the Grand Canyon. But then again, neither was the amount of water that created Burlingame Canyon anywhere near the amount of water that created the Grand Canyon. Obviously the Grand Canyon is much larger because the amount of water left over from the flood catastrophe that formed it was much larger. The general rule of thumb here is simply this: The more water, the more the damage, and the quicker that it will come about.

Another example of rapid canyon formation in recent times is what some have dubbed "Little Grand Canyon," also in the state of Washington, near Mt. St. Helens. On March 19, 1982, a small eruption of Mt. St. Helens took place which melted the snow that had accumulated in its crater during the winter. The resulting flow of mud, as it poured into the Toutle River, eroded a canyon system which dug up to 140 feet deep into the Earth. Here again we have a good demonstration of what the power of water is capable of doing in a short time. Thus we should be able to see, once again, that neither the Grand Canyon nor the Scablands necessarily required millions of years to form, as long as we take into account the tremendous amount of rushing water draining out to sea in the fairly early post-flood era.

Regarding the Scablands, specifically, even mainstream scientists have acknowledged that this formation most likely came about in a rather short time-period, toward the end of the Ice Age. In fact, the official website of the Washington State Tourism Agency had this to say about the swift formation of the Scablands: "Explore the other-worldly scene of a flood of biblical proportions that scoured the Channeled Scablands of eastern Washington. Learn about the unfathomable forces and the brave assertions of a lone geologist who envisioned a two-day torrent of water ten times greater in volume than all the rivers of the world combined.

"In the 1920s a young geologist named J. Harlen Bretz from the U.S. Geological Survey spent a summer studying northeastern Washington. He was awed and mystified by the surreal landscape of dramatic, dry canyons and high, gravel hills. When he proposed that a giant flood created the strange formations, his colleagues scoffed. In 1927, six leading geologists even challenged Bretz to debate the merits of his 'outrageous' ideas. Basically, the size of the floods necessary to carve the coulees and rippled complexion of the Colombia Plateau defied the imagination.

"Years later, Bretz and his theory would be vindicated when the colossal source of the water was identified and further remnants of the flooding were documented. Between 13 and 15 thousand years ago [according to their unreliable dating methods], glaciers dammed an enormous body of water in western Montana, Glacial Lake Missoula. This inland sea covered 3,000 square miles and was over 2,000 feet deep at the edge of the dam.

"Eventually, as the lake grew large enough to float the wall of ice, it undercut the dam and collapsed a section 100 miles wide. In two days, Lake Missoula emptied through the breach, raging westward toward the sea in a towering torrent several hundred feet high, traveling at speeds between 50 and 90 miles per hour. Boulders weighing many hundreds of tons were carried almost to the coast of Oregon, and some were dropped, as if from the sky, in the middle of the plains of eastern Washington.

"Repeated floods scarred the land with deep channels and created the largest cataracts in the history of the planet. To peer out and ponder the forces unleashed here is both eerie and awesome for the sheer scale and beauty. Over three thousand square miles of eastern Washington were flooded. From space, these Channeled Scablands resemble the channeled regions of Mars. See the monumental sculpture carved into the columnar basalt of the high desert. It's a short trip to another planet"(56).

Let us now ponder this question: If mainstream scientists can fathom the Scablands being formed in a somewhat short time, why can't they do the same for the Grand Canyon?

Before closing this particular discussion, there's an alternative viewpoint about how formations like the Grand Canyon and the Scablands came into existence that does not involve water at all, which deserves consideration. According to the expanded Earth hypothesis, all such chasms (or most of them, anyway) are the result of stretching and tearing of the upper crust when the Earth expanded, and the fact that water now flows in them has nothing to do with how they came about. By no means is this an unreasonable or unlikely proposition. It may be that both water flow and expansion played a role in the creation of such formations. There is even another mechanism proposed in Appendix D, which also has great merit, and may have worked in tandem with water flow and Earth expansion. Perhaps we'll never know for sure.

A great wind

A number of ancient records reveal that the flood catastrophe was accompanied by a tremendously powerful wind, which added to the multiplied horrors of this event. Such a wind would be expected from the shock waves produced by the asteroid impact. But some of these ancient records portray the great wind event as having been ongoing, occurring all during the flood, while others place it after the rains had ceased, at a time when the engulfing waters began to recede (and some of these records claim that this great wind had actually assisted the receding process of the flood waters).

The differences in the timing and duration of the wind event in these ancient records may simply be due to there having been great winds that occurred all throughout the flood catastrophe--at its beginning, while the rains were falling, and after they had stopped. Perhaps there was a series of enormous, powerful hurricanes. In any case, it is significant that many ancient records do associate the flood catastrophe with significant pneumatic activity.

Let us now turn our attention to some of these very old reckonings:

- The Hebrew flood account states: "...and God made a wind to pass over the earth, and the waters assuaged [receded]; The fountains also of the deep and the windows of heaven were stopped, and the rain from heaven was restrained." Genesis 8:1, 2.

- Ancient Babylon's flood legend declares: "As soon as something of dawn shown in the sky, a black cloud in the foundation of heaven came up. Inside it the god Adad thundered....The whirlwind of Adad swept up to heaven. Every gleam of light was turned into darkness....Swiftly [a mighty flood] mounted up...[the water] reached to the mountains [and] attacked the people like a battle. Brother saw not brother....The gods were terrified at the cyclone....The wind, the storm raged, and the cyclone overwhelmed the land. When the seventh day came the cyclone ceased, the storm and battle which had fought like an enemy. The sea became quiet, the grievous wind went down, the cyclone ceased. I looked on the day and vices were stilled, and all mankind were turned into mud, the land had been laid flat like a terrace"(47).

- Aztec folklore tells a similar story. In his book *The Rise and Fall of Maya Civilization*, Eric S. Thompson wrote: "The Aztecs believed that the world had been destroyed four times, the present age being the fifth [age]. Each age had been brought to a violent end, the agents being respectively ferocious jaguars, a hurricane, volcanic eruptions, and a flood"(57).

Minus the mention of the jaguars, could this legend's references to a hurricane, volcanic eruptions, and a flood be faded recollections of three of the major phases of the asteroid impact?

The Aztecs also believed and taught that in ancient times "there fell a rain of fire; all which existed burned; and there fell a rain of gravel....Now in this day, in which men were lost and destroyed in a rain of fire...the sun itself was on fire...everything was consumed.

"...A tremendous hurricane...carried away trees, mounds, houses and the largest edifices....All this time they were in darkness, without seeing the light of the sun, nor the moon..."(17).

- A Toltec legend states: "...a furious hurricane [swept]...over the entire world....Whirlwinds and cyclones swept over the world, picking up sand, stones, rocks, waters and finally trees, houses and human beings. The snowy tops of the mountain peaks were whisked away....Nowhere could the panic-stricken humans find safety..."(195).

- The Buddhist book, *Visiddih-Magga*, has this to say: "There are three destructions: the destruction by water, the destruction by fire, and the destruction by wind....After a long time had elapsed, after the rain had stopped, a second sun appeared...there was no difference between day and night...and incessant heat burned down on the world"(25).

Could these "three destructions" of water, fire, and wind also be a reference to three of the major phases of the asteroid impact mentioned in Aztec folklore?

- Another Buddhist text declares: "There arises a wind to destroy the world cycle...it rains a fine dust...boulders as large...as mighty trees on the hilltops...[it] turns the ground upside down...[when] worlds clash with worlds"(17).

Whatever the nature or cause of this wind (or winds), there is actual physical evidence that a great atmospheric disruption did indeed accompany the flood catastrophe. Here is a good example: A large number of fossil insects, according to *When the Earth Nearly Died*, have been found in Geiseltal, Germany. But what's so interesting about them, in the context of our "great wind" discussion, is the fact that "The vast majority have been torn apart--and suddenly at that, because the process of fossilization of all surviving parts with silica invading the tissues must have been virtually instantaneous...and was undoubtedly responsible for preserving the membranes and original colours of these insects so marvelously....

"Only some tremendous hurricane acting in concert with vast surging masses of water could have transported and accumulated these remains at Geiseltal--a hurricane and a deluge operating on a scale and with a ferocity far beyond that of which modern humanity has any experience"(Ibid.).

There exists physical evidence of at least one other such wind, which occurred sometime after the flood, perhaps several centuries later, during the "Ice Age." This evidence is expounded upon thusly in *When the Earth Nearly Died*: "[W]hat agency other than extremely violent winds could have brought together in one place [the California tar pits] such dissimilar avian species as the following? They included: Grebes, herons, bitterns, storks, wood ibises, spoonbills, swans, various geese (including the snow geese), ducks, American vultures, kites, many kinds of hawks, falcons, eagles, caracaras, the Teratornis, quails, cranes, partridges, turkeys, rails, gallinules, parrots, coots, plovers, turnstones, woodcock, snipes, surfscooters, stilts, sandpipers, barn owls, seven other owl species, flycatchers, woodpeckers, swallows, jays, crowns, magpies, titmice, chickadees, ravens, mockingbirds, waxwings, thrashers, meadowlarks, shrikes, two species of blackbird, redwings, orioles, finches, sparrows and buntings. Remains of all these birds were discovered in...tar-seeps at McKittrick in California, and the asphalt pits at Rancho La Brea in the same state."

This same book then went on to mention how such finds like this are not uncommon in the "Ice Age" fossil record. It described, for example, the San Pedro deposits, also in California, and stated that most of the species found here and in the two locations cited above do not occupy the same habitat, and so were thus obviously "brought together from various directions involuntarily by irresistible winds and buried in a common grave by, it would appear, catastrophic agencies."

Section I: From heaven to hell
Part 2: A history of Mars

Did Mars have a global flood, too?

As mentioned earlier, Mars exhibits many classic signs of having suffered a flood catastrophe in its distant past, many of which parallel the signs here on Earth of its own colossal flood. It is to these Martian signs that we will now turn our attention.

Former continents and oceans

When we look at geological maps of Mars, the global landscape reveals raised areas that are reminiscent of Earth's continents, along with large lowland regions that resemble Earth's ocean basins, which may in fact have once contained Martian oceans. These oceans would have averaged 1,850 feet deep, reaching up to 10,000 feet at their greatest depths (Utopia Planitia).

Stretched across the surface of the whole planet, these waters today would cover Mars 330 feet deep, if the entire surface was leveled flat. So much liquid water did Mars once appear to have, in fact, that, in proportion to its size as a planet (as compared to Earth), it actually had more water than Earth does today(58).

Incidentally, the red color of Mars may be due to iron oxidation (rust) in surface basalts, resulting from past contact with its once-abundant global water supply. By comparison, the bulk of Earth's ocean floors, interestingly enough, are covered with red clay.

I must also draw attention to the fact that the highland areas (upper crust) of Mars are heavily cratered, whereas the lowland areas (lower crust) are almost devoid of craters. This is the same type of pattern that we find, as you may recall, on Earth, the Moon, Jupiter's Ganymede, and several other worlds in the solar system that show signs of expansion. And, in addition to this, the few craters in the lowland areas, where the oceans once resided, are often surrounded by a splash pattern, indicating that such regions are (or at least were) extremely muddy in nature. By contrast, the craters in the highland regions do not exhibit such a feature.

Possible indication of certain mountains being formed by buckling

Mars has a most curious semi-circular mountain range, called Thaumasia, which is located in the Tharsis region of the planet, just southwest of Valles Marineris (both of which we will discuss shortly). This mountain range closely resembles so-called island-arc forms on Earth, such as the Himalayas and Japan(58), which most scientists associate with continental drift (however, with the Earth expansion scenario, the cause would have been the buckling that resulted from expansion). So, does the Thaumasia mountain chain further hint at a prehistoric expansion of the planet Mars? Quite possibly so.

Martian "stretch marks"?

NASA's "Space Science News" Internet site reported on April 29, 1999: "NASA's Mars Global Surveyor has discovered surprising new evidence of past movement of the Martian crust, suggesting that ancient Mars was a more dynamic, Earth-like planet than it is today.

"Scientists using the spacecraft's magnetometer have found banded patterns of magnetic fields on the Martian surface. The adjacent magnetic bands point in opposite directions, giving these invisible stripes a striking similarity to patterns seen in the crust of Earth's sea floors [i.e. fracture zones, or stretch marks]"(140). Did these result from Mars expanding long ago, because of internal heating?

Mars also has visible stretch marks in several areas, just like Earth and Venus have (as we saw earlier).

Decaying magnetic field

Like Earth, Mars has a decaying magnetic field. So badly has Mars' field decayed, in fact, that it is almost non-existent, and thus very difficult to detect. If this decay factor of both of these magnetic fields is due somehow to the catastrophe that spawned the global floods of these two planets, the fact that Mars' field is more severely diminished than Earth's may simply be the result of Mars being a significantly smaller planet than Earth, which means it obviously had a much weaker magnetic field to begin with.

Incidentally, judging from the limited "ghost" remnants of Mars' ancient magnetic field, it appears that the once-stronger field that Mars once had was tilted out of alignment with its rotational axis, just like Earth's field is today, except that Mars' appears to have been tilted more significantly--about 40 degrees or so.

A Martian antediluvian subterranean water chamber

The original source for much of the waters of the flood on Mars appears to have been a planet-wide underground aquatic chamber (underneath its upper crust layer), just as the bulk of the Earth's flood waters came from the "fountains of the great deep" (below its top crust layer). On this matter, one Mars specialist, William Hartmann, wrote: "Where did the Martian water come from? The Martian floods probably originated not in surface lakes but...underground..."(58). Elaborating further, a February 20, 2008 Yahoo News article stated: "Fan-shaped deltas at the edge of huge basins scattered across Mars were probably formed by a titanic influx of water, gushing from the bowels of the Red Planet....

"...Mars [appears to have] once had [an]...underground water system that may have latticed the entire planet"(198)[Endnote 41].

To have had so much liquid water flooding its surface in the remote past, Mars must have once had a much warmer climate (like Earth did). But did Mars truly have a warmer climate in ancient times?

A pre-flood Martian greenhouse

In much the same way that Earth was a relative paradise before its flood catastrophe, it would seem that Mars was also once a paradise compared to its current inhospitable conditions of extreme cold temperatures and a very thin atmosphere, both of which make liquid water on Mars virtually impossible today [Endnote 42].

Mars, in fact, may have once had a universally warm "greenhouse" climate, like Earth seems to have had. In this regard, Hartmann wrote: "[A] thicker...atmosphere of carbon dioxide would have produced a greenhouse effect, which warmed the early Martian climate"(58).

While Hartmann here proposes that higher levels of carbon dioxide may have caused a planet-wide Martian greenhouse effect, this same effect could have instead resulted from the presence of an ancient water canopy above Mars' atmosphere, similar to what Earth might have had. Though there's no way to prove this, it's impossible to resist the temptation to speculate about it.

Rotational axis tilt

Today Mars' rotational axis is tilted at 25 degrees, very similar to the 23.5 degree tilt of Earth. These similar axial tilts of Mars and Earth would seem to suggest that both of these worlds may have been struck by debris coming from the same direction and angle, if they were in close proximity to each other (in their orbital positions around the Sun) at the time of their proposed impacts, during the Great Bombardment period.

<u>Rotational axis wobble</u>

Mars' rotational axis, like Earth's, is known to have a slight wobble to it (precession)(142). Could this have also been a result of its asteroid impact/flood catastrophe, just as the Earth's axis wobble appears to have resulted from its own related catastrophe? If so, it would seem safe to conclude that Mars, like Earth, has not yet stabilized itself from the massive, flood-producing calamity that offset its axial orientation.

<u>An altered length of the Martian day</u>

We talked earlier about the possibility of the length of the Earth's day having been altered at the time of the catastrophe--being either sped up or slowed down, depending on where and from what direction the cosmic projectile hit. Well, the same appears to have been the case with Mars--that its day-length was altered as a result of its asteroid impact (in this case, its day appears to have been lengthened). As the authors of *When the Earth Nearly Died* explain: "[C]alculations made of the mass and angular momentum of Mars, when compared with other members of the solar system, result in an ideal axial rotation of 8 hours"(17).

Let us bear in mind, however, that these calculations were made with the assumption that Mars was always the same size it is today. But if Mars expanded (as the available evidence strongly suggests), then these calculations are rendered null and void. The actual rotational rate of Mars could therefore very well have once been less than 8 hours.

<u>Huge channels and tributary systems</u>

Features like these, obviously cut by running water, can be seen all over the surface of Mars. At first there was a great debate among scientists as to what carved these river-like structures across the surface of this planet. But as time went on, nearly all were finally forced to concede that water was indeed the culprit, since all other suggested mechanisms could not account for all the observable phenomena. There is almost no more opposition, in fact, to the idea of water having created Mars' channels and related scars. As Hartmann wrote: "After the outflow channels were discovered, an argument broke out among

Martian researchers. Were these features really carved by water? What about some other fluid? Could lava have done the job, or liquid carbon dioxide? Or horrific winds, whistling through dusty canyons? One by one, the alternatives were dismissed. Lava creates modest channels, but tends to solidify as it flows downstream, so that lava channels often dwindle to nothing, whereas the Martian channels get wider and deeper downstream, matching the properties of large terrestrial rivers. Carbon dioxide is plentiful on Mars, comprising 96% of Mars' scanty atmosphere, but its liquid form requires five times the pressure of air found on Earth's surface, so it is not a good candidate to flow hundreds of kilometers under the extremely thin air of Mars. Wind has never been seen to carve the kinds of valleys found on Mars. Many Martian channels have tributary systems, streamlined islands, and other features of earthly river channels, especially those associated with flood systems. The same features can be seen in terrestrial water flow systems, from the Mississippi to the dribble of water flow from your garden hose across a bare patch of dirt. Most researchers finally agreed that most Martian channels must have come from outflows of water"(58).

Hartmann then went on to make the astonishing claim that "many of the large Martian channels were carved not during millennia of leisurely flow, but during catastrophic, sporadic flood events"(Ibid.). Another Mars specialist, Michael Carr, further wrote in his book *Water On Mars:* "Enormous floods occurred episodically throughout Martian history. Many of the floods had discharge rates that ranged as high as...100 times the peak discharges of the largest known terrestrial floods and 100,000 times the average discharge rates of the largest terrestrial rivers"(98).

While both of these scientists believe that the flood features on Mars didn't all form at once (perhaps some didn't, but I believe that most of them did), nevertheless, they both agree that Mars has indeed endured massive catastrophic flooding in the past.

Ares Vallis is one particular outflow channel, or riverbed, that is worth noting. It averages about 1.5 miles wide, with a depth ranging between 100 and 1,000 yards. It has been estimated that this river once flowed between 50 and 2,000 times the flow rate of the Mississippi(58).

Clearly, Mars had a great quantity of water on its surface long ago, which inflicted a tremendous amount of damage during a catastrophic flood disaster, just like on Earth. But the next questions is: Where did all this water go? A large portion of it still actually exists on Mars, but it is frozen at the poles [Endnote 43], frozen deep underground [Endnote 44], and chemically locked away in hydrated minerals. Some may have also evaporated into space(Ibid.). It is also believed that a good deal of it may be hidden in plain sight--frozen right on the surface, but buried underneath a thin layer of wind-blown sand and loose pebbles(97).

Sedimentation

One could not scan the surface of Mars for very long without running into layered rocks. The red planet is literally plastered with sedimentary formations, and a great many of them show clear signs of having been deposited by water. Hartmann tells us that there are "many places on Mars where massive layers of sediments fill low areas such as craters. Some of this material may merely be layered lava flows, but much of it appears to form weakly resistant layers, like loosely consolidated sedimentary rock. Some of the layers may be deposited by wind, but others by water. This makes Mars more Earthlike than a merely igneous planet, like the Moon"(58).

The PhysicsWeb.org website made these relevant remarks on December 5, 2000: "New images of the surface of Mars [taken by the Mars Global Surveyor spacecraft] have revealed numerous layered outcrops similar to sedimentary rocks on Earth, strongly suggesting that liquid water was present on Mars at an early stage in its history....Some of the images show hundreds of identically thick layers, which is almost impossible to have without water"(96).

Signs of ancient glacial flows

Did Mars once have glaciers as well? To answer this question, we will again quote from Hartmann: "Many researchers...began as early as the 1970s and '80s to point out various features that they interpreted as remnants of...Martian glaciers. These include soil lineations and possible eskers and moraines, or winding, ridge-like deposits carved by streams running under glaciers or left as glaciers melt and drop loads of gravel that have been picked up and accumulated in the moving ice"(58).

Was there once a Martian equivalent of a post-flood Ice Age? The evidence certainly seems to suggest so.

The Tharsis Bulge

The Tharsis Bulge is a 2,486 mile-wide area on Mars where the crust literally bulges out about 6 miles above the rest of the planet's surface. This Martian landmass is topped by 4 of Mars' largest volcanoes and an enormous crack, or "canyon" system (Valles Marineris), which probably formed when this area was pushed up by cataclysmic volcanic activity at the time of the catastrophe.

It should be borne in mind that, on the opposite side of Mars from the Tharsis Bulge (and its accompanying huge volcanoes and canyon system), there lies an enormous crater known as Hellas. It is nearly 6 miles deep and 1,300 miles in diameter, and its basin is surrounded by a ring of material that is 1.25 miles above the surrounding terrain, stretching out to about 2,500 miles from the basin's center. So vast is this ring of material, in fact, that it would cover the entire continental United States with a two-mile-thick blanket of dirt and rock. This material was undoubtedly blasted out and thrown to its current location by the impact of the asteroid that formed this crater.

There are two other large craters in the same hemisphere as Hellas. One is called Isidis, which is about 690 miles across, and the other is called Argyre, with a diameter of roughly 120 miles. These two craters, along with Hellas, are among the largest impact basins in the entire solar system.

Seeing how massive the shockwaves from these Martian impacts must have been, it was probably these same collisions, which most likely happened at about the same time, that kicked off the Martian global catastrophe. It was certainly these impacts and their resulting shockwaves that caused, or contributed to, the formation of the Tharsis Bulge and its related scars on the exact opposite side of Mars.

There is another, lesser bulge located just north of Tharsis, known as the Elysium Bulge (also having several large volcanoes associated with it)[Endnote 45], which probably formed at the same time and by the same three impacts that created the Tharsis Bulge. And, it was most likely these three impacts that were also responsible for the deformation of Mars' orbital path, making it one of the most extremely elliptical in all the solar system. Earth's orbit, as mentioned earlier, is also an ellipse, but to a much lesser degree, most likely because it is significantly larger in size and was thus not moved as much when impacted (not to mention the fact that Mars appears to have been struck by larger objects than Earth was).

It should also be mentioned at this point that the two moons of Mars, Phobos and Deimos, may also have originated at the time of the catastrophe. Their shapes and sizes make them look as though they are captured asteroids (broken pieces of the exploded planet). Or, as some scientists have postulated, they may represent the remains of a larger moon that once orbited Mars, but was broken up during the cataclysm(143).

One final point about the Hellas basin: This impact crater looks rather deformed, and its perimeter seems somewhat ill-defined. Most scientists have attributed this to erosion. However, it may very well be that this crater was once quite a bit smaller, but got stretched out when the planet expanded. If this be the case--that the original crater was significantly smaller--the impact that created it would still have been powerful enough (along with the impacts that created its other two accompanying craters) to have produced the Tharsis Bulge and the Valles Marineris crustal spread on the other side of Mars, since the entire planet Mars itself would have been proportionately smaller. What's so significant about these impact basins is not so much their current diameters, but their size ratios to that of the planet as a whole.

By the way, the same can be said for some of the roughly circular-shaped lunar lowlands. They were probably once impact craters that got stretched out and slightly deformed in shape when the Moon expanded.

Can you not see how the expansion theory explains so many phenomena throughout the solar system that are otherwise difficult to make sense of?

Another point to bring out here is that Mars, like Earth, has an equatorial bulge (an entirely separate feature from its Tharsis and Elysium bulges), which must surely have come about at the time of its catastrophe, just as the Earth's bulge probably resulted from its own catastrophe. And as we should come to expect from Mars by now, its bulge is significantly more prominent than Earth's (its equatorial diameter is 42 miles larger than its polar diameter). Compare this with Earth's difference of only 27 miles.

Olympus Mons, or Olympia Mountain

This colossal feature, the biggest of the 4 extinct volcanoes on the Tharsis Bulge (in fact, it is the biggest volcano in the entire solar system), measures 15 miles high and 340 miles in diameter, and is rimmed by a scarp (a line of cliffs produced by faulting or erosion) that is 4 miles high. By contrast, Earth's largest volcano, Mauna Loa, is only 6 miles high and 75 miles across. Earth's tallest mountain, Everest, is almost 3 times lower in height than Olympus Mons. This is particularly strange when we consider the small size of Mars as compared to Earth (4,200 miles and 8,000 miles, respectively). But then again, if the asteroid that hit Mars was of comparable size to the one that hit Earth, we would expect a greater degree of damage done to this smaller planet.

Valles Marineris, or Mariner Valley

This formation, of truly mammoth proportions, clearly points to a catastrophe for the explanation of its existence. It is a canyon-like system over 2,700 miles long and up to 6 miles deep. Hartmann, noting some curious points about this enormous feature, wrote: "The closest terrestrial analogs are [the] great rift valleys...on Earth. One example lies in North America, where Baja California is being split off the mainland of Mexico, creating the Gulf of California. The notorious San Andreas Fault near Los Angeles is part of the same fracture system that is separating the California coast from the mainland and widening the Gulf of California. Indeed, the Gulf of California is about the same size and shape as Valles Marineris. Also the same size and shape is the Red Sea, of biblical fame; it, too, was created by plate-tectonic fracturing as the Arabian peninsula separated from Africa. The fractures along these faults allow magma to reach Earth's surface, and both the Gulf of California and the Red Sea are dotted with volcanoes, supporting the connection between the Valles Marineris and the volcanoes of Tharsis"(97).

Mariner Valley was probably formed by a combination of Mars' three largest impacts (as described above) and the general expansion of the planet itself, which tore the outer crust to shreds, as has happened on so many other planets and moons throughout the solar system.

====================

Ancient mythology and the Martian catastrophe

Our ancient ancestors were obsessed with the planet Mars, often making reference to it in their mythologies. But Mars wasn't looked upon with much favor. Instead, it was frequently depicted as a malignant agent of war, pestilence, and cataclysmic disaster. This was the case among peoples such as the Greeks, Romans, Babylonians, Indians (of India), Chinese, Mayas, Aztecs, etc.

So great was the fear of this bloody "god of war" in certain ancient cultures, in fact, that in order to appease his wrath, they offered human sacrifices to him.

What was it about Mars that could have prompted such morbid conceptions and grizzly acts to be associated therewith? And how can we explain so many ancient cultures around the world having shared such similar beliefs and practices that revolved around this planet-god? Could it be that they (or their ancestors) had witnessed the Hellas impact on Mars?

This impact was so significant that it is highly probable that its effects would have been seen from Earth, with the naked eye. Between the massive explosion from the impact itself, and then the resulting eruptions of the enormous Tharsis volcanoes on the other side of the planet, it's difficult to imagine how it could have escaped the notice of Earth-bound observers gazing at the dark night sky.

If the ancients truly did witness the Martian catastrophe, they probably did so several months before the Earth faced its own catastrophic asteroidal impact, since it would have taken debris from the exploded trans-Martian planet several months longer to reach here than it took for such debris to reach Mars (assuming that Earth and Mars were relatively close to each other, in the direction of the oncoming debris, at that time). Also, it isn't likely that anyone would have been able to witness the Mars impact after the Earth suffered its own catastrophe, since the skies here would have been darkened for a considerable amount of time thereafter, due to volcanic and impact debris in the upper atmosphere, as discussed earlier.

Thus, if the perception the ancients had of Mars as a symbol of destruction did stem from eye witness accounts of the Hellas event, these accounts must have been transmitted over from before Earth's flood catastrophe.

There does indeed seem to be such accounts recorded in ancient documents from around the world. We will now cite one of the best surviving examples.

In the *Iliad*, Homer says that Mars (Ares) was thrust with a spear (debris from the exploded planet?) "mightily against his nethermost belly" (the Hellas impact?). Next, said Homer, "brazen Ares bellowed loud as nine thousand warriors...join[ed] in the strife of the War-god" (perhaps a reference to the multitudinous flaming chunks of debris that were seen blowing off of Mars as a result of the near-fatal impact).

Because of his victory in this violent battle, the personified and deified Mars was hailed in hymn after hymn as a fearful and undefeatable war hero, especially by the Babylonians and Assyrians, both of which called this planet-god Nergal.

Esarhaddon, son of Semnacherib, said of Mars: "Nergal, the almighty among the gods, fear, terror, awe-inspiring splendor."

Shamash-shum-ukin, king of Babylonia and grandson of Semnacherib, wrote these words about this conquering deity: "Nergal, the most violent among the gods."

Another grandson of Semnacherib, Assurbanipal, king of Assyria, contributed this salutation to Mars: "Nergal, the perfect warrior, the most powerful one among the gods, the pre-eminent hero, the nightly lord, king of battle, lord of power and might, lord of the storm, who brings defeat"(12).

An intriguing point to reference here is the fact that the ancient Greeks seem to have been aware that Mars had two moons. In their mythology, they made mention of Ares having two companions--two "blazing steeds," Phobos and Deimos. Hesiod wrote thusly of this triune Martian arrangement: "Ares..., sceptred King of manliness, who whirls your fiery sphere among the planets in their sevenfold courses through the aether wherein your blazing steeds Deimos and Phobos ever bear you above the third firmament of heaven."

Phobos and Deimos, of course, are now the actual names of Mars' moons, assigned to them in 1877 when they were discovered (or rediscovered). In keeping with the image of Mars being a mighty and victorious warrior, the meaning of the name Phobos is fear, and that of Deimos is "great fear" or "great awe."

Just as the Hellas impact on Mars was apparently witnessed, the collision that destroyed the trans-Martian planet must also have been witnessed by the ancients. In fact, it must have been a far more brilliant site to behold than the Martian impact, since it blew the whole planet apart. So why, you may ask, wasn't this planet made a "god of war," and why weren't sacrifices offered to it? The answer to this is quite obvious--because this planet lost the battle, and "died."

An ancient Chinese chronicle appears to have captured an eyewitness account of the destruction of the trans-Martian planet. It talks about "two suns" (the trans-Martian planet and the object that struck and destroyed it) battling in the heavens, which caused a disruption of the "five planets" (Mercury, Venus, Earth, Mars, and perhaps the Moon). The text states that, during the reign of Emperor Kwei (Koei-Kie), "the two suns were seen to battle in the sky. The five planets were agitated by unusual movements. A part of Mount Tai-chan fell down"(Ibid.).

Mars is not the only planet associated with a catastrophic spectacle in the heavens that was viewed and recorded by the ancients. So, too, was Venus, which was also a highly-esteemed deity in many ancient pantheons. And so it is to some of these ancient eyewitness accounts that we will now turn our attention.

The *Iliad* states that Pallas Athene (Venus) "darted down to earth a gleaming star" with sparks springing from it. Was this perhaps a description of what Venus looked like when it was hit by a large piece of debris from the trans-Martian planet? Did it appear as though it "darted down toward Earth" because of the "flying sparks" that were seen shooting out from it? Seeing that Venus never strays far from the horizon, either at dawn or dusk, any "sparks" shooting from it when it was impacted would have indeed made Venus look as though it was darting "down to earth."

By comparison, the Persian Mithra (Venus) is said to have descended from the heavens and "let a stream of fire flow toward the earth" and "filled our world with its devouring heat." Here it would seem that the flood-causing chunk of debris that hit the Earth, disseminating great heat (as we discussed earlier), may have been thought to have originated from Venus.

The planet Venus was seen as the cause of Earth's catastrophe by the ancient Babylonians as well, at least at one point in their history. Or perhaps they only saw it as contributing thereto. In any case, the Babylonian psalms record Ishtar (Venus) as saying:

> By causing the heavens to tremble and the earth to quake,
> By the gleam which lightens in the sky,
> By the blazing fire which rains upon the hostile land,
> I am Ishtar.
> Ishtar am I by the light that arises in heaven,
> Ishtar the queen of heaven am I by the light that arises in heaven.
>
> I am Ishtar; on high I journey...
> The heavens I cause to quake, the earth I cause to shake,
> That is my fame.

In ancient India, Venus was referred to in the *Vedas* as having hurled "fire upon earth and heaven," precisely paralleling the above accounts. The *Vedas* also dubbed Venus with the title "the bull," which is more than likely the origin of the cow and bull being viewed as sacred by the Hindus.

But why did the ancient Hindus liken Venus to a bull? Probably because a bull has two horns, and Venus, when in crescent phase, looks like the two horns of a bull (as does the moon when in its own crescent phase, which was likened to a bull by the Babylonians).

I must further point out that the Mayas made reference to the "horns of Venus"(18), and that the Babylonian Ishtar and the Egyptian Isis (which was that culture's name for Venus) were both depicted as having two horns (the ancients would have been able to see the crescent shape of Venus, which cannot be resolved with the naked eye, for reasons that will be explained later on).

Even though Venus is closer in toward the Sun than Earth, and thus would have technically been further away than Earth from the oncoming debris of the exploded trans-Martian planet, it's quite possible that Venus may have been more readily in line with a large piece of this debris before Earth was, if the Earth was on the other side of the Sun when the debris was approaching. Thus Venus may have been struck first, before Earth, enabling the ancients to have witnessed this event and transmit accounts of it to post-flood generations, as was obviously the case with the eyewitness accounts of the Mars impact, as discussed above.

======================

Having now completed our examination of the physical and ancient literary bodies of evidence that reveal a radically different history of the Earth and its planetary neighborhood, we are now ready to take a look at some physical and ancient literary pieces of evidence that paint a profoundly different picture of the history of ancient man.

Section II: Human History: Man in the post-flood world
Part 1: Intellect and technology

Those "primitive" cave dwellers

As the flood catastrophe came to a close and the Earth began to somewhat settle down to a relative calm, the few human survivors, stripped of any conveniences that their pre-flood civilization had provided them, were desperate to find ways to continue surviving in a totally hostile and broken-down world. The alien environment of less oxygen, extremes of cold and hot through seasonal changes, the lack of shelter and societal organization, scarcity of foodstuffs, occasional volcanic and tectonic "aftershocks" from the catastrophe, rotting vegetation and corpses of flood victims laying all about, polluted water, etc., made daily life a grim and burdensome affair, being totally devoid of leisure or luxury.

All of this drove these new progenitors of the human race to settle for living in caves and underground hovels to protect themselves from the merciless elements. Any housing structures that were built by human hands in this early post-flood world had to be constructed out of stone, as there were, at first, no trees to cut down for wood. And because of this lack of wood, even starting a fire must have been a very difficult task. Any food gathered at this time would have had to be eaten quickly, as no methods of preservation were readily available. We could thus imagine malnutrition and disease being widespread during this period.

Many of those who didn't die from these conditions probably met a homicidal fate, as other, less-scrupulous flood survivors (and early generations of their offspring) would have killed them to obtain their food and other vital necessities. It was obviously a very difficult time to live. The survival of the human race must have literally hung in the balance as critically at this time as it did during the flood catastrophe itself.

Numerous ancient legends from cultures around the world make reference to this era of desperation and deprivation that immediately followed the flood.

The Mayas recall "the great pain [the people] went through," and how "there was nothing to eat, nothing to feed on."

In their *Atra-Hasis*, the Babylonians similarly testified that, after the flood, "the womb of the earth did not bear...vegetation did not sprout...the broad plain was choked with salt"(17).

Likewise, the Chinese wrote about the plight of their post-flood ancestors thusly: "[They] dwelt in caves and desert places, eating raw flesh [Endnote 46] and drinking blood"(146).

In British Columbia, the Kwakiult Indians state that "when the flood had subsided there was no fresh water to be found. They were nigh on dying from thirst....But then a great rain came, very heavy and very long, which filled the valleys with fresh water lakes and rivers, which have remained there ever since."

An ancient prayer addressed by the Aztecs to their god Tezcatlipoca, evidently written way back during this time of struggle for survival, states: "Know, O Lord...the men have no garments, nor the women, to cover themselves with, but only certain rags rent in every part, that allow the air and the cold to pass [through] everywhere....With great toil and weariness they scrape together enough for each day, going by mountain and wilderness seeking their food; so faint and enfeebled are they that their bowels cleave to their ribs, and all their body re-echoes with hollowness, and they walk as people affrighted, the face and body in likeness of death....[T]hey draw a rag over them at night, and so sleep; there they throw their bodies, and the bodies of the children thou hast given them. For the misery that they grow up in, for the filth of their food, for the lack of covering, their faces are yellow, and all their bodies of the color of the earth. They tremble with cold, and for leanness they stagger in walking. They go weeping and sighing, and full of sadness...though they stay by a fire, they find little heat"(17).

This life-and-death struggle had remained the plight of these flood survivors and their offspring for several generations, until they could eventually repopulate, reorganize, and recivilize themselves by starting to rebuild a structured society. It was this period of fear and uncertainty that is commonly called the Stone Age. It was a "Stone Age," not because the people ("cavemen") who lived at that time were intellectually incapable of manufacturing more sophisticated instruments than stone tools, but because stone, as mentioned above, was for a time the only material available. Since the whole surface of the Earth had been completely reworked by the flood, it took a while for metals and other deposits of important natural resources to be discovered and mined [Endnote 47].

No discussion of "primitive" cavemen would be complete without mentioning Neanderthals. All kinds of crazy theories have arisen to try and explain what fossils of Neanderthal Man represent. Most of these theories are based around the fact that, while Neanderthals were obviously humanoid, they nonetheless possessed features that were distinctly different from modern man, i.e. irregularities in their bone structure. But this could easily be explained (for the most part, anyway) as being a consequence of the above-described excruciating living conditions they had to endure in the early post-flood era. As Harold Coffin wrote: "More than a hundred years ago [as of 1983], Rudolph Virchow, a distinguished German biologist, examined Neanderthal skeletal material and prepared a careful report in which he credited many of the skeletal characteristics to rickets, a Vitamin D deficiency disease." After mentioning osteoarthritis as another possible explanation for Neanderthal bone abnormalities, this author then went on to say: "In addition to rickets and osteoarthritis, the suggestion has arisen that Neanderthal Man suffered from syphilis. The disease produces bone deformity similar to that of rickets--indeed, they often occur together. Perhaps the low social and cultural life style envisioned for Neanderthal Man favored the progress of both diseases"(34).

The truth is, Neanderthal Man was not really that much different from modern man. As authors Alan and Sally Landsburg wrote in their book *In Search of Ancient Mysteries*: "Pioneer paleontologists pictured Neanderthals as barrel-chested, hairy, slow-moving, chinless, and brutish. But we now deduce that they had no more body hair than we do...and that they weren't always dull brutes.

"Today's experts are more inclined to think that if we gave a Neanderthal a shave and haircut, and dressed him in well-fitting clothes, he might walk down New York's Fifth Avenue without getting many second glances"(59).

As far as the shape of Neanderthal skulls goes, exhibiting a somewhat forward-protruding upper and lower jaw, it is important to point out that all modern ethnic groups have quite differently-shaped skulls. Neanderthal might simply represent an extinct ethnic group that did not survive to modern times--or perhaps it did. According to the following quote from *Insight* magazine, there is precious little difference between the skull shape of Neanderthals and modern Europeans: "For years, many anthropologists have maintained that Neanderthal man disappeared from Earth approximately 35,000 years ago. 'But the fact is, the West European Neanderthals are today's West Europeans,' says C. Loring Brace, curator of the physical anthropology division at the University of Michigan's Museum of Anthropology....'In every respect, the shape of the West European skull is closer to the shape of the classic Neanderthal cranium than to that of any other modern group in other parts of the world,' he says, adding that there is no evidence Neanderthals walked any less erect than modern man and from the neck down, the only difference between Neanderthals and modern man is the 'indication of generally greater ruggedness in Neanderthal joints and muscles'"(60).

Incidentally, the same holds true for Cro-Magnon Man and other supposedly "primitive" human remains that scientists have found--the differences between them and modern man can either be attributed to non-extant ethnic groups, or simple bone deformities caused by rampant disease conditions that prevailed after the flood.

As time went by after the catastrophe, most of these "primitive" cavemen managed to recivilize themselves and had gone on to build great cities and establish far-reaching trade networks. However, there were some who, for various reasons, retained their primitive existence, and in some cases had even degenerated down to the status of brutish beasts, remaining in this condition for centuries and even millennia after the flood. There are actually several ancient records that talk of these ancient dregs of society.

One of these records is found in the Babylonian story of Gilgamesh, which mentions a certain Enkidu, who, as a youth, had lived like an animal among the animals. His hair was long and his nails and teeth were developed for gathering and eating herbs, and he was devoid of speech. This was his plight until, one day, one of his civilized contemporaries took him in and taught him the ways of urban life. The ancient text makes it clear that Enkidu's lot was not uncommon. The only unique aspect of his story is that he was one of the rare examples of his kind that leaned to adapt to a more "modernized" existence.

India's *Ramayana* contains a story similar to this one. It describes a race of "ape-men" who aided the noble Rama in a war against the Ravana kingdom of Ceylon. The most famous among this group was Hanuman. His appearance was that of an ape, yet he was endowed with a good sense of humor, was able to speak, and had great bravery. He also possessed a great knowledge of edible and curative wild plants, and he was intimately familiar with the remote hill and forest country.

The Chinese likewise have had an ancient tale of such a primitive band of people bequeathed to them by their remote ancestors, except that their version indicates that this breed was a menace to society. These brutish people were called Mao-tse, and were described in the *Shu King* as "an ancient and perverted race who in olden days retired to live in rocky caves, and the descendants of whom are still to be found in the vicinity of Canton." They were said to have "troubled the earth, which became full of their robberies." Emperor Shi Huangdi, so disgusted with their behavior, ordered his generals Tchang and Lhy to exterminate them. It should be pointed out here that it was just outside of Canton that the remains of "gigantopithecus blacki" were discovered--a race of people, as you may recall, that were between 10 and 12 feet tall. It's also important to draw attention to the fact that this group of people, according to archaeologists, disappeared suddenly, without a trace. Was this because they were exterminated?

Finally, the biblical book of Job relates the existence of a similar race of men in the region of Mesopotamia. They lived in solitude in the wilderness, eating wild grasses and leaves, and often resorted to stealing food and whatever else they could get their hands on, just like the Chinese Mao-tse. Furthermore, they lived in the rocks and cliffs like animals, and had no intelligent speech. Job denounced them all as "a scourge to the land" and "children of fools"(9). Was this the community from whence Enkidu came, as mentioned in the Babylonian story of Gilgamesh?

I want to emphasize that these unintelligent degenerates were relatively rare--the exception, and not the rule. Yet modern science wants us to think in reverse terms. They pawn ALL cavemen off on us as endemically dumb, primitive, brutish thugs that were entirely incapable of advancing. But why? Why would they apply the "dumb brute" paradigm to all prehistoric cave dwellers, as though they truly were incapable of progress? The answer is twofold.

The first part of the answer is a simple matter of pride. Modern man loves to extol himself in thinking that his current status constitutes the pinnacle of what is said to have been an ongoing process of scientific and technological development, and that no period in history has ever achieved, or even been capable of achieving, the recent advancements that have been made in these areas. But is there any truth to these assertions? Is there any justification for such pride? Not according to a fascinating Time-Life book called *Feats and Wisdom of the Ancients*, from which we now quote: "Smug in the certainty that we stand in the vanguard of history's forward march, modern humans tend to regard the past as little more than a backdrop for current achievements. Progress, we think, has been a steady, linear advance, leading inexorably from superstition to enlightenment, from primitivism to sophistication, from ignorance to knowledge. This comfortable assumption is being challenged, however, as archaeologists, historians, and other scholars point out evidence hinting that humankind's progress has been less a lockstep march than a lurching dance, with steps forward and steps back, with bursts of genius ahead of its time and lost knowledge that even now rests unretrieved--or is gone forever....How much more did our ancestors know? In our own ignorance, we are just beginning to learn the answer"(62).

The second part of the answer to the question of why early man is deemed to have been inferior to modern man is the dogma of evolution. Since evolution teaches that all plants and animals, including man himself, evolved from simpler to more complex forms, it is held that early man, supposedly having emerged from the ape family, must have been decreasingly intelligent the further back in time we look. The lack of a written language for early cave dwellers is often cited as a good rationale for drawing this conclusion. But this is utterly ridiculous. It's like saying that modern secluded cultures in the remote areas of Africa and South America, which are also devoid of a written language, are lower (or unevolved) forms of humanity.

========================

Is there a correlation between a written language and an advanced society?

The truth is, there have been many highly-advanced cultures in the past that did not develop a system of writing. One prime example of this was the people who built the advanced and complex city of Teotihuacan, in modern-day Mexico City. Here's what the official website of the Metropolitan Museum of Art has to say about this ancient necropolis: "Teotihuacan, located in the highlands of central Mexico, is one of the world's most impressive archaeological sites. Between 100,000 and 200,000 people lived there at its peak around 600 A.D., making it one of the ancient world's largest cities with an urban core covering some twenty square kilometers. Settlement began about 200 B.C. and the basic layout of the city was complete by the mid-second century A.D. Most of the major construction was accomplished within the next hundred years. In plan, Teotihuacan is a complex urban grid filled with single-and multi-floor apartment compounds. This grid, unique in Mesoamerica in its scale and organization, implies a high degree of social control. Presumably an elite group of nobles directed the building projects and coordinated trade and tribute relations with far-flung corners of Mesoamerica"(120).

If such sophistication could be developed without writing, it should be quite clear that the practice of writing is not a prerequisite for an advanced civilization. This is one of the great misconceptions of modern anthropology.

Another advanced ancient people that did not possess a written language was the Anasazi Indians of the Southwest United States. This culture built the famous Pueblo Bonito compound, the largest of 14 Anasazi "houses" constructed in Chaco Canyon between A.D. 900 and 1150. The D-shaped floor plan of this set of ruins included five stories, 800 rooms, and a central plaza with 37 kivas (ceremonial chambers). It was designed and oriented in such a way that it would be heated internally by the Sun. It is also aligned with the four cardinal points, like many other ancient constructions(121).

The Anasazi were obviously able to develop, without the use of writing, a highly-advanced civilization, and had managed to pass on their collective knowledge for centuries, again, without the aid of writing.

Evidence shows that the Anasazi had constant contact with other cultures, like the Aztecs, who did have writing. So it is clear that the Anasazi were aware of the concept of writing, but found no need for developing their own version thereof (or adopting anyone else's). It simply wasn't something they felt was necessary, and apparently they were correct.

Still another ancient people that had no written language, and yet were undeniably highly advanced, were the builders of the ancient Bolivian city of Tiahuanaco, located high in the Andes Mountains, not far from Lake Titicaca. It has got to be the most impressive city ever built by any ancient culture--including the ones that did have a written language.

Some of the blocks that make up this complex weigh as much as 440 tons, and were quarried over 10 miles away. They were also hauled up the 12,500-foot mountain upon which they now rest. Such a monumental task could not be accomplished even today. Yet we're supposed to believe that we, with our written languages and heavy-duty construction equipment, are smarter than they were?(122)

While today's "experts" would have us believe that a written language is required of a civilization if it is to be labeled as advanced, there was at least one very early Egyptian king who held a totally opposite view. He saw writing, instead, as a sign of degradation of civilization.

In his book *Phaedrus*, the Greek philosopher Plato talked of how the Egyptians believed their hieroglyphic system of writing to have been invented by Thoth, the god of learning. Thoth, as the story goes, took his invention to the pharaoh of the day, Thamus, and claimed that it would help increase wisdom. But the king was not so optimistic, expressing his concern that the reverse would happen. He argued that writing would result in people, even the educated, forgetting information and becoming too dependent upon their personal libraries for the retention of information(61).

========================

One of the best ways to expose the fallacy of the presumption that earlier generations of man were less sophisticated than his modern descendants is to look at the inventions and discoveries of Classical Greece, Rome, and other contemporary cultures. Doing so will also help us to see that our modern advancements are not the cumulative result of a continuous, unbroken, uphill climb throughout the millennia, and that a good many of them are not unique to our era.

Advanced developments in Classical Greece, Rome, and other contemporary cultures

The following brief catalog of discoveries and inventions from Classical Greece, Rome, and other cultures of roughly the same time-period focuses mostly on innovations that were long forgotten and/or suppressed during the Dark Ages, only to be reinvented or rediscovered at the time of the Renaissance and the centuries that have followed. As you read through this list, ask yourself this question: If the Classical Greeks, Romans, and various other cultures of roughly 2,000 years ago weren't less intelligent than us today, why should we believe that those of any previous era were less intelligent?

We will begin with Pythagoras, a noted Greek philosopher who founded the famous Pythagorean school--a man who deduced that the Earth was round back in the sixth century B.C. Contrast this with what was taught in medieval Europe just four short centuries ago--that the Earth was flat, and that one who taught otherwise was a "heretic"(62).

Eratosthenes, a Greek mathematician, geographer, and astronomer, measured the Earth's circumference fairly accurately, as well as the relative sizes of the Sun and Moon, in the third century B.C. The accuracy of such measurements was not matched or surpassed until the Renaissance(Ibid.).

The Greek scholar Aristarchus, also of the third century B.C., knew that it was the Earth that went around the Sun, and not the other way around. He even correctly stated that the orbit of Earth was oblique (elliptical, or non-circular). In addition, he determined that the day/night cycle was due to the rotation of the Earth on its axis(Ibid.). But even earlier than this, Anaximander (610-547 B.C.) stated: "The earth is round and it revolves around the sun." Heraclides (fourth century B.C.) and Seleucus of Erythrea (second century B.C.), probably copying Anaximander, both said these exact same things(61). Yet many centuries later, during the Dark Ages, not only was it "heresy" to deny that the Earth was flat, but it was also forbidden to teach that it orbited the Sun. So forbidden, in fact, that it was a crime punishable by death.

Democritus, a Greek philosopher from Abdera in Thrace who lived in the fifth century B.C., stated: "Space is filled with myriads of stars and the Milky Way is but a vast conglomeration of distant stars." He also once said: "There are more planets than the ones we can see"(Ibid.). These were amazing insights that did not begin to be rediscovered until the seventeenth century.

The Greeks were also quite the ingenious inventors of various types of sophisticated mechanisms. Probably the best example of this is the famous Antikythera device. In 1900, sponge fishermen off the coast of Antikythera found an ancient sunken Greek vessel. The most impressive item recovered from its cargo was a badly-corroded bronze artifact that later caught the attention of archaeologists working at the Greek National Archaeological Museum in Athens. This object turned out to be a mechanical gadget made up of a series of gears, looking much like a giant watch. The device dated to about the first century B.C., and appeared to be some kind of navigational tool--perhaps an early astrolabe, an instrument used to determine the positions of celestial bodies. Later, in the 1950s, x-rays and gamma-rays were used to probe the artifact's internal structure, which revealed that it contained layer upon layer of various-sized gears. Upon further investigation, it was determined that this instrument was indeed able to calculate the movements of the Sun, Moon, and planets. In other words, this object was a 2,000-year-old analog computer. Time-Life described it like this: "The whole system of gears required the turning of a handle once a day to keep it running continuously and accurately, tracking the movements of heavenly bodies. The gears were linked by a toothed turntable that acted as a differential gear train, permitting two shafts to rotate at different speeds. Operating on the same principle that allows the traction wheels on modern automobiles to turn at different rates on curves, the differential gear had long been assumed to be an invention of the seventeenth century....[The] Greeks, it seems, had a mechanical genius after all"(62).

This was apparently not the only device of this sort in ancient Greece. *Nature* magazine's news website reported on November 29, 2006: "Almost everyone who has studied the [Antikythera] mechanism agrees it couldn't have been a one-off--it would have taken practice, perhaps over several generations, to achieve such expertise. Indeed, Cicero wrote of a similar mechanism that was said to have been built by Archimedes. That one was purportedly stolen in 212 BC by the Roman general Marcellus when Archimedes was killed in the sacking of the Sicilian city of Syracuse. The device was kept as an heirloom in Marcellus' family: as a friend of the family, Cicero may indeed have seen it"(164).

It was not just the Greeks who created "computers" like these. The Chinese had produced a similar device, although not until several centuries later. Speaking of Tao Hung Chin (452-536 A.D.), the device's inventor, Andrew Tomas, author of *We Are Not the First*, wrote: "He also built a celestial globe which was about one meter high. The earth was in the middle and remained stationary while the heavens revolved about it. The twenty-eight stellar mansions thus fulfilled their periods, and the Seven Bright Ones [Sun, Moon, and five planets] pursued their courses. The stars were luminous in the dark and faded in the light. The globe was constantly revolving by a mechanical device, and the whole thing agreed with the actual motion of the heavens"(61).

Coming back to the Classical Greeks, let us consider their development of the screw pump, which is still in use today. This ingenious device of interacting levers, pulleys, and grips was primarily used for lifting very heavy weights. One specific application of it was to grab, tilt, and sink enemy ships(63).

The Greeks, and later the Romans, had built powerful tripod cranes that utilized a series of gears and pulleys. These mechanisms were used for lifting very heavy stones into place during construction projects(64).

Henry Hodges, in his book *Technology In the Ancient World*, informs us of another innovative Greek invention: "Mechanized milling of corn is known from written sources to have been developed by the first century B.C. in northern Greece and western Anatolia. The type of mill used appears to have been essentially a turbine in which water from fast mountain streams was directed through a chute on to the blades of the water wheel, so rotating the upper millstone. This device seems to have spread rapidly throughout the Roman Empire, and in basically the same form is to be seen in many mountainous areas today from the Middle East to northern Europe"(Ibid.).

There were other Greek innovations, which, though no less ingenious, did not seem to be put to much practical use. For instance, in the fourth century B.C., the Greeks constructed a vault with "magnetic stones" so that idols could be suspended in mid-air, dazzling onlookers(63). Perhaps the only practical application here was the exploitation of gullible patrons.

Another apparently useless but brilliant Greek creation is described for us in *Technology of the Gods*, by David Childress: "The ancient Greeks built steam boilers that worked, but they [appear to have] used them only as gadgets and toys rather than as practical sources of power. One such toy was a sphere that was spun by two steam jets, and was 'invented' in Ptolemaic Greek Egypt circa 200 B.C."(15)

Speaking of Ptolemaic Egypt, Childress went on to expound upon a rather modern-sounding invention: "[I]n the second century before the Christian era, Egyptian temples had slot machines for holy water. The quantity of water which flowed from the tap was in direct relation to the weight of the coin thrown into the slot. The Temple of Zeus in Athens had a similar, automatically-controlled holy water dispenser. A coin was dropped into a sealed vessel which made a small plunger pull up, which then allowed a measured quantity of fluid to be dispensed. The famous Greek/Egyptian inventor Hero of Alexandria invented one such well-known device in 120 B.C. It is evident from this example that temples and priests were involved in technology at an early time"(15).

Moving on to the Romans, we find that Lucretius, a poet and deep thinker of the first century B.C., believed that matter is composed of atoms, an idea that he no doubt inherited from the Greeks (particularly Democritus). But he expanded on it, saying that even energy, such as lightning, is composed of infinitesimally-small particles. He said that lightning was a "rarified fire...composed of minute and mobile particles to which absolutely nothing can bar the way." He also noted that thunder and lightning occurred simultaneously, even though lightning is most commonly seen before its accompanying thunder is heard, and properly explained that the reason for this is that light travels faster than sound(65).

The Romans had produced quite an assortment of high technology inventions. This is not surprising, considering the legacy left to them by the Greeks. Nevertheless, some of their technological achievements were totally unique.

The first example we will consider might not seem so impressive at first. However, we need to realize that it had represented at that time a tremendous leap forward, revolutionizing the world of architecture. As Hodges explains: "One of the few original contributions of Rome to building technology was the introduction of cement. Apart from its use as a bonding material, it was also employed in the making of concrete, which, combined with a brick facing, allowed the construction of solid arches, so eliminating the need for buttressing the supports. Arches of this type were to be seen in their most dramatic form in the building of aqueducts. Indeed, the Romans devoted a great deal of time and money to public sanitation, creating a water supply, drainage and other forms of sanitation often as good as, if not better than, those to be found in many parts of Europe today"(64).

We generally think of our modern luxury cruise ships as being endemic to our technologically advanced society, and that such things didn't exist--or couldn't have existed--in ancient times. But back in the 1920s, two Roman ships were found at the bottom of Lake Nemi in Italy, which revealed that there was precious little difference in the accommodations afforded passengers on large ships between then and now. These Roman vessels were made to carry 120 passengers in 30 cabins, plus the crew, which had their own separate quarters. Each boat was adorned with mosaic floors that depicted scenes from the *Iliad*, walls of cypress paneling, paintings in the lounges, and even a library. There were also sun dials in the ceiling for telling the time, along with musicians for entertainment. The sterns contained restaurants and a kitchen. Copper heaters provided hot water for the baths, utilizing modern-looking brass water pipes(61).

It was perhaps in the medical field where the Romans were most proficient and resourceful. They built large military hospitals and performed surgeries of all types, and with the most delicate, finely-crafted instruments. Most impressive was their skill in removing cataracts, as we see from the book *Ancient Inventions*: "The medical writings of Cornelius Celcus, who lived under the Emperor Tiberius (A.D. 14-37), include a detailed description of a cataract operation. Great care was needed, particularly in the preparation, as Celcus stressed: 'The patient...is to be seated opposite the surgeon in a light room facing the light, while the surgeon sits on a slightly higher seat; the assistant from behind holds the head so that the patient does not move; for vision can be destroyed permanently by slight movement. In order also that the eye to be treated may be held more still, wool is put over the other eye and bandaged on. Further, the left eye should be operated on with the right hand, and the right eye with the left hand'"(65).

For the less-skilled surgeon, Celcus recommended leaving the cataract in the eye, but simply pushing it out of the way so as not to obstruct the patient's vision. He wrote: "The needle used is to be sharp enough to penetrate, yet not too fine; and this to be inserted straight through...at a spot between the pupil of the eye and the angle adjacent to the temple, away from the middle of the cataract, in such a way that no vein is wounded. The needle, however, should not be inserted timidly. When the spot is reached, the needle is to be sloped against the colored area [lens] itself and rotated gently, guiding it little by little below the pupil; when the cataract has passed below the pupil, it is pressed upon more firmly in order that it may settle below....After this the needle is drawn straight out; and soft wool soaked in white of egg is to be put on, and above this something to check inflammation; and then bandages"(Ibid.).

Trepanation--a very delicate head surgery that involved removing a section of the skull to expose the brain (probably done to relieve pressure caused by an injury, or maybe even for removing tumors)--was practiced by many ancient cultures, including the Romans. The Romans began implementing this procedure in the second century B.C., which involved the use of a bronze surgical drill kit. These operations were obviously quite successful, by the way, since the skulls found of those who underwent them show signs of healing, as well as indications that the patients lived long after these surgical procedures were performed.

Anesthetics were employed during such operations. In fact, the Aztecs used the same type of anesthetic as the Romans, except for a less wholesome purpose. As *Ancient Inventions* tells us: "The most powerful anesthetic known to the Romans was the mandrake, a plant with an awesome reputation. Though it is difficult to be sure that all references to 'mandrake' in ancient literature are actually to the same plant, it is reasonably certain that the mandrake used in medicine was datura (also known as jimson weed or thorn apple). Datura contains atropine and hyoscine, drugs that slow down the heart rate and, in the right doses, can completely deaden pain as well as reduce the trauma experienced by a patient undergoing surgery. Datura tea was one of the substances fed to their sacrificial victims by the Aztecs to keep them drowsy and oblivious to their eventual end. Around A.D. 75 Pliny described how it was put to more constructive use by Roman doctors: 'When the mandrake is used as a sleeping draught the quantity administered should be proportional to the strength of the patient, a moderate dose being one cyathus [about three tablespoonfuls]. It is also taken in drink for snakebite, and before surgical operations and punctures to produce anesthesia. For this purpose some find it enough to put themselves to sleep by the smell'"(Ibid.).

Gynecology was developed to a very advanced stage in ancient Rome, on a par with modern medical science. Speaking on this matter, Childress wrote: "[A]ccording to the October 20, 1900 issue of *Scientific American*, excavations at Pompeii revealed gynecology to be but a 'reinvention in the world of surgery.' Instruments buried in the Temple of Vestal Virgins since the eruption of Vesuvius in A.D. 79 were found to demonstrate that 'gynecology was a science flourishing in its perfection long before [our time]....[I]n every instance the instruments are almost in the minutest particulars exact duplicates of those in use by the most approved modern science of today....The workmanship is as fine as anything to be produced in this line in [modern times]....The instruments are hand wrought...and capable of delicate manipulation as anything to be found in today's achievements"(15).

Physicians in India were no less competent than those in Rome at about this same time. In the last few centuries B.C., Indian surgeons performed plastic surgery, particularly on those who lost their earlobes (perhaps from earrings being torn off during battle). The Hindu work *Sushruta Samhita* described the procedure this way: "A surgeon well versed in knowledge of surgery should slice off a patch of living flesh from the cheek of a person devoid of earlobes in a manner so as to have one of its ends attached to its former seat [i.e. the cheek]. Then the part where the artificial earlobe is to be made should be slightly scarified [with a scalpel] and the living flesh, full of blood and sliced off as previously directed, should be stuck to it"(65).

India also produced some adept astronomers several centuries later. The Hindu astronomical text *Surya Siddhanta*, dating to about 400 A.D., states that the Earth is "a globe in space." Later on in this same text, it states: "Everywhere on the sphere [Earth] men think their own place to be on top. But since it is a sphere in the void, why should there be an above and an underneath?"(18).

Returning to our discussion of the medical profession, we find that the Etruscans, in the sixth century B.C., were skilled in the art of dentistry. They were able to make braces out of gold, and had replaced missing teeth with human or oxen teeth, riveting them to a gold band fastened to the base of the other surrounding teeth in the mouth(62).

We couldn't end this discussion without mentioning some of the astonishing medical practices of the Chinese during approximately this same period.

In the late fifth century B.C., for example, Chinese surgeons performed heart transplants, utilizing techniques that were similar to those of today. While laying on an operating table, the patient had his heart removed from his chest as tubes fed him infusions. After the new heart was introduced into the chest cavity, two surgeons closed up the arteries and then stitched the opening in the chest(63).

Hou Han Shu, an ancient Chinese chronicle, discusses the surgical methods of Hua To, who had employed anesthetics during the late Han Dynasty (25-220 A.D.). Speaking of him and his techniques, Andrew Tomas wrote: "He first made the patient swallow hemp-bubble-powder mixed with wine, and as soon as intoxication and unconsciousness supervened, he made an incision in the belly or the back and cut out any morbid growth. If the stomach or intestine was the part infected, he thoroughly cleansed these organs after the use of the knife, and removed the contaminating matter which had caused the infection. He would then stitch up the wound, and apply a marvelous ointment which caused it to heal in four or five days, and within a month the patient was completely restored to health"(61).

* * * * * * *

The Classical Greeks, Romans, and other nearly contemporaneous cultures had indeed made incredible advancements over two millennia ago. And yet, not long after this period of progress had passed, nearly all these advances were forgotten, and mankind as a whole soon slipped into the Dark Ages. Clearly, this sequence of events does not coincide with the notion of mankind having undergone continuous, progressive stages of evolutionary intellectual and technological development. We see, instead, a long era of *devolution* that followed a surge of advancements.

Just imagine if the pattern of progress begun by the Classical Greeks had not been interrupted. Think where we would be today if mankind had continued to build on that foundation. How much further advanced do you suppose we would be today? And we probably don't even know the half of what they had discovered and invented. How many more things did they know than what we give them credit for? How many other advances did they make that are now perhaps lost forever? But the real questions we should ask are these: How did many of their advancements become lost, and why didn't their progress continue, without any interruptions or setbacks, into our day?

Advancements are not so easily handed down to posterity

Two of the main reasons why so much of the progress of the Classical Greeks and their contemporaries has been lost to us, and why that pattern of progress did not continue in a steady, upward climb, are: a). wanton destruction resulting from senseless wars of conquest, and b). national internal strife, both of which have ever been the ceaseless plights of mankind. As one country overthrew another, or as one political faction vied for control against another within a given nation, the casualties were devastating, not only in the loss of human lives, but in the loss of the libraries that housed their accumulated knowledge. It is to some of the more significant examples of these irreplaceable literary losses that we now turn our attention.

Andrew Tomas documented quite a few of these incidents. He wrote: "The famous collection of Pisastratus [Pisander] in Athens (sixth century B.C.), was ravaged. Fortunately the poems of Homer somehow survived. The papyri of the Temple of Ptah in Memphis were totally destroyed. The same fate befell 200,000 volumes in the library of Pergamus in Asia Minor. The city of Carthage, razed by the Romans in a seven-day fire in 146 B.C., is said to have possessed a library of half a million volumes. But the greatest blow to history was the burning of the Alexandrian library in the Egyptian campaign of Julius Caesar, during which 700,000 priceless scrolls were irretrievably lost"(61).

The Alexandrian library recovered, only to fall victim to two more great burnings. One was done under the direction of the Roman emperor Theodosius I, around 390 A.D. In his day, Roman Christianity was the official religion of the empire, so he deemed it necessary to destroy all the knowledge of what he called "pagans." The last burning followed the Arab conquest of Egypt in 641 A.D. For six months, millions of scrolls were used to heat the city's bath water. Caliph Omar, the Arab leader, decreed: "The contents of these books are in conformity with the Koran or they are not. If they are, the Koran is sufficient without them; if they are not, they are pernicious. Let them therefore be destroyed."

The fate of libraries in other times and places was hardly any better. For example, China's first Q'in emperor, Shi Huangdi, ordered the burning of most Chinese books in 212 B.C.

Emperor Leo III sent 300,000 books to the furnaces of Constantinople during the iconoclastic controversy of the eighth century.

Literary losses were particularly significant during the Dark Ages. As Tomas wrote: "The number of manuscripts annihilated by the Inquisition...in the Middle Ages can hardly be estimated. Because of these tragedies we have to depend on disconnected fragments, casual passages, and meager accounts"(Ibid.).

And then there was the destruction that took place in the New World. The Mayan chronicles, for instance, were annihilated by Bishop Diego de Landa in the sixteenth century. Along with the loss of these ancient volumes was the key to translating the Mayan system of writing, which we know almost nothing about today(18). Speaking in regards to this literary demolition, de Landa wrote: "We found great numbers of books [written on deer skin] but as they contained nothing but superstitions and falsehoods of the devil we burned them all, which the natives took most grievously, and which gave them great pain"(10).

We could go on to cite more instances like this. However, enough has been said to illustrate the point of just how fragile man's intellectual and technological achievements really are. What can take centuries, and even millennia, of observation, research, testing, experimentation, and development can be irretrievably lost in an alarmingly short time. Not only can such destruction be wrought at the hand of man, but also by natural catastrophes (see Appendix B), which can be even less merciful than man, as was the case with the flood catastrophe.

But even on the occasions when records of ancient knowledge have escaped both manmade and natural calamities, the hand of fate has often dealt out another deadly blow, which is no less harsh--disintegration. There is absolutely no material known to man, whether organic or inorganic, that is immune to this process. Everything, given enough time and enough exposure to the elements, will eventually wear away and break down into tiny specks of dust.

So it's quite probable, due to all of the above-mentioned destructive factors, that there were indeed many other great advancements made by the Classical Greeks, Romans, and other cultures of that period, which have not been able to leave for us even a faint trace of their former existence.

We therefore arrive at a most provocative question: If so little remains of the knowledge of the Classical period, which was only about 2,000 years ago, what about the millennia that preceded that time? Were there not advanced cultures that existed long before the Classical Greeks, whose knowledge was lost to these Greeks and their contemporaries, leaving it up to them to rediscover and reinvent at least some of what was already known long before their era? This seems to be exactly what the ancient Hebrew ruler, King Solomon, was alluding to when he wrote in the biblical book of Ecclesiastes the following fascinating statements: "The thing that hath been, it is that which shall be; and that which is done is that which shall be done: and there is no new thing under the sun. Is there any thing whereof it may be said, See, this is new? it hath been already of old time, which was before us. There is no remembrance of former things; neither shall there be any remembrance of things that are to come with those that shall come after." Ecclesiastes 1:9-11.

A somewhat parallel declaration to this is found in Plato's *Timaeus*: "One of the [Egyptian] priests, who was of a very great age, said: 'O Solon, Solon, you Hellenes [Greeks] are never anything but children, and there is not an old man among you.' Solon in return asked him what he meant. 'I mean to say,' he replied, 'there is no old opinion handed down among you by ancient tradition, not any science which is hoary with age.'" In other words, the priest was informing Solon that Egyptian wisdom and science had a very long tradition that dated back to archaic times, whereas the Greeks, in comparison, were relative newcomers to the realm of scientific inquiry.

The first-century Jewish historian Josephus was even more frank when he wrote on this wise: "[Y]ou will find all things among the Greeks to be recent, having come into existence, as one might say, yesterday or the day before; I mean the foundation of their cities, and their invention of the arts, and the registration of their laws....

"[A]ll with one voice acknowledge that...the Greeks...got their learning from Egyptians and Chaldeans [Babylonians]..."(110).

Shortly we shall see that there were definitely advanced cultures that long preceded the Classical Greek period, which had apparently been even more highly developed than the Greeks and their cohorts. But first let us turn the clock back even further, to before the flood catastrophe. We will endeavor to answer the question, Is it possible that pre-flood man may also have been intellectually and technologically advanced, but the flood disaster prevented him from passing on to future generations much of his discoveries and inventions? Plato certainly gives us this impression. He stated in *Timaeus* that the Egyptian priests told Solon: "[W]hen the stream from heaven [the flood catastrophe] descends like a pestilence and leaves only those of you who are destitute of letters and education...you have to begin all over again as children and know nothing about what happened in ancient times..."(29).

Pre-flood man's superior brain function and consequent advanced intellectual and technological achievements

The notion that ancient man, the further back in time we look, was decreasingly mentally sophisticated is nothing but pure nonsense. If anything, the evidence suggests that early man, especially before the flood catastrophe, was potentially *more* mentally developed than his modern descendants. The very environment of the pre-flood Earth, which we discussed earlier, may have facilitated the development of a higher level of brain performance. Here are three reasons why:

- It is a known fact that the brain is highly dependent upon oxygen for its proper and optimal functioning. It would make sense to say, then, that since the Earth's antediluvian atmosphere contained more oxygen than it does at present, the functioning capacity of the brain had likely been of a higher order than it is today.

- If the increased oxygen content of the antediluvian atmosphere did indeed create, as we discussed earlier, a larger growth size for many plants and animals, including man himself, this would have resulted in a larger human brain size, and thus would have further contributed to an expanded brain functioning capability.

- As we also saw earlier, the increased oxygen in the pre-flood atmosphere may have precipitated extended lifespans for all members of the plant and animal kingdoms. Just think of what the scientific and technological implications of this may have been for man at that early date. Scientists and inventors would have had a much longer stretch of time to develop and perfect their inventions and refine their theories. To put this in a more modern perspective, imagine how much more Einstein could have accomplished if he had lived for several hundred more years(66)[Endnote 48].

If pre-flood man truly did possess a more advanced functioning brain, would it not stand to reason that he may also have possessed an advanced technology, perhaps even superior to our own? While it is obvious that these antediluvians did not require a superior brain in order to develop an advanced technology, it is clear that having such a superior mind would have increased the chances of such a technology being developed. But is there any actual evidence that pre-flood man achieved a high level of technological development? Indeed, there is.

Looking first to the realm of ancient records, we find some fascinating accounts of this very thing--a technologically advanced civilization that predated the great flood catastrophe. Commenting on this issue, catastrophist Richard Mooney wrote: "Most legends say that before the flood lived gods, or god-men with strange powers...whose knowledge [was] great....Are the legends of...the Golden Ages of many races echoes of the world as it once was, before a great catastrophe occurred?

"The concept of an older, vanished civilization exists as far back as written records exist. Much of religious tradition and literature points, even if obscurely, to such a concept"(5).

What follows are a few examples of ancient legends of this variety:

- Plato, in his *Timaeus*, mentioned the "great and marvelous actions of [the Atlantean] ancestors of the Athenians, which had passed into oblivion through time and the destruction of the human race"(29).

- Josephus, in the first century, apparently relying on then-available ancient documents, wrote about the survivors of the flood catastrophe, saying that they did not desire to be "deprived of those good things which they enjoyed before the flood"(67).

- Lao Tzu, born in China in 604 B.C., wrote: "The Ancient Masters were subtle, mysterious, profound, responsive. The depth of their knowledge is unfathomable"(68).

Though no specific reference is made here to the pre-flood world, the mention of these "Ancient Masters" having a knowledge that was "unfathomable" seems to suggest a people who, at a very remote point in history, possessed superior brain functions.

- The Mayan *Popol Vuh* talks about the people of "the first race," who were "capable of all knowledge"(18).

Could this not also be a reference to pre-flood man, who was "capable of all knowledge" because of his superior brain capacities? Surely the mention of the "first race" must be referring to this very early point in human history [Endnote 49].

- Josephus talked about a desire on the part of pre-flood man to preserve all antediluvian technological discoveries for post-flood generations. He wrote: "And that their inventions might not be lost--upon Adam's prediction that the world was to be destroyed...by the violence and quantity of water--they made two pillars, one of brick, the other of stone: they inscribed their discoveries upon them both, that in case the pillar of brick should be destroyed by the flood, the pillar of stone might remain and exhibit these discoveries to mankind..."(20).

Since we only know about the discoveries of the Classical Greeks, Romans, and others of that period from a few records and artifacts that managed to survive to our day, we shouldn't be surprised that there is virtually no written record remaining of the technology that appears to have pre-existed the flood catastrophe, except for a few tantalizing hints from dusty old manuscripts that record faded memories thereof, such as the five examples cited above. But what about physical evidence? Here, too, we find the traces to be almost non-existent. Yet there have been some curious artifact finds made over the years, like the ones described below, which plainly point to a pre-flood origin, since they have been found encased in solid sedimentary rock that undoubtedly formed during the time of the flood. Though these artifacts don't all necessarily point to a highly advanced technology, per se, they nevertheless do show that the people who made them were quite industrious, and that they had a considerable degree of intelligence that at least equaled our own, if not having outright surpassed it.

The first intriguing artifact we will look at was discovered on February 13, 1961, when three rock collectors from the LM and V Gem and Gift Shop in Olancha, California, went rock hunting in the Coso Mountains just outside of Olancha. Near the top of a 4,300-foot peak they found an unusual round rock with fossil shells encrusted on the outside. Upon sawing this rock open, it was found to contain an object that resembled a spark plug. In the center of this object was a metal core about 2 mm in diameter, encased in a cylinder-shaped ceramic material which was further encased in a hexagonal sleeve of wood that had become petrified. It appeared that some copper wiring was also present at one time, which had long since corroded away. An x-ray of the object revealed what looked like a screw and a washer within the rock as well(9, 62).

Beginning in 1991, on the banks of the Narada, Kozim, and Balbanyu rivers in the eastern Ural Mountains, gold prospectors have been finding tiny metal objects, deep underground, that are obviously of human manufacture. Made of copper, tungsten, and molybdenum, they have mostly spiral shapes and range in size between 1.2 inches and .003 millimeters long. They look, in fact, like products of modern nano-technology, yet they have been found in upper Pleistocene deposits that are claimed to be 100,000 years old(124).

Over the past several decades, beginning back in the 1970s, miners at Klerksdorp in South Africa have uncovered hundreds of metal spheres in a stratum of earth supposedly 2.8 billion years old. The spheres measure approximately one inch in diameter, and contain very finely-etched grooves which specialists conclude could not have been created by any natural process. Two types of these spheres have been noted: one is composed of a solid bluish metal with flecks of white; the other is hollowed out and filled with a spongy white substance(69). While it is not known what function these objects may have served, it is obvious that they were made by human hands.

Inside a supposedly 65-million-year-old Cretaceous chalk bed at a quarry in Saint-Jean de Livet, France, several semi-ovoid metallic tubes of identical shape but varying size were found back in 1968(39). Though there's no way of knowing what purpose these tubes were used for, they strongly hint at an advanced culture that clearly existed before the flood, when their encasing chalk bed was formed.

Back in 1885, in an Austrian foundry owned by Isador Braun of Vocklabruck, a block of coal was broken open revealing a small metal cube inside. Braun's son, astonished by this discovery and realizing that it must represent something very significant, brought this mysterious cube to the Salzburg Museum, where it was closely inspected by Karl Gurls, an Austrian physicist. His tests indicated that the object was composed of a steel and nickel alloy. Its measurements were 2.64 by 2.64 by 1.85 inches, it weighed 1.73 pounds, and it had a specific gravity of 7.75. The edges of this unusual cube were perfectly straight and sharp. Four of its sides were flat, while the other two sides, opposite each other, were convex. A fairly deep groove was cut all the way around the cube, about halfway up its height. The entire piece bore the marks of having been machine-made, and it looked as though it was part of a much bigger mechanism of some sort(9).

Back in June of 1934, in the town of London, Texas, Emma Hahn and her family found a chunk of limestone with a piece of carbonized wood protruding from it. Once they split the rock open, later at home, it was discovered that the piece of wood was the splintered remains of a handle to an iron-headed hammer, which was still remarkably preserved inside the stone, with no sign of oxidation whatsoever(62).

A fascinating find was made on June 9, 1891, by Mrs. S.W. Culp of Morrisonville, Illinois. While shoveling a load of coal into her kitchen stove, she noticed that one piece had broken in two, exposing a gold chain entombed within that was beautifully crafted. The *Morrisonville Times* of June 11 reported: "Mrs. Culp thought the chain had been dropped accidentally in the coal, but as she undertook to lift the chain up, the idea of its having been recently dropped was shown to be fallacious, for as the lump of coal broke, it separated almost in the middle, and the circular position of the chain placed the two ends near to each other; and as the lump separated, the middle of the chain became loosened while each end remained fastened to the coal....This is a study for the students of archaeology who love to puzzle their brains out over the geological construction of the Earth from whose ancient depth the curious are always dropping out"(9).

In 1912, while heaving coal mined from Wilberton, Oklahoma, into furnaces in the Municipal Electric Plant in the nearby town of Thomas, two employees, unable to handle one chunk of coal because it was too large, had taken a sledge hammer to it. Once the piece was broken open, the workers discovered that an iron pot had been encased within. This piece of coal, by the way, was from a formation that supposedly dated between 300 and 325 million years ago(70).

The June 1851 issue of *Scientific American* (Vol. 7) carried a report, reprinted from the *Boston Transcript*, about a metallic vase that had been dynamited out of solid rock on Meeting House Hill in Dorchester, Massachusetts. It stated: "On putting the two parts together it formed a bell-shaped vessel, 4 ½ inches high, 6 ½ inches at the base, 2 ½ inches at the top and about an eighth of an inch in thickness. The body of this vessel resembles zinc in color, or a composition metal in which there is a considerable portion of silver. On the sides there are six figures of a flower, a bouquet, beautifully inlaid with pure silver, and around the lower part of the vessel, a vine, or wreath, inlaid also with silver. The chasing, carving and inlaying are exquisitely done by the art of some cunning craftsman. This curious and unknown vessel was blown out of...solid pudding stone [conglomerate], fifteen feet below the surface"(9). This artifact has since been lost, but at least a picture of it still remains.

Such artifacts as these are quite a tease, providing only a faint glimpse of what pre-flood man had accomplished technologically. We can only guess at the level of development that these early humans had actually reached. Perhaps we will never know for sure. However, we can gain a better insight into this matter by looking at some early post-flood technological advances. For if a highly-progressed technology did indeed exist before the flood, it would make sense that at least some of that technology would have trickled over to the post-flood period, especially if the flood survivors still possessed, at least for a time, some of the superior brain functions that antediluvian man seems to have had, making it easier to retain a memory and workable knowledge of this technology, at least in part.

A post-flood resurgence

We have already seen that many of our modern inventions and discoveries are not unique to our time, but are actually repeats of what the Classical Greeks, Romans, and others had stumbled upon much earlier. We have also touched upon the idea that these Classical civilizations themselves were unable to lay claim to uniqueness for many of their discoveries and inventions. Indeed, as we shall now see, these Classical civilizations merely reinvented or rediscovered most of their advancements, which were known long before them, going back to the early post-flood era (which in turn were evidently known even further back, in the antediluvian world, as we have just discussed). But before we take a look at the technological and intellectual developments of pre-Classical cultures, let us first address another mystery from history that can now finally be explained.

When we look at the earliest known ancient civilizations, like Sumer and Egypt, for example, we find that all of them seem to have sprung up overnight, without any sign of earlier stages of development leading up to them. This has been, and still remains, a great mystery to most archaeologists, anthropologists, and historians. But if we take into account an advanced pre-flood society that passed on the knowledge of (or at least a partial knowledge of) building an organized civilization, it would make sense for centers like Sumer and Egypt to have emerged in the rather sudden manner that they did. Note the following quotes regarding this matter:

- "No more surprising fact has been discovered by recent excavation, than the suddenness with which civilization appeared in the world. It was expected that the more ancient the period, the more primitive would excavators find it to be, until traces of civilization ceased altogether and aboriginal man appeared. Neither in Babylonia [Sumer] nor Egypt, the lands of the oldest known habitations of man, has this been the case." - P. J. Wiseman, *New Discoveries in Babylonia*(73).

- "The archeological evidence suggest[s] that rather than developing slowly and painfully, as is normal with human societies, the civilization of ancient Egypt...emerged all at once and fully formed. Indeed, the period of transition from primitive to advanced society appears to have been so short that it makes no kind of historical sense. Technological skills that should have taken hundreds or even thousands of years to evolve were brought into use almost overnight--and with no apparent antecedents whatever. What is remarkable is that there are no traces of evolution from simple to sophisticated, and the same is true of mathematics, medicine, astronomy and architecture and of Egypt's amazingly rich and convoluted religio-mythological system..." - Graham Hancock, *Fingerprints of the Gods*(20).

- "Egyptian science, medicine, mathematics and astronomy were all of an exponentially higher order of refinement and sophistication than modern scholars will acknowledge. The whole of Egyptian civilization was based upon a complete and precise understanding of universal laws....[E]very aspect of Egyptian knowledge seems to have been complete at the very beginning. The sciences, artistic and architectural techniques and the hieroglyphic system show virtually no signs of a period of "development;" indeed, many of the achievements of the earliest dynasties were never surpassed, or even equaled later on. This astonishing fact was readily admitted by orthodox Egyptologists, but the magnitude of the mystery it poses is skillfully understated, while its many implications go unmentioned. How does a civilization spring full-blown into being? Look at a 1905 automobile and compare it to a modern one. There is no mistaking the process of "development," but in Egypt there are no parallels. Everything is right there at the start. The answer to the mystery is of course obvious, but because it is repelling to the prevailing cast of modern thinking, it is seldom seriously considered. Egyptian civilization was not a "development," but a legacy." - John Anthony West, *The Serpent in the Sky*(72).

- "A recent series of stunning archaeological finds in South America has revealed that the Incas were merely the final act in an Andean civilization that was far older and far more sophisticated than ever imagined. New excavations have turned up huge stone pyramids and other monuments that date back nearly 5,000 years, to about the time when the Great Pyramids were being constructed in Egypt." - *U.S. News & World Report*, April 2, 1990(74).

- "[I]n all main culture centers from Mexico to Peru--that is, wherever high culture flourished in ancient America--no trace of gradual evolution from primitive society to civilization has been discovered anywhere within the New World. Wherever archaeologists have dug, they have found that civilization appeared suddenly in America, in full bloom, superimposed upon a primitive, archaic society. The sudden flourishing of civilization begins at a peak and shows a decline rather than progress through the centuries leading to the arrival of the first Europeans. The Incas of Peru astonished the Spaniards with their high degree of culture, yet archaeology has shown that the Incas had borrowed most of their cultural elements from the earlier Tiahuanaco and Mochica cultures, which in many respects had had an even more sophisticated and impressive civilization that suddenly appeared without traceable background in the highlands of the Andes and on the desert coast of Peru. Modern archaeology has established that contact took place between these early, pre-Inca civilizations and the contemporaneous civilizations of Mexico and Central America. In Mexico, correspondingly, the great cultures of the Aztecs, Toltecs, and Mayas had drawn their basic lessons from the highly advanced civilization of the Olmecs, an unknown people who suddenly established Mexico's earliest civilization--with script, calendar system, pyramid building, etc. fully developed..." - Geoffrey Ashe, Thor Heyerdahl, et al, *The Quest for America*(80).

Can you not see from these quotes that the earliest ancient civilizations must have inherited their advanced developments from a still older civilization, the traces of which have not been found because they were destroyed by the flood catastrophe? [Endnote 50]

This is exactly what is implied in the following quote from the ancient Egyptian Edfu Building Texts: "An ancient world...was destroyed [by the flood], and as a dead world it came to be the basis of a new period of creation which at first was the re-creation and resurrection of what once had existed in the past"(209).

Let us now take a look at some of the amazing intellectual and technological accomplishments of the early post-flood period, long before the age of Classical Greece and Rome.

The early ancestors of the Incas, mentioned a moment ago, who settled in the Andes region probably a short time after the flood catastrophe, built enormous structures out of gargantuan stone blocks that rival anything erected by the Romans, the Greeks, and even the Egyptians. Elaborating on such amazing feats of construction, Rene Noorbergen wrote: "[In] Ollantaitambo, Peru, is [a] pre-Inca fortress, with rock walls of tightly-fitted blocks weighing between 150 and 250 tons each. Most of the blocks consist of very hard andesite, the quarries for which are situated on a mountaintop seven miles away. Somehow, at an altitude of 10,000 feet, the unknown builders of Ollantaitambo carved and dressed the stone (using tools, the nature of which we can only guess, that could penetrate such hard rock), lowered the 200-ton rocks down the mountainside, crossed a river canyon with 1,000 foot sheer rock walls, then raised the blocks up another mountainside and placed them in the fortress complex. As South American antiquarian Hyatt Verrill notes, no number of men--Indian or otherwise--could duplicate this feat with only stone implements or crude metal tools, ropes, rollers and muscle power. 'It is not a question of skill, patience and time,' Verrill explains. 'It is a human impossibility.'...

"One of the most impressive 'mystery fortresses' of the Andes is Sacsahuaman, located on the outskirts of the ancient Inca capital of Cuzco. It rests on an artificially leveled mountaintop at an altitude of 12,000 feet, and consists of three outer lines of gargantuan walls, 1,500 feet long and 54 feet wide, surrounding a paved area containing a circular stone structure believed to be a solar calendar. The ruins also include a 50,000-gallon water reservoir, storage cisterns, ramps, citadels and underground chambers.

"What is truly remarkable about Sacsahuaman is the stonework. Here extremely skilled stonemasons fit blocks weighing from 50 to 300 tons into intricate patterns. A block in one of the outer walls, for example, has faces cut to fit perfectly with twelve other blocks. Other blocks were cut with as many as 10, 12, and even 36 sides. Yet all the blocks were fit together so precisely that a mechanic's thickness gauge could not be inserted between them. And even more extraordinary is the fact that the entire Sacsahuaman complex was built without cement.

"As with the other mystery fortresses [in the Andes], the question of how the stones of Sacsahuaman were transported remains unanswered. The quarries from which the stones for Sacsahuaman were brought are located 20 miles away, on the other side of a mountain range and a deep river gorge. How the massive stones were moved across such hopeless terrain is anyone's guess.

"Sacsahuaman poses many mysteries, yet it possesses one more which few orthodox historians are willing to recognize or study because of its 'impossibility.' Within a few hundred yards of the Sacsahuaman complex is a single stone block that was carved from the mountainside and moved some distance before it was abandoned. An earthquake apparently interrupted the progress of the movers, for the stone was turned upside down and is damaged in several places. It contains steps, platforms, holes and other depressions--a masterpiece of precision cutting and dressing, clearly intended to become a part of the fortification. What is truly impossible about the block is that it is the size of a five-story house and weighs an estimated 20,000 tons! We have no combination of machinery today that could dislodge such a weight, let alone move it any distance. The fact that the builders of Sacsahuaman could and did move this block shows their mastery of a technology which we as yet have not attained"(9).

Another enormous stone construction of unknown antiquity is found in the ancient city of Baalbek, near modern-day Beirut. Commenting on this structure, the science-frontiers.com website states: "The city of Baalbek, called Heliopolis by the ancient Greeks, lies some 50 miles northeast of Beirut. Here are ruins of the greatest temple the Romans ever tried to construct. However, we must focus not on mundane Roman temples but upon a great assemblage of precisely cut and fitted stones, called the Temple today, which the Romans found ready-made for them when they arrived at Baalbek. It was upon this Temple, or stone foundation, that the Romans reared their Temple of Jupiter. No one knows the purpose of the much older Temple underneath the Roman work.

"...Being 2,500 feet long on each side, the Temple is one of the largest stone structures in the world. Some 26 feet above the structure's base are found three of the largest stones ever employed by man. Each of these stones measures 10 feet thick, 13 feet high, and...over 60 feet long....Some people with impressive engineering skills cut, dressed, and moved these immense stone blocks from a quarry 3/4 of a mile away.

"A walk to this quarry introduces the observer to the Monolith, an even larger block of limestone: 13 feet, 5 inches; 15 feet, 6 inches; and 69 feet, 11 inches. The Monolith weighs in at over 2,000,000 pounds. In comparison, the largest stones used in the Great Pyramid tip the scales at only 400,000 pounds. Not until NASA moved the giant Saturn V rocket to its launch pad on a giant tracked vehicle has man transported such a large object.

"Today, one sees no evidence of a road connecting the quarry and Temple. Even if a road existed, logs employed as rollers would have been crushed to a pulp. But, obviously, someone way back then knew how to transport million-pound stones. Just how, we do not know"(144).

Since reference was just made to the Great Pyramid of Cheops (or Khufu), let's take a moment to ponder this amazing structure. It was built as early as 2700 B.C., probably not too long after the flood. According to Mooney, it is "composed of 2.6 million blocks, each weighing between two and a half and ten tons, with slabs lining passageways weighing over fifty tons, and eighty-four weight-relieving beams, weighing eighty-seven tons each, placed one above the other over the main chamber"(5). "Wikipedia" further tells us that "For four millennia it was the world's tallest building, unsurpassed until the 160-metre tall spire of Lincoln Cathedral was completed c. 1300 A.D. The accuracy of [the] Pyramid's workmanship is such that the four sides of the base have a mean error of only 50 mm in length, and 12 seconds in angle from a perfect square. The sides of the square are closely aligned to the four ordinal compass points to within 3 minutes of arc and is based not on magnetic north, but true north"(123).

Not only were these massive structures built so that they would endure for centuries (and millennia) as reliable calendars, but also, presumably, so that they might be durable enough to survive another possible flood catastrophe.

The next early post-flood advancement we will look at is in the area of metallurgy. Noorbergen tells us that "In 1968, Dr. Koriun Megurtchian of the [former] Soviet Union unearthed what is considered to be the oldest large-scale metallurgical factory in the world, at Medzamor, in Soviet Armenia. Here, 4,500 years ago, an unknown prehistoric people worked with over 200 furnaces, producing an assortment of vases, knives, spearheads, rings, bracelets, etc. The Medzamor craftsmen wore mouth-filters and gloves while they labored and fashioned their wares of copper, lead, zinc, iron, gold, tin, manganese and fourteen kinds of bronze. The smelters also produced an assortment of metallic paints, ceramics and glass. But the most out-of-place discovery was several pairs of tweezers made of steel....The steel was later found to be of exceptionally high grade, and the discovery was verified by scientific organizations in the Soviet Union, the United States, Britain, France and Germany.

"French journalist Jean Vidal, reporting in *Science et vie* of July 1969, expressed the belief that these finds point to an unknown period of technological development. 'Medzamor,' he wrote, 'was founded by the wise men of earlier civilizations. They possessed knowledge they had acquired during a remote age unknown to us that deserves to be called scientific and industrial'"(9).

Please understand that this site dates to nearly two full millennia before the onset of the so-called Bronze Age. Thus it is obvious that the Bronze Age smiths were by no means the world's first highly-skilled metal workers. So does the Medzamor site truly represent one of the early signs of civilization being redeveloped after the flood? Interestingly, the Genesis record tells us of a skilled early post-flood metal worker named Tubalcain, describing him as "an instructor of every artificer in brass and iron." Genesis 4:22. Did he develop, or help to develop, the metalworking technology that was utilized at the Medzamor site?

Complex and efficient waterworks systems appeared very early in human history. For example, the pyramid temple of Sahure at Abusir, Egypt, had a network of copper piping installed within it over 4,500 years ago(65).

At about the same time, the cities of Harappa and Mohenjo Daro, located in India and Pakistan, were constructed with the same planning and efficiency as any modern city, particularly regarding their waterworks assemblages. They had a proficient water supply and drainage system, as well as rubbish chutes. There were also public swimming pools, and many of the residences contained bathrooms(61).

Going back even further, to roughly 5,000 years ago, we find that the residents of Tell Asmar, in the Tigris Valley near Baghdad in Iraq, had homes and temples with elaborate arrangements for sanitation. One temple was found to have six toilets and five bathrooms, with most of the plumbing equipment connected to drains that emptied into a main sewer one meter high and 50 meters long. The Babylonians employed earth-ware pipes that were connected to one another in the same fashion as modern pipes, where one end of a pipe section, being more narrow than the other end, would fit snugly into the wider end of an adjacent pipe section(15, 76).

Coming up to a more fairly recent age, yet still many centuries before the Classical Greeks, we see that the Persians were also highly innovative with waterworks systems, especially for crop irrigation purposes. *Scientific American* informs us that "Some 3,000 years ago the ancient Persians discovered a method of digging underground aqueducts that would bring mountain ground water to their arid plains. Still extant and functional, the system of irrigation provides 75 percent of the water used in Iran today"(15, 75). Was this where the Romans got their aqueduct technology from? And did the Persians themselves get such a technology from an earlier civilization?

The so-called "primitive," late "Stone Age" peoples of England (about 4,000 years ago) performed trepanation, the same delicate brain operation carried out by the Romans centuries later, which we mentioned before. The ancient British methods of this practice were likewise very successful, since the cut areas in the skulls that have been dug up show clear signs of healing. And the patients who underwent this procedure, just like the Roman patients of a much later date, had continued to live for many years thereafter, indicating an extraordinarily high level of surgical skills on the part of those who conducted these early operations.

It so happened that this same medical procedure was commonly carried out the world over during prehistoric times, and was equally successful from one culture to the next(65). Another good example of how well-developed it was is expounded upon by Noorbergen in this next quote: "Professor Andronik Jagharian, anthropologist and director of operative surgery at the Erivan Medical Institute in Soviet Armenia, examined a number of skulls from the ancient site of Ishtukunuy, located near Lake Sevan. The site was inhabited by a prehistoric people called the Khurits who settled the area prior to 2000 B.C.

"Two of the skulls examined by Professor Jagharian revealed extraordinary skill in head surgery. The first is the skull of a woman who died at approximately thirty-five years of age. In her youth she had suffered a head injury which made a hole one-quarter inch in size in her skull. This accident must certainly have left brain tissue exposed, and a considerable amount of blood must have been lost. The prehistoric surgeons skillfully inserted a plug of animal bone, and the woman survived the delicate operation. This could be seen from the woman's skull, as her own cranial bone grew around the plug before she eventually died years later.

"The second Khurits skull shows even more evidence of complicated surgery. The skull is of another woman, who was approximately forty years old when she died. A blow to the head had caused a blunt object about an inch in diameter to puncture the skull, splintering the inner layers of cranial bone. The surgeons of 4,000 years ago carefully cut a larger hole around the puncture in order to remove the splinters that had penetrated into the brain. Even by modern standards, such an operation would be considered extremely difficult; yet the prehistoric operation was successful. Evidence shows that the woman survived the surgery for fifteen years"(9).

As was the case with the Romans, the Babylonians, many centuries before them, had performed delicate eye operations for cataracts in the eighteenth century B.C., according to the code of Hammurabi. These Babylonian surgeons must have had a considerable degree of skill, since the Hammurabi Code called for a physician's hand to be cut off if an incompetent slip of the hand resulted in the patient's blindness(65). Perhaps the Romans inherited this technology from them.

India's early inhabitants were also very knowledgeable and skillful in the medical field. The *Sactya Grantham*, a Brahmin book compiled about 1500 B.C., contains the following passage, giving instructions on the smallpox vaccine: "Take on the tip of a knife the contents of the inflammation, inject it into the arm of a man, mixing it with his blood. A fever will follow but the malady will pass very easily and will create no complications"(61).

The Dogons of Chad and the Sudan knew, probably from back in the early days of ancient Egypt, that a dark star orbited Sirius (the "dog star") every 50 years--a fact that was only fairly recently confirmed, in 1862(77).

Mooney tells us that "Babylonian priests and astronomers had charts that showed four of the largest satellites of Jupiter"(5).

An ancient Assyrian cylinder seal depicts the planet Saturn encircled by a ring(188).

The Mayas also knew, in ages long gone by, about the existence of Jupiter's 4 major satellites, as well as Saturn's 7 major ones(18).

In 1857, Heinrich Karl Brugsch discovered a star chart on a coffin lid from a tomb in Thebes, Egypt. It showed the sky goddess Nut surrounded by the 12 zodiacal signs, the Sun, Moon, and all the planets, including, most astonishingly, Uranus and Neptune.

Even more astounding, the ancient Assyrians stated that the planet beyond Saturn (Uranus) is tipped on its axis, sometimes presenting its poles toward the Sun(63).

In addition to all of this, recall from a discussion in Section I of this work how the ancients were also able to see the Moons of Mars and the crescent shape of Venus.

Just how were these people of so long ago able to see all these things that cannot be resolved with the naked eye? Did they have telescopes? It would appear so. For lenses made of rock crystal have been found in ancient Sumer and Babylon(5), as well as in Egypt and other places around the world.

Additional evidence of telescope technology in the ancient world comes to us in the form of written records.

One example of such records is from China, where reference is made to Emperor Chan, circa 2283 B.C., who is said to have arranged two powerful magnifying glasses in such a way as to facilitate detailed observations of the planets(207).

Another example, although dating to a later time (the Classical Greek period), comes from Archimedes, a scientist and inventor, who wrote about an instrument he had invented "to manifest to the eye the largeness of the sun." Democritus mentioned another instrument--or perhaps the same one--that was used "for studying sky details"(63).

It would also appear that at least two ancient civilizations had developed the microscope as well, or at least a powerful magnifier, as we see from the following two quotes:

- "In the Museum of the American Indian in New York there are tiny particles of gold on display [found in Ecuador]. They appear to be natural grains--until they are viewed with a magnifying glass. These "particles" are perfectly wrought beads, some smaller than the head of a straight pin, elaborately engraved, some composed of almost invisible pieces welded together.. All are pierced." - Richard Mooney, *Colony: Earth*(5).

- "One of the most puzzling [giant etchings on the desert floor of the Nazca Valley in Peru] is the picture of a spider, 150 feet long, drawn with a single continuous line half a mile in length. What is so peculiar about the spider is that one of its legs is deliberately lengthened and extended, and at the tip there is a small cleared area. There is only one spider known that uses the tip of its third leg in precisely the manner depicted in the desert drawing, and that is the *Ricinulei*, which lives in caves deep in the Amazon jungle, a thousand miles from Nazca. Known to scientists for its unique method of copulation, for which the spider uses that extended leg in the described manner, the *Ricinulei* is extremely rare. Its mode of reproduction can be observed only with the aid of a microscope." - Rene Noorbergen, *Secrets of the Lost Races*(9).

While there have not been any ancient lenses found in South America, magnifying mirrors were excavated there that had been ground to the precision of powerful lenses. So we can safely assume that lenses must have also been made there, which were surely used to incise the tiny Ecuadorian gold beads, and to see the microscopic extension on the *Ricinulei* spider's leg that is represented on the Nazca plain. After all, how else could these phenomena be accounted for?

Shu-Ching (circa 2200 B.C.), an ancient Chinese document, states that matter is composed of distinct, separate elements. This same document also hypothesizes that sunbeams are made of particles(71).

Pythagoras's theorem, which states that the sum of the squares of the two perpendicular sides of a right triangle is equal to the square of the hypotenuse, was actually first proclaimed by the Babylonians at least fifteen hundred years before Pythagoras was born(Ibid.).

The Greeks did not originate the heliocentric view of the solar system, that is, the idea that all the planets orbit the Sun. This concept also dates back to the Babylonians, long before them, which was actually acknowledged by Diodorus, who wrote: "[T]he Chaldeans named the planets....[I]n the center of their system was the Sun, the greatest light, of which the planets were 'offspring,' reflecting the Sun's position and shine"(171).

* * * * * * *

These ancient advancements were truly remarkable, primarily because of their great antiquity--dating back to a time-period when man, according to mainstream modern science, was supposed to have been too "primitive" for such developments. Yet, the truth is, as amazing as these achievements are, they don't even scratch the surface of just how sophisticated the ancients really were. In the next few subsections, we will be delving into some fascinating legendary and archaeological revelations about ancient highly-advanced technology, spanning all regions of the globe and all periods of ancient history, beginning just after the flood and following right up to the early centuries A.D. As we proceed, let us continue to embrace both the courage to investigate uncharted territory, and the humility to admit that we may have been more wrong than we thought about ancient man and his capabilities.

Electricity in the ancient world

Once we recognize that ancient man was just as intelligent and competent as modern man (if not more so), then the idea of electricity having been discovered and utilized in the distant past should not be so surprising. Let us now direct our attention to some amazing and irrefutable evidence that early man did indeed reach such a technological height.

Several civilizations produced electroplated jewelry and other objects in ancient times. Noorbergen tells us that "[E]lectroplated objects...have been discovered in Iraq in Babylonian ruins dating back to 2000 B.C. [Such objects] were also found in Egypt by the famous nineteenth-century French archaeologist Auguste Mariette. Excavating in the area of the Sphinx of Gizeh, Mariette came upon a number of artifacts at a depth of 60 feet. In the *Grand Dictionaire Universal du 19th Siecle*, he described the artifacts as 'pieces of gold jewelry whose thinness and lightness make one believe they had been produced by electroplating, an industrial technique that we have been using for only two or three years'"(9). The Chimu culture of South America also appears to have carried out gold plating with results that today can only be achieved by electroplating(5).

Can we assume that the ancients were able to somehow produce electrical currents in order to have accomplished the task of electroplating? Did they know how to make and use batteries? We can answer both of these questions in the affirmative, with no room left for doubt. The evidence comes to us, as we have seen so often throughout this work, in the forms of both ancient records and archaeological artifacts. We will first look at the evidence in the realm of ancient records.

An ancient document from India, known as *Agastya Sanhita* and preserved in the Indian Princes' Library at Ujjain, contains these instructions for making a battery: "Place a well-cleaned copper plate in an earthenware vessel. Cover it first by copper sulfate and then by moist sawdust. After that put a mercury-amalgamated-zinc sheet on top of the sawdust to avoid polarization. The contact will produce an energy known by the twin name of Mitra-Varuma [positive and negative charges?]. Water will be split [electrolysis?] by this current into Pranavayu and Udanavayu [hydrogen and oxygen?]. A chain of one hundred jars is said to give a very active and effective force"(61).

Archaeological evidence for the existence of batteries in the ancient world is even more compelling. Over the years since 1936, there have been many actual ancient batteries found in the Baghdad area of Iraq. Speaking on this matter, Time-Life had this to say: "While excavating ruins of a 2,000-year-old village near Baghdad, Iraq, in 1936, workers discovered a mystifying artifact. It was a small earthenware vase in which was set a soldered sheet-copper tube about one inch wide and four inches long. The bottom of the tube was sealed with a crimped copper disk; an iron rod, seemingly corroded by acid, projected through an asphalt plug at the top. German archaeologist Wilhelm Konig, then living in Iraq, examined the object and reached a startling conclusion: If the tube had been filled with an acidic solution, it would have served as a rudimentary electric battery. Such batteries, he speculated, may have been used by ancient artisans to electroplate metals. He pointed out that similar objects had been found at other sites in the region, along with thin copper and iron rods that might have been used to link an array of such batteries"(62). This book then went on to tell of how each battery was estimated to have been capable of generating a half volt of electricity, but when linked together with other batteries, a much more substantial charge was capable of being produced. To put these discoveries into historical perspective, the first modern battery was not invented until 1799, by Alessandro Volta.

The designers and builders of the Aswan Valley Temple in Egypt left a fascinating sign of a superior technology that must have involved the use of electricity. Here we find drill holes that were made in extremely hard pink granite blocks. These drill holes have spiral grooves on their insides that tell us a lot about the drills that made them--that they were capable of cutting into very hard surfaces at the rate of a quarter inch for every full turn of the drill. This had to involve a power tool of some sort.

Turning to the Far East, we find possible physical evidence of electricity having been in use there as well, in ancient times. In the summer of 1938, a Chinese archaeological expedition undertook the effort of mapping a series of interlocking caves in a part of the Himalayas called the Baian Kara Ula Mountains, along the border of China and Tibet. The expedition, led by Dr. Chi Pu Tei, stumbled across a remarkable find in some of the caves--neat lines of graves of a strange people averaging about 4 feet in height, having long spindly limbs and round, oversized heads. Not very far away, a researcher stumbled across the most interesting find of all--a strange, hand-crafted stone disc. It was approximately 9 inches in diameter, with a 3/4-inch, perfectly-round hole in the center.

This disc was found to have a groove spiraling outward from the center, tightly wound like a phonograph record. However, upon further inspection, the groove was revealed to be no groove at all, but rather a continuous line of written characters, unlike any ever seen before. The writing was almost microscopic in size. In all, 716 of these discs have been recovered over the years, ranging in size from several inches to several feet in diameter. Collectively they are known as

Dropa Stones, or Dropa Discs.

The stones contain cobalt and other metals, which make one wonder how a so-called primitive people could have inscribed the minute hieroglyphics on such a hard substance. A Russian scientist named W. Saitsew tested some of these discs in an oscillograph that produced some unusual results. A surprising oscillation rhythm was recorded, as though the discs were once electrically charged or had functioned as electrical conductors. Today these extraordinary artifacts sit in the basement of Beijing University, awaiting further study.

If a knowledge of electricity existed in remote history, we shouldn't find it so unusual to discover that electric light bulbs were also in use. And when looking to ancient Egypt, we discover what may very well be evidence of just that. On a stone wall of the Temple of Hathor at Dendera, Egypt, there are human figures carved that are holding the bases of what appear to be five-foot-long light bulbs with elongated "serpent" filaments inside. The other end of each "bulb" is held up by what look like high-tension insulators. Connected to the base of each "bulb" are braided cables that seem as though they are plugged into a transformer. Interestingly, the snake filaments inside these "bulbs" are referred to in an adjacent text as "seref," which means "to glow"(29, 130).

As to whether or not these unusual carvings actually depict light bulbs remains to be seen. But if this is what they are, it would certainly explain why there are so many subterranean tombs and temple corridors in ancient Egypt that are elaborately decorated with highly-detailed carvings and paintings. There's simply no way that such spectacular works of art could have been executed without bright lighting. Forget the idea that oil lamps or torches were used, since there is no trace of oil smudges or smoke stains on any of the ceilings. And it couldn't have been a series of mirrors that were used either, to reflect light from above ground, since many of the subterranean corridors and chambers in question are located within complex networks of winding passageways. If any light could have reached these remote areas through reflection at all, it would have been way too dim to have been of any use to the carvers and painters anyway.

It so happens that ancient India may have had electric light fixtures as well, which would account for the archaic Indian legend of the magic, ever-burning lamps of the Nagas--the serpent gods and goddesses who lived in the underground abodes of the Himalayas(61).

A great many other ancient cultures seemingly had electric lighting.

For instance, Numa Pompilius, the second emperor of Rome, is claimed to have had a "perpetual light" shining in the dome of his temple.

In Hierapolis, Syria, during the second century A.D., there was, we are told, a "shining jewel" in the forehead of the goddess Hera that brilliantly illuminated the whole temple at night.

Around 70 A.D., in Lebanon, a lamp in the temple of Minerva is alleged to have shined continually for about a year.

Colombia has an old legend that tells of how the distant forefathers of the people of this land "made fire and light by strange means."

Lastly, the Maya and Aztecs of Mexico preserve legends of ancient cities that were lit up by lights that never went out by day or night(63).

In addition to electroplating and possible electric-powered lighting, there may have been other uses that the ancients put electricity to. Perhaps we have one such example--a possible x-ray machine--in the following Noorbergen quote: "An Indian contemporary of Buddha, a physician named Jivaka, was given the title King of Doctors about 500 B.C. Records tell us that he had a 'gem' which he used for diagnosis, and that when a patient was placed before the gem it 'illuminated his body as a lamp lights up all objects in a house, and so revealed the nature of his malady.'

"Jivaka's magic gem disappeared in history, but three centuries later there was [another ancient record written which tells us that there was] discovered in the palace of Hien-Yang in Shensi, a 'precious mirror that illuminates the bones of the body.' The mirror was rectangular--[the equivalent of] 4 by 5 feet--and gave off a strange light on both sides. The view of the organs of the body that the mirror gave could not be obstructed by any obstacle, which would be typical by the penetration power of x-rays"(9).

There seems to have been many other devices invented in the ancient world that were powered by electricity, which parallel modern gadgets.

In the apocryphal book of Enoch, for example, we find that a certain Azaziel taught men to make "magic mirrors" whereby distant people and scenes could clearly be seen. Were these archaic equivalents of modern video monitors that received signals from remote television cameras?

The Romans presumably had the same thing--"magic mirrors," or looking glasses, of enormous dimensions lying over a shallow well. Whoever descended into the well, it was claimed, could hear and see all things that went on in cities throughout the Roman world.

In China, during the first millennium B.C., we also find mention of "magic mirrors" that could transmit images across great distances.

An ancient Brazilian Native legend tells of "magic stones" that were used to look into the distance and see cities, rivers, hills, and lakes(63).

Also equipped with this technology, apparently, were the distant ancestors of the Mayas. For the *Popol Vuh* speaks of them thusly: "They were endowed with intelligence; they...could see instantly far, they succeeded in seeing, they succeeded in knowing all that there is in the world. When they looked, instantly they saw all around them....The things hidden [in the distance] they saw all, without first having to move; at once they saw the world, and so too, from where they were they saw it. Great was their wisdom..."(213).

Another electronic gadget in the ancient world appears to have been the two-way communications radio, as the following few examples will illustrate.

The Hsing Nu civilization of Tibet was, according to legend, able to convey speech over large distances.

Phoenician and Carthaginian records make reference to "speaking stones" and "animated stones" (transistors?) that, it would seem, facilitated communication both in the twelfth century B.C. and the second century A.D.

We are told that Eusebius, a Christian historian, carried just such a "speaking stone" on his chest, which would respond to questions he asked with a soft voice. The response sounded like a "low whistling." We are further told that Arnabius, of the same era, also made use of a "speaking stone"(Ibid.).

If the ancients truly were endowed with a competent understanding of electromagnetism, and knew how to utilize it to their technological advantage, then how much of a stretch would it be to imagine them having mastered the force of gravity (in the form of antigravity devices)? Examining the proceeding pieces of evidence, gathered from ancient written records, will make it difficult to deny this possibility.

The priests of ancient Babylon, legend has it, were able to raise heavy rocks into the air that 1,000 men could not budge, simply by the means of sound(61). Is this how the massive stone structures around the world were constructed--levitation of enormous stones through vibrational energy? Did the ancients know something about the potential of soundwaves that modern science has yet to discover or master?

Most attempts by the ancients at defying gravity appear to have involved the repulsive force of magnetism, which we will now cite several examples of.

As was mentioned earlier, in fourth-century Greece a vault was constructed with "magnetic stones" that were utilized to suspend idols in mid-air.

In Alexandria, Egypt, around 400 A.D., a sun disk was levitated in the air "by magnetism" in the temple of Serapis.

During the second century A.D., in Syria, an image of a deity was also raised in the air, presumably by the same method, although the record doesn't say exactly how it was done.

Similar things were reported being accomplished during the early centuries A.D., in places like Tibet, Medina, and Abyssinia(Ibid.).

Looking to the Americas, we discover the same technology having been employed in the remote past.

The Natives of the Andes region, for example, claim that the great stone cities of the area were built by godlike heroes who caused the great stones of the ruins to fly into place from their quarries.

A parallel assertion is made by the locals on Ponape, in the Carolina Islands, about the Cyclopean ruins of Nan Madol that reside there.

According to the ancient traditions of Easter Island, the great statues there were lifted and placed in their current resting places by a mysterious vibrating energy of "mana."

Similar statements have been made by the ancient Druidic people of England--that methods of this sort were used by their ancestors to move the massive stones used in the construction of Stonehenge(63).

The development of antigravity technology is something that is currently being pursued with great enthusiasm, through various methodologies, and with frustratingly vacillating degrees of successes. But because of the progress that has been made in recent decades, the notion of this goal being reached is no longer viewed by mainstream scientists as an unattainable fantasy.

Here are several examples of such progressive steps that have been taken in contemporary times:

- Some modern techniques employed to achieve antigravity involve the use of magnetism, as was the case with many of the ancient techniques mentioned above. Several successful experiments along this line have been performed at the Nijmegen High Field Magnet Laboratory in the Netherlands. Examples of objects that have been levitated in magnetic fields at this facility include simple metals (bismuth and antimony), liquids (propanol, acetone and liquid nitrogen), and living things such as plants, frogs, fish, and mice(180).

- We made reference a moment ago to an ancient Babylonian legend that discusses how the power of sound was used to raise heavy rocks off the ground. It turns out that this exact same technology is currently under development. As the Wikipedia website reports: "Acoustic [ultrasonic] levitation is a method for suspending matter in a medium by using acoustic radiation pressure from intense sound waves in the medium. Acoustic levitation is possible because of the non-linear effects of intense sound waves.

"Some methods can levitate objects without creating sound heard by the human ear such as the one demonstrated at Otsuka Lab, while others produce some audible sound. There are many ways of creating this effect, from creating a wave underneath the object and reflecting it back to its source, to using an acrylic glass tank to create a large acoustic field....

"There is no known limit to what acoustic levitation can lift given enough vibratory sound, but currently the maximum amount that can be lifted by this force is a few kilograms. Acoustic levitators are used mostly in industry and for researchers of anti-gravity effects such as NASA; however some are commercially available to the public"(211).

- The space.com website reported on February 15, 2006: "An 'antigravity' propulsion system was proposed at the Space Technology and Applications International Forum (STAIF) in Albuquerque on February 14 by Dr. Franklin Felber. His new exact solution to Einstein's gravitational field equation gives hope to space enthusiasts that it might be possible to accelerate space craft to speeds approaching that of light without crushing the contents of the craft. If it works, it could be even better than apergy, as described by science fiction writer Percy Greg in 1880.

"Dr. Felber's paper states that a mass moving faster than 57.7 percent of the speed of light will gravitationally repel other masses lying within a narrow 'antigravity beam' in front of it. This 'beam' intensifies as the speed of the mass approaches that of light"(181).

- A Canadian independent research scientist by the name of John Hutchison has developed a method of levitating objects that employs a very complex scalar-wave interaction with matter, which he refers to as the Hutchison Effect. He discovered this phenomenon as a result of his studies into the research of Nikola Tesla. Like the Nijmegen High Field Magnet Laboratory, he has been able to cause many different types of objects, non-living and living alike, to hover in mid-air(182).

These modern technological innovations should serve to dispel any misconceptions we might have that the ancient antigravity legends cited above are merely the products of ripe imaginations.

========================

Was ancient alchemy a science?

Many people in modern times have laughed at the ancient notion of alchemy--the professed ability to transmute one element into another, such as changing copper into gold. Some have claimed that ancient practitioners of this craft were either scam artists or evil magicians who performed their feats by the "power of the devil." But the truth is, modern science has shown that alchemy is indeed possible, and has itself successfully executed transmutations of many elements. All one has to do to change one element into another is to add or subtract protons from the nuclei of the atoms of the element that is being worked on. The problem is, it is a very involving and enormously expensive undertaking, at least in the way that it is done today.

The first person to successfully undertake alchemy in modern times, in the way it's understood today, was the English physicist Ernest Rutherford, in 1919. He transmuted nitrogen into oxygen and hydrogen by bombarding it with helium. But it would appear that the ancients were able to accomplish this task with relative ease, and in a far more cost-effective manner. Were they privy to a knowledge of physics or chemistry that is lost to us today?

So successful was this practice in ancient times, in fact, that there were occasions when it became viewed by political leaders as an economic menace. The Roman emperor Diocletian, for example, issued an edict in Egypt, around 300 A.D., demanding that all books on "the art of making gold and silver" be burned. One would have to ask why such an edict would have been issued in the first place, unless alchemy was indeed a reality.

According to the alchemist Zosimus, from this same period (300 A.D.), the Temple of Ptah at Memphis had huge furnaces that were allegedly the means of transmuting base metals into precious metals.

Speaking of Egypt and alchemy, the word alchemy itself, along with related words such as chemistry, chemist, etc., are all derived from the ancient name for Egypt--Khemt.

In the 8th century, the Arab Jabir (Geber) adapted and expanded upon the Egyptian knowledge of alchemy, and thus became known as the "Father of Alchemy." In his writings, he described, not only the laboratory equipment necessary for carrying out this endeavor, but even the moral and mental attitude needed to be an acceptable apprentice. He wrote: "The artificer of this work ought to be well skilled and perfected in the sciences of natural philosophy...lest he happen not to find the art, and be left in misery." He went on to assure his students that "copper may be changed into gold," and that "by our artifice we easily make silver."

Alchemy was practiced universally in ancient times. In China, around 133 B.C., there emerged the story of Chia and the alchemist Chen, which tells of how whenever Chia needed money, Chen would rub a black stone on a tile or a brick and transmute these common articles into precious silver. Was this just a fairy tale? Or was it a simplified version of a more complicated, scientific process known to Chen for changing the chemical properties of one element into another? Perhaps this "black rock" was cadmium or palladium, both of which are only one proton away from silver.

Another ancient Chinese reference to alchemy--a seemingly far more ancient one--is in the biography of Chang Tao-Ling, who studied at the Imperial Academy in Peking. It makes reference to the *Elixir Refined in Nine Cauldrons*, which he said he found in a cavern, allegedly written by the Yellow Emperor (26th century B.C.).

The basic ingredient of alchemy in ancient China was cinnabar, or mercuric sulfide, used in transmutation. "You may transmute cinnabar into pure gold," was the assurance given in the historical text *Shih Chi*, written in the first century B.C.

In addition to the ancient Chinese using cinnabar for alchemic purposes, they also used it in the preparation of what was called "gold juice," or the elixir of youth--something else they were very interested in, associating it with alchemy.

As was the case with ancient Rome, China, in 175 B.C., also forbade the counterfeiting of gold by alchemical means, indicating that this art, by that time, must have become widespread.

India also dabbled into alchemy. And just like the other parts of the world that practiced it--Europe, Egypt, China, etc.--the Indian practitioners believed that sulfur and mercury were the primary elements for accomplishing transmutation. They attributed a positive polarity to mercury and a negative polarity to sulfur. Like the Chinese, they were also interested in finding an elixir of immortality, which they also associated with alchemy. But as far as alchemy goes, the Indians were particularly keen on utilizing it for gold-making purposes.

Interestingly, the atomic number of mercury is 80, whereas that of gold is 79, so these metals are actually very close cousins.

It was because of the writings of Arab alchemists like Jabir, as well as Al Razi, Farabi, and Avicenna, who lived from the eighth to the eleventh centuries, that the discipline of alchemy arose in medieval Europe, and quite heavily so. During this time, expensive, hand-copied books were carried from city to city. Some of these writings were very methodic and scientific, talking about funnels and furnaces, including charts and diagrams, while others were poor copies that were carelessly written, and thus of little practical value. Of course, the distribution of these books had to be done with extreme caution, since the authorities, as in ancient China and Rome, were not very fond of this economic destabilization practice.

Because of fear of government authorities, as well as worries over potential competitors stealing his secrets, medieval alchemist Sir George Ripley, in his book *Compound of Alchemy* (1471), was compelled to advise his students "to keep thy secrets in store unto thyself for wise men say store is no sore." He is said to have amassed tremendous wealth from his alchemic skills.

One of the pioneers of modern science, Albertus Magnus (1206-1280), was a strong believer in alchemy, and gave similar advice about keeping the secrets of this discipline private. His recommendation was "to carefully avoid association with princes and nobles and to cultivate discretion and silence."

There were many famous personages from medieval Europe who were accomplished alchemists. Another one was Roger Bacon (1214-1294). He left a manuscript that contained an encoded message on how to make copper.

Paracelsus (1493-1541) was the discoverer of zinc. But his real fame in his day stemmed from his expertise as an alchemist. In fact, his grave was once opened because it was rumored that it contained his secret of the craft. However, nothing was ever found therein. His secret had died with him.

Nicolas Flamel (1330-1418), who was a Paris notary, was another renowned alchemist. He claimed to have learned the secrets of this science from an age-old book that he translated. So profitable was his alchemical business that he personally funded the building of many hospitals and churches in Paris.

During the thirteenth and fourteenth centuries, alchemy had become so widespread in Europe that it captured the attention of the Vatican. In fact, a bull issued by Pope John XXII explicitly forbade the practice in the year 1317. This document, *Spondent Pariter*, called for the exile and heavy fining of anyone caught engaging in transmutations. Not surprisingly, this same pope later wound up pursuing alchemy himself. In fact, he wrote a book on it called *Ars Transmutatoria*, in which he boasted that he had created 200 bars of gold weighing one quintal, or 100 kg.

Henry IV of England issued an act similar to that of Pope John XXII, in 1404, which stated that the "multiplying of metals" was a crime against the Crown. This occurred during the 100 Years' War and the Peasants' Revolt--not a likely time to issue a decree forbidding a non-existent threat.

Though the British Crown generally viewed alchemy with disfavor, it made exceptions when this craft could be utilized for its own benefit. Thus King Henry VI granted permits to John Cobbe and John Mistelden to practice "the philosophic art of the conversion of metals." Parliament also approved of these licenses, and the gold produced was used in minting coins.

The ban on alchemy by Henry IV was later officially repealed by William and Mary in 1688. This decree stated: "And whereas since the making of the said statute, diverse persons have by their study, industry and learning, arrived to great skill and perfection in the art of melting and refining of metals, and otherwise improving and multiplying them."

This act of repeal also declared that from the reign of Henry IV, many Englishmen went to foreign countries "to exercise the said art" to the great detriment of the kingdom. But now, said this act, "all the gold and silver that shall be extracted by the aforesaid art" was to be turned over to the Royal Mint, where it would be purchased at full market value.

This same king and queen later went on to issue a decree that mentioned the desirability of studying alchemy. Can we not infer from this, as with the previously-mentioned declarations in other countries, that alchemy must indeed have been a legitimate science?

Johann Helvetius (1625-1709), a physician to the Prince of Orange, was an alchemist well known for his ability to transmutate base metals into gold. He was also known for being able to make already-existing gold items contain more grains of gold than they originally had(61).

The most famous scientist of all, from this period, who dabbled into alchemy was Sir Isaac Newton. In his personal notes he made this intriguing statement, which parallels what we have seen so many other alchemists say about this craft down through the millennia: "The substance which ye first take in hand is mineral, composed of mercury and crude sulphur"(206). He also expressed, like so many before him, the importance of keeping secret the methods of transmutation. In a letter, he urged a fellow alchemist, Robert Boyle, to keep "high silence" when publicly discussing the principles of alchemy.

The British Museum has what may be an actual piece of transmuted gold. If this is indeed what it is, it would mean that the secret of alchemy was known and practiced as late as the early nineteenth century. The artifact is a gold bullet. Its identification tag reads: "Gold made by an alchemist from a leaden bullet in the presence of Colonel MacDonald and Dr. Colquhoun at Bupora in the month of October, 1814."

This is not the only artifact of this nature. Nor is the British Museum the only museum that houses such artifacts. In fact, the Kunsthistorisches Museum in Vienna contains extraordinary evidence of alchemy. For example, one artifact found there, known as the Alchimistisches Medaillon, an oval medal 40 by 37 cm in diameter, and weighing 7 kg, is made of pure gold except for the upper third part, which is made of silver. The implication is that the alchemist who worked on the piece had not finished his task of completely transmuting the silver into gold. The transmutation of this disc, we are told, involved a mysterious red powder. Was it perhaps cinnabar?

The most recent example on record of the practice of this "lost art" of transmutation occurred during the last few years of the 19th century. In 1897, Dr.

Stephen H. Emmens, a British physician working in New York, claimed to discover a method of transmuting silver into gold. Between April of 1897 and August of 1898, more than $10,000 worth of gold was sold by him to the U.S. Assay Office on Wall Street. The New York *Herald* printed this headline in reference to him: "This Man Makes Gold and Sells it to the United States Mint"(61).

So, can there be any doubt that alchemy was a real science that is now apparently lost to us, at least in the comparatively simple and cheap manner in which it was formerly practiced?

========================

Airflight, too?

Assuming that we have been able to accept the notion that the discovery and usage of electricity is not unique to modern times, we are now ready to go to the next level and ask, Did ancient man possess the technology to fly? Let us look at the evidence--both legendary and archaeological--beginning with ancient legends.

Noorbergen tells us that "One of the earliest records of flight is found in a Babylonian document known as the Halkatha. In this document is found the following statement: 'To operate a flying machine is a great privilege. Knowledge of flying is most ancient, a gift of the gods of old for saving lives.' An ancient Chaldean work known as *Sifr'ala* dates back to about 3000 B.C. It gives a detailed account on how to build and fly an aircraft. The text speaks of various parts such as vibrating spheres, graphite rods, and copper coils. And when speaking of flying the craft, the author discusses the subject of wind resistance, gliding, and stability. Unfortunately, many key lines of the text are missing, making any attempt at reconstructing the craft impossible"(9).

Ancient tales of flying craft are also common throughout Asia.

For instance, in the fourth century A.D., a flying craft appeared in Ceylon (Sri Lanka), where, according to legend, the Buddhist monk Gunarvarman used it to fly to the island of Java--a distance of 2,000 miles(Ibid.).

In China, sometime between 340 and 278 B.C., a poem entitled "Li Sao" was written by Chu Yuan, in which he described an aerial voyage he made at a very great height, in the direction of the Kun Lun Mountains. This journey, which took place in a jade-colored craft, was said to have been unaffected by the wind and dust below(63).

Feats and Wisdom of the Ancients tells us that "Records inscribed on bamboo tablets some 2,000 years old tell of how the emperor Shunh [of China], who reigned between 2258 and 2208 B.C., built a flying craft to escape his parents' plot to kill him. Not only did the emperor effect a successful flight, the story goes, but he also tested a primitive parachute, landing unharmed after his getaway. Four and a half centuries later, Emperor Cheng Tang is said to have ordered his engineer, Ki-Kung-Shi, to design a 'flying chariot,' the term still used by the Chinese to describe an airplane. According to legend, the emperor ordered the machine destroyed after a successful flight so that its secrets would not fall into the wrong hands. In the fourth century A.D., the scribe Ko-Hung wrote of a 'flying car,' fashioned from the wood of a jujube tree, that used 'ox leather straps fastened to rotating blades to set the machine in motion.'...Ko-Hung's description calls to mind a helicopter"(62).

Please notice the mention above of the suppression of airflight technology, and the concern of it falling into the wrong hands. This is a point that pops up in many ancient legends that talk of airflight. Another example comes from a Tibetan document, which states that the knowledge of flying machines "is secret and not for the masses"(63). In the next subsection, we will see precisely why such suppression existed.

In the ancient Chinese chronicle, *Records of the Scholars*, it is revealed that the Han Dynasty astronomer and engineer, Chang Heng, constructed a "wooden bird" that had a mechanism in its interior which enabled it to fly a distance that is roughly the equivalent to our modern mile(15).

At a grave in the province of Shantung, dating to about 147 A.D., a stone carving can be seen that depicts a dragon chariot flying high above the clouds(61).

According to Jonathan Gray, in his book *Dead Men's Secrets*, a Tibetan text of unknown antiquity contains a description of "an enormous flying wagon made of a black metal with an iron base, not drawn by horses or elephants but by machines as large as those animals"(63).

The ancient Israelis quite possibly possessed the technology of airflight as well. As Gray wrote: "Familiar to us all is the Bible story of the visit to King Solomon by a queen from the south. Now comes the discovery of the Ethiopian epic *Kebra Nagast* (c. 850 B.C.), which tells the story from the other side. It records that King Solomon of Israel lavished on a visiting Ethiopian queen enormous riches and gifts 'and a vessel wherein one could traverse the air.' Carrying a cargo of animals as well as men, via Egypt, it 'traveled in one day a distance which (usually) took three months to traverse'"(Ibid.).

The Greeks were certainly no exception to having records of airflight in the ancient world. Gray tells us that "a Greek legend relates that Daedalus constructed wings for himself and his son Icarus, whom he advised neither to fly high, lest the glue should melt in the sun and the wings should drop off, nor fly too near the sea, lest the pinions be detached by the damp"(Ibid.).

Looking again to *Feats and Wisdom of the Ancients*, we are informed of a couple additional ancient Greek references to airflight: "In 468 B.C., the Greek mathematician Archytas made a wooden pigeon that actually flew. Hailed as a wonder of the ancient world, this pigeon was powered by an internal mechanism of balanced weights and a mysterious, unknown propulsive agent....[C]enturies later...the Greek scientist Hero...designed a rotating boiler that showed the propulsive effect of a jet of steam. Suspended above a fire, his device spun about as steam shot from its four nozzles"(62).

The early inhabitants of Britain weren't exempt from tales of flying craft either. Druid legends speak of "magical machines capable of traveling on land, sea and air"(63).

In the South Pacific, the natives on the island of Ponape tell of how their antiquated ancestors were visited by learned, light-skinned men from the west who came in "shining boats" that "flew above the sea." These guests were said to perform many great "magical works" (advanced technological feats?)(177).

On Mangareva, another South Pacific island (the largest of the Gambier Islands), the aboriginal inhabitants speak of how, in archaic times, they were visited by "flying canoes" with "great wings clasped tightly to the sides." These aerial craft are said to have been able to fly as far as the Hawaiian Islands, over 2,500 miles away(Ibid.).

On Easter Island, the Natives there tell the story of an ancient god-king, Hotu Matua, who descended from heaven to Earth in a great "ship." The legend states: "He came down from heaven to earth....He came in the ship of his youngest son...from heaven"(209).

Of all ancient cultures, none talked more about airflight than the people of India. Their descriptions of their aerial vehicles--how they were made, how they were flown, and what they were used for--were far more detailed and realistic-sounding than any others that have survived to our day.

The first example we will look at is a document known as the *Valmiki*, which dates to the second millennium B.C., and proclaims: "The sky chariot...is gilded and lustrous throughout...it leaps above the hill and the wooded valley, winged like lightning...covered in smoke and flowing lamps, speedy and round of prow"(177).

The next example we'll examine is elaborated on thusly by Jonathan Gray: "The *Ramayana*, the great Indian epic poem dating from the third century B.C., describes a double-deck circular aircraft with portholes and a dome....Fueled by a strange yellowish white liquid, the craft was said to travel at the 'speed of wind,' attain heights that made the ocean look like 'a small pool of water,' and stop and hover motionless in the sky"(62).

This same document, the *Ramayana*, further states of these ancient sky vehicles that they were "furnished with window compartments and excellent seats." One person named Bhima, according to the text, "flew along in his car, resplendent as the sun and loud as thunder....The flying chariot shone like a flame in the night sky of summer." These craft, we are told, were able to rise "vertically into the air with a whole family on board, and with a tremendous noise." Pilots, the text states, had to be well trained before ever being able to sit at the controls(63).

Among the old Hindu sacred books that deal with flying, we also find the *Samaranga Sutradhara*, a collection of texts compiled in the eleventh century

A.D., but dating back to a very remote period in India's history. It contains 230 stanzas that describe, in detail, every aspect of flying, from how the craft were powered to the proper clothing and diet of the pilots. What follows are a few excerpts from this text: "The aircraft which can go by its own force like a bird, on the earth or water, or through the air, is called a vimana. That which can travel in the sky from place to place is called a vimana by the sages of old....The body must be strong and durable and built of light wood, shaped like a bird in flight, with wings outstretched. Within must be placed the mercury engine, with its heating apparatus made of iron underneath....In the larger craft, because it is built heavier, four strong containers of mercury must be built into the interior. When these are heated by controlled fire from the iron containers, the vimana possesses thunder power through the mercury. The iron engine must have properly welded joints to be filled with mercury, and when fire is conducted to the upper part, it develops power with the roar of a lion. By means of the energy latent in mercury, the driving whirlwind is set in motion, and the traveler, sitting inside the vimana, may travel in the air, to such a distance as to look like a pearl in the sky"(9).

Though this text gives some tantalizing information on these vimana aircraft, it deliberately left out specifics on how to construct such a vehicle. Later on in the text, the reason for this is given: "Any person not initiated in the art of building machines of flight will cause mischief"(Ibid.). Once again we find the suppression of airflight technology in ancient literature, which we will, as already stated, be taking up in considerable detail in the next subsection.

In the last-cited Hindu document, we read about the "energy latent in mercury." What could this be referring to? Actually, it is not known today what the exact nature is of this reference to latent mercury energy. But it is quite obviously indicative of a real technology that has been lost to us--that the ancients knew something about mercury that we have yet to rediscover [Endnote 51]. Notice what Isaac Newton once wrote about this: "Because [of] the way by which mercury may be impregnated, it has been thought fit to be concealed by others that have known it, and therefore may possibly be an inlet to something more noble, not to be communicated without immense danger to the world"(9).

Of particular interest here is the fact that there have been several recorded instances of pure liquid mercury being found in various ancient ruins around the world.

For example, in the early 1970s, writes Noorbergen, "Soviet explorers excavating a cave near Tashkent in the Uzbek S.S.R. discovered a number of conical ceramic pots, each carefully sealed and each containing a single drop of mercury. A description and illustrations of the mysterious pots were published in the Soviet periodical *The Modern Technologist*. There is no clue to what these mercury containers were used for, but they must have been highly treasured and used for something that is apparently beyond our present understanding and technology. It was a secret that was found, used and preserved by a select few--only to be lost again, perhaps forever."

Another key location where liquid mercury has been found while excavating ancient ruins is Lamanai, a Mayan site located in Belize(Ibid.).

Returning to our coverage of ancient Indian records of airflight, we find the eighth century A.D. document, *Mahavira of Bhavabhuti*, recording that "an aerial chariot, the Pushpaka, conveys many people to the ancient capital of Ayodhya. The sky is full of stupendous flying machines, dark as night, but picked out by lights with a yellowish glare"(63).

Nearby Nepal had its share of tales relating to flying in its early history. Noorbergen reveals that "References to flight also appear in the *Budhasvamin Brihat Katha Shlokasamgraha* of Nepal, a twelfth-century [A.D.] written version of an oral tradition of unknown age. It was first published in Europe in Felix Lacote's French translation in 1908. The *Brihat Katha* tells the story of Rumanvit, the servant of a king who desired to travel about the earth in a flying vehicle. In order to satisfy his master, Rumanvit commanded the court designers to construct the needed flying apparatus, but they informed him that they were unable to do so. They knew the workings of many machines, they declared, but the secret of flying machines was known only to the 'Yavanas.'

"...The story of Rumanvit ends with the appearance...of [the Yavanas] from the west who fulfill the monarch's wish to see the world from the air, but without revealing to him the mechanics of flying. There appears to have been a conscious effort on the part of the high civilization centers not to proliferate advanced technology...but rather to keep that technology for their own use and power"(9). Yet again we find the suppression of airflight technology being referenced in an ancient document, the reason for which will be elaborated on shortly.

Coming back to the *Ramayana*, we find that mention is made of a seven-storey vimana--the mother of all ancient aircraft--which was known as a Puspakavimana. The text refers to it as a "great aerial mansion" that was "decorated with pearls and diamonds." It was also said to have been "featured with artistic windows made of refined gold"(177). This last comment is of particular interest, seeing that NASA has made the visors of astronaut helmets with a fine, transparent gold coating. Now how do you suppose the early inhabitants of India could have been aware that gold was able to be rendered transparent? Does this not tell us that they must have had the technology to do this?

Can you not see that, because there are so many references--and quite detailed ones at that--to airflight in ancient literary works, it is difficult to write them off as mere mythical fantasy? But there are also some tantalizing pieces of physical evidence of this ancient technology, upon which we will now focus our attention.

We find on a ceiling beam of a 3000-year-old temple at Abydos several unusual carved features, which look like modern air and sea craft. The aircraft featured on the beam look remarkably like modern, aerodynamically-designed vehicles. There's a helicopter (complete with propeller), a blimp, and a glider, all three of which contain a tail fin. The sea craft looks very much like a submarine.

Back in 1969 in the basement of the Cairo Museum, while rummaging through boxes of old artifacts discovered in the late nineteenth century, Egyptologist Khalil Messiha noticed one particular item that stood out from the rest. In a box marked "bird objects," Messiha found Special Register No. 6347, which did not look quite so much like a bird. With straight wings and a sleek, tapered body, along with a vertical tail fin, this curious object had more the appearance of a scale model of a glider plane. Dating from about 200 B.C. and found in a tomb near Saqara, it was 7 inches long and was constructed of light sycamore wood, weighing in at only 1.11 ounces. Its construction incorporated sophisticated principles of aircraft design that took modern engineers decades to discover and perfect. This glider, incidentally, after nearly 2,300 years, was still able to fly. And, furthermore, there is an interesting jagged edge on the base of the tail which has led some to conclude that this artifact may have once contained some type of propulsion mechanism(62).

An important point that needs to be brought out here is the fact that the ancient Egyptians often made scale model replicas of real-life objects, which they placed in their tombs for use in the "afterlife." Could this model airplane have been a representation of a real plane from the ancient Egyptian world? Regardless, this artifact at least demonstrates that the people of ancient Egypt had a working knowledge of the science of aerodynamics.

Certain pyramid texts in Egypt seem to provide us with further confirmation that the people of this land knew how to fly in ancient times. While speaking of a particular king, or pharaoh, one text states: "...the doors of the sky are thrown open to you, the doors of the starry firmament are thrown open for you." Another one reads: "The king is a flame, moving before the wind to the end of the sky and to the end of the earth...there is brought to him a way of ascent to the sky..." A few lines later, this same text asks the king: "What fiery boat shall be brought to you?" And then it adds this line, which is apparently the king's response: "Bring me [that which] flies and alights." Next, the king goes on to say: “I have ascended in a blast of fire having turned myself about.” Finally, the inscription ends by asking: “Wherewith can the king be made to fly up?” And to this it replies: “There shall be brought to you the hnw-bark [“hnw” is untranslatable]...and the [text missing]...of the hn-bird [“hn” is also untranslatable]. You shall fly up therewith....You shall fly up and alight”(20).

The pre-Incan Sinu culture of Colombia, dating somewhere between the sixth and ninth centuries A.D., made several gold pendants that look strikingly like modern aircraft. These “planes” have a perfect aerodynamic design, complete with deltoid wings and upright tails(62).

Not too far from where these pendants were found, there exists a most fascinating system of grand-scale ground markings in the Nazca Valley of Peru, which we touched upon earlier. These markings were created in ancient times (perhaps in the early centuries A.D.) by removing pebbles embedded in the desert floor, exposing the hard, compacted sand underneath. These etchings have survived the many centuries since their creation because this area receives almost no rain whatsoever, which otherwise would have washed them away a long time ago. For centuries, since the Spaniards first visited this area, it was thought that these etchings were some sort of road system, except that they didn’t appear to lead anywhere. It wasn’t until a pilot flew over the area in the 1940s that their true nature was realized--that these enormous carvings are actually, for the most part, giant pictoglyphs that cover an area over 60 miles long. Sometimes these lines form very recognizable patterns--animals, such as a monkey, a condor, or a spider--and other times they appear simply as straight lines that stretch for miles across the desert floor. What was their function? And, better yet, how were they made with such precision, on such a large scale?

Even with our modern surveying equipment, reproducing these precise, colossal forms would be impossible without an aerial perspective. Understand that these patterns can’t even be discerned except from a great height, directly above. So how could the ancients have appreciated them, let alone having created them in the first place, without airflight technology? Did the ancient creators of these patterns truly have the ability to fly? Did their craft look like the small, gold “airplane” pendants found in nearby Colombia, as described above? Or could it be, as some have suggested, that the ancient Nazca artists employed hot air balloons to aid them in forming these enormous etchings? Along this line, take note of the following insightful quote from *Dead Men’s Secrets*: “Four pieces of Nazcan fabric from tombs were examined under a microscope. They proved to have a finer weave than present-day parachute material and a tighter weave than that used by modern hot-air balloonists--205 X 110 threads per square inch, compared with 160 X 90.” Many of these Nazca textiles, incidentally, depict flying men(63).

Interestingly, near to the Nazca plain, according to Gray, there circulates "[a]n ancient Incan legend [which] tells of a boy named Antarqui who [somehow] flew behind the enemy lines and reported their positions--thus helping the Incas in battle"(Ibid.).

There exists an ancient Assyrian cylinder seal, now housed in the British Museum, which portrays a winged disk hovering in the air, carrying three seated human occupants. Is this a representation, albeit obviously simplified, of another ancient aerial craft? Another Assyrian work of art has been found that depicts a single man sitting in an aerial craft with wings. Can we really write such portrayals off as visual myths?

Then we have what appear to be eyewitness accounts of people in the ancient world who spotted flying craft, which some have interpreted to be of an extraterrestrial nature. But could they not simply have been glimpses caught of ancient terrestrial air vehicles, like the vimanas of India, for instance, the nature of which was not understood by those who viewed them? Here are some examples of sightings of this sort:

- During the reign of Pharaoh Thutmose III, around 1450 B.C., there were seen multiple "circles of fire" hovering over the skies of Egypt, measuring about 5 metres in size. This sighting is known to us from an ancient papyrus, which states: "In the year 22 of the 3rd month of winter, sixth hour of the day...the scribes of the House of Life found it was a circle of fire that was coming in the sky....It had no head, the breath of its mouth had a foul odor. Its body [was] one rod long and one rod wide. It had no voice. Their hearts [the hearts of the scribes] became confused through it; then they laid themselves on their bellies...they went to the Pharaoh...to report it. His Majesty ordered...[an examination of] all which is written in the papyrus rolls of the House of Life. His majesty was meditating upon what happened. Now after some days had passed, these things became more numerous in the skies than ever. They shone more in the sky than the brightness of the sun, and extended to the limits of four supports of the heavens....Powerful was the position of the fire circles. The army of the Pharaoh looked on with him in their midst. It was after supper. Thereupon, these fire circles ascended higher in the sky towards the south....The Pharaoh caused incense to be brought to make peace on the hearth...and what happened was ordered by the Pharaoh to be written in the annals of the House of Life...so that it be remembered for ever"(168).

Regarding this text's mention of these "circles of fire" shining "more...than the brightness of the sun," let us recall how the *Ramayana* talks of certain flying craft that were "resplendent as the sun."

- Alexander the Great's army, while crossing a river into India in 329 B.C., saw "two silver shields" in the sky that dove repeatedly towards them, causing great panic. Later on, in 322 B.C., when Alexander was besieging Tyre in Phoenicia, another "flying shield," moving in triangular formation with smaller "shields," had appeared. Supposedly the larger object shot beams of light at the city, shattering its walls and other defenses, at which time Alexander's army, quickly taking advantage of the situation, seized the city. The objects then disappeared(167).

These "shields" could very well have been vimanas, since the ancient Indian text, the *Mahabharata*, describes one model of this line of craft as being spherical in shape, which would have looked like a metallic shield (or shields, in this case) to ground-based observers.

The likelihood that Alexander and his army did indeed see vimanas is especially probable in the case of the first sighting mentioned above, which took place in India--the home of the vimanas.

- Julius Obsequens, a Roman author believed to have lived in the fourth century A.D., drew on Livy, as well as other sources of his time, to compile his book *Prodigorium Liber*. In this work, he described three historical sightings of flying craft, which are cited below:

216 B.C.: "Things like ships were seen in the sky over Italy....At Arpi [east of Rome, in Apulia] a 'round shield' was seen in the sky....At Capua, the sky was all on fire, and one saw figures like ships..."

99 B.C.: "When C. Murius and L. Valerius were consuls, in Tarquinia, there fell in different places...a thing like a flaming torch, and it came suddenly from the sky. Towards sunset, a round object like a globe, or round or circular shield took its path in the sky, from west to east."

90 B.C.: "In the territory of Spoletium [north of Rome, in Umbria] a globe of fire, of golden color, fell to the earth, gyrating. It then seemed to increase in size, rose from the earth, and ascended into the sky where it obscured the disc of the sun with its brilliance. It revolved towards the eastern quadrant of the sky"(170).

- In more recent times--the thirteenth century A.D.--Friar Roger Bacon made an intriguing statement indicating that flying craft were being made (and apparently were witnessed flying about) in his day, and that these flying contraptions were known from ancient times. Here are his exact words: "Flying machines as these were of old, and are made even in our days"(15).

Worthy of our attention here are two high-tech-looking flying craft that are pictured in a fresco above the altar at the Visoki Decani Monastery in Kosovo, Yugoslavia, which was painted in the year 1350. Were these examples of the flying craft that Roger Bacon made reference to a century earlier?

Such flying craft were apparently witnessed in Europe into the seventeenth century. For we find that a French coin from this period contains the image of a round flying craft in the sky, hovering over the French landscape. It contains the Latin words "Opportunus Adest," which mean "It is here at an opportune time."

* * * * * * *

If the ancients truly did have airflight technology, what, you may ask, happened to their flying machines? Why do we not find any traces of them in the archaeological record? Well, the materials used to make these craft, as described above--mostly wood and iron--would have disintegrated a long time ago. There's also the issue, described above as well, of how many of these craft were intentionally destroyed to prevent the dissemination of airflight technology into the wrong hands. Keep in mind, too, that the ancient world surely did not have as many flying craft as we do in our time. Thus, the less craft there were in existence, the less chance of one surviving, in any recognizable form, into modern times.

But as far as the suppression of ancient airflight technology goes, let us now direct our attention toward some of the sinister uses this technology was put to, in order to see why there was such a strong desire to keep it secret.

Ancient aerial warfare

It was in 1903 that the Wright brothers first began tests of their flying craft, which was just a little over a decade before such contraptions were used to inflict massive devastation during World War I. It didn't take long at all to turn this great scientific advancement into a tool of destruction. And so it was in ancient times. Though ancient flying craft were initially said to have been designed "for saving lives," inevitably this technology was utilized to put lives to an end.

Not much changes throughout history. As we shall see throughout the course of this subsection, nearly all the ancient legends that talk of airflight also mention how this innovation was employed in time of war. And, as we shall also see on the pages that follow, the devastation wrought from the air was often so thorough and pervasive that it equaled modern man's destructive capabilities, hence the drive to prevent the broad dispersal of airflight technology in ancient times.

Let us open with a review of some ancient records that talk of the use of somewhat conventional weapons in aerial warfare, beginning with a few citations from Gray's *Dead Men's Secrets*: "The Piute Indians tell of the 'Hav-Musuvs' who traveled from the Gulf of California to Death Valley (when it was still green and fertile)....They flew in large silent craft, which had weapons....

"The Hopi Indians of the southwest United States have a...tradition [known as Kuskurza] of a people who made a 'patuwvota' which soared through the sky. On this, many of them flew to attack a great city. Soon others from many nations were making 'patuwvotas' and flew to attack one another....

"G.R. Josyer, director of the International Academy of Sanskrit Research in Mysore, has translated into English the 3,000-year-old *Vymanika Shastra*, meaning 'the Science of Aeronautics.' It has eight chapters (6,000 lines)...on the construction of three types of aircraft." The information covered includes the design of a helicopter-type cargo plane and passenger planes for as much as 500 people; plans for a craft that flew in the air, traveled under water, or floated platoonlike on the water's surface; the qualifications and training of pilots; the inclusion on the aircraft of what can only be described as cameras, radios, and a type of radar; instructions on how to make the aircraft invisible to enemies (a radar jammer or cloaking device?)[Endnote 52], how to paralyze other aircraft, how to zigzag in the sky like a snake, how to see inside an enemy's aircraft, and how to spy on "all activities going on down below on the ground"; also given was the correct proportions of certain chemicals which, when discharged outside the aircraft, would give the appearance of a cloud (for camouflage purposes?)(63).

In one episode of India's national epic, the *Mahabharata* (cited earlier), the Vrishnis, a tribe whose warriors included a hero named Krishna, were beset by the forces of a leader named Salva. From there, the cruel Salva had arrived on the scene, riding on the Saubha chariot, and from it he killed many Vrishni youths and devastated all the parks in the city.

The story goes on to tell of how the Vrishni heroes were well equipped to launch a counter attack against Salva. At one point, a Vrishni warrior was about to finish Salva off with a special weapon, but then decided against it because "not a man in battle is safe from this arrow."

Soon Krishna took to the skies in pursuit of Salva, but Salva's Saubha "clung to the sky at a league's length....He threw at him (Krishna) rockets, missiles, spears, spikes, battle axes, 3-bladed javelins, flame throwers, without pausing [Endnote 53]....The sky...seemed to hold a hundred suns, a hundred moons...and a hundred myriad stars. Neither day nor night could be made out, or the points of compass."

Krishna then warded off Salva's attack with the equivalent of anti-ballistic missiles, destroying Salva's rockets, missiles, and spears in mid flight, before they impacted: "Krishna warded them off, as they loomed towards him, with his swift-striking shafts, as they flashed through the sky. Krishna cut them into pieces...there was a great din in the sky above."

Nonetheless, Krishna was still in distress, for Salva's Saubha somehow became invisible [again, was this a cloaking device being used?]. Krishna then loaded a special weapon. It appears to have been an ancient version of a smart bomb: "Krishna quickly laid on an arrow, which killed by seeking out sound....All of...Salva's army...lay dead, killed by the blazing sunlike arrows that were triggered by sound"(9).

In light of this fascinating account, it is important to point out that the meaning of the biblical name Methuselah, a person who lived before the flood catastrophe, is "man of the flying dart, man of war"(Ibid.). Could we perhaps gather from this that advanced military weaponry also existed before the time of the flood? If so, this might possibly explain, at least in part, what the ancient Hebrew scriptures meant when they talked of how, before the flood, "the wickedness of man was great in the earth, and that every imagination of the thoughts of his heart was only evil continually....The earth also was corrupt...and...was <u>filled with violence</u>." Genesis 6:5, 11.

Leaving our discussion of the use of somewhat conventional weapons in ancient aerial warfare, let us continue on to unconventional weapons--weapons of mass destruction. Were the ancients also in possession of grand-scale, death-inflicting munitions, such as atomic bombs?

Well, let me first point out that, in modern times, less than four decades passed between the development of the first airplane and the creation and usage of the first atomic bomb. Could the same speed of development have occurred in ancient times? Irregardless how long it may have taken the ancients to create atomic weapons, it would appear that they truly did have such innovations, deploying them from their flying craft, as we shall see from the pieces of legendary and archaeological evidence that we will now consider.

Turning again to the *Mahabharata* (Book 16, *Mausala Parva*), we find astonishing accounts therein of what can only be described as devastating atomic warfare. One such account tells of an 18-day war between the Kauravas and the Pandavas, who inhabited the upper regions of the Ganges. Not long after this war, a second war was waged against the Vrishnis and the Andhakas in the same area. In both conflicts, vimanas, described in this particular text as "celestial cars" and "aerial chariots with sides of iron clad with wings," were used to launch a weapon of terrible destructive power.

The first of these two wars fought in these aircraft is described as follows: "The valiant Adwattan, remaining steadfast in his vimana, landed upon the water and from there unleashed the agneya weapon, incapable of being resisted by the very gods. Taking careful aim against his foes,...[he] let loose the blazing missile of smokeless fire with tremendous force. Dense arrows of flame, like a great shower, issued forth upon creation, encompassing the enemy. Meteors flashed down from the sky. A thick gloom quickly settled upon the Pandava hosts. All points of the compass were lost in darkness. Fierce winds began to blow. Clouds roared upward, showering dust and gravel. Birds croaked madly, and beasts shuddered from the destruction. The very elements seemed disturbed. The sun seemed to waver in the heavens. The earth shook, scorched by the terrible violent heat of this weapon. Elephants burst into flame and ran to and fro in a frenzy, seeking protection from the terror. Over a vast area, other animals crumpled to the ground and died. The waters boiled, and the creatures residing therein also died. From all points of the compass the arrows of flame rained continuously and fiercely. The missile of Adwattan burst with the power of thunder, and the hostile warriors collapsed like trees burnt in a raging fire. Thousands of war vehicles fell down on all sides"(9).

The description of the second battle that took place is no less frightening or realistic-sounding than the first: "Gurkha, flying in his swift and powerful vimana, hurled against the three cities of the Vrishnis and [the] Andhakas a single projectile charged with all the power of the universe. An incandescent column of smoke and fire, as brilliant as ten thousand suns, rose in all its splendor. It was the unknown weapon, the iron thunderbolt, a gigantic messenger of death which reduced to ashes the entire race of the Vrishnis and [the] Andhakas. The corpses were so burnt that they were no longer recognizable. Hair and nails fell out [Endnote 54]. Pottery broke without cause. Birds, disturbed, circled in the air and were turned white. Foodstuffs were poisoned. To escape, the warriors threw themselves in streams to wash themselves and their equipment"(Ibid.).

The above descriptions of the resulting devastation from these battles are far too detailed to be written off as the products of ancient fertile imaginations. Foodstuffs being poisoned, corpses burnt beyond recognition, hair and nails falling out, fierce winds, a darkening sky, clouds roaring upward, warriors running to streams to wash themselves, etc.--all of these things are precisely what would result from radioactive fallout. But there is also physical evidence for ancient atomic warfare in India.

Precisely in the area where these conflicts are reported to have taken place (in the vicinity of Harappa and Mohenjo-Daro), archaeologists have discovered numerous burnt ruins which have yet to be fully explored and excavated. The observations that have thus far been made indicate that these ruins were not burned by ordinary fire. In fact, no natural burning flame or volcanic eruption could have produced the types of phenomena observed in these ruins. Only the heat released through atomic energy could have done this damage, namely, stone walls whose outer surfaces have been melted (vitrified) and split.

In these same ruins there have also been found charred, radioactive human bones scattered about in the streets. Speaking on this matter, Gray wrote: "Skeletons in Mohenjo-Daro and Harappa are extremely radioactive. The ruins of these ancient Indus Valley cities are immense. They are thought to have contained well over a million people each. Practically nothing is known of their histories, except that both were destroyed suddenly. In Mohenjo-Daro, in an epicenter 150 feet wide, everything was crystallized, fused or melted; 180 feet from the center, the bricks are melted on one side, indicating a blast.

"Ancient Indian texts speak of a city's people being given 7 days to get out--a clear warning of total destruction. Excavations down to the street level revealed forty-four scattered skeletons, as if doom had come so suddenly they could not get into their houses. All the skeletons were flattened to the ground. A father, mother and child were found flattened in the street, face down and still holding hands.

"The skeletons, after thousands of years, are still among the most radioactive that have ever been found, on a par with those of Hiroshima and Nagasaki"(63).

Furthermore, according to researcher David Childress, "at Mohenjo-Daro,...the streets were littered with 'black lumps of glass.' These globs of glass were discovered to be clay pots that had melted under intense heat!"(169)

There is even evidence that the ancient people of India not only knew about the existence of the atom, like the Greeks and Romans after them, but actually had a working knowledge of atomic structure, which, of course, they would have needed in order to develop atomic weapons. In the fourth-century Brahmin treatises, *Valsesika* and *Nyaya*, the Yoga Vasishta says: "There are vast worlds within the hollows of each atom, multifarious as the specks in a sunbeam"(61).

Though this text dates to a much later time than the records that discuss the ancient atomic battles, it may simply be that this understanding of atomic structure had been handed down from an earlier age.

A further piece of evidence of a working knowledge of the nature of the atom in ancient India is provided by Noorbergen: "The use of atomic weapons in India 4,000 years ago presupposes a knowledge of nuclear physics rivaling our own. There is evidence of such a knowledge preserved among the ancient Hindu records. Several Sanskrit books, for example, contain references to divisions of time that cover a very wide range. At one extreme, according to Hindu texts dealing with cosmology, is the *kalpa* or 'Day of Brahma,' a period of 4.32 billion years. At the other, as described in the *Bihath Sathaka*, we find reference to the *kashta*, equivalent to three one-hundred-millionths of a second. Modern Sanskrit scholars have no idea why such large and such minuscule time divisions were necessary in antiquity. All they know is that they were used in the past, and they are obliged to preserve the tradition.

"Time divisions of any kind, however, imply that the duration of something has been measured. The only phenomena in nature that can be measured in billions of years or in millionths of a second are the disintegration rates of radioisotopes. These rates range from those of elements like uranium 238, with a half-life of 4.51 billion years, to subatomic particles such as K mesons and hyperons, with mean half-lives measured in the hundred-millionths, billionths, trillionths, and even smaller fractions of a second. The ranges of ancient Hindu time division and of radioisotope disintegration thus partially coincide, and the former could have been used to measure the latter"(9).

A most curious point of interest that is worthy of note here is the fact that Robert Oppenheimer, the "father of the atomic bomb," was also a Sanskrit scholar. On one occasion, when speaking of the first atomic test conducted near Alamogordo, New Mexico, on July 16, 1945, he quoted from the *Mahabharata*, saying, "I have unleashed the power of the Universe. Now I have become the destroyer of worlds." When asked during an interview at Rochester University, seven years after this first atomic test, whether it was the first such bomb ever to be detonated, he replied by saying, "Well, yes, in modern history"(116). Obviously, then, Oppenheimer believed that the ancient people of India had beaten modern man in the development of atomic warfare technology.

Still another record from India of ancient atomic warfare is found in the *Bhagavata Purana*, which states: "The flames of the Brahmastra-charged missiles mingled with each other and surrounded by fiery arrows they covered the earth, heaven and space between and increased the conflagration like the fire and the Sun at the end of the world....All beings who were scorched by the Brahmastras, and saw the terrible fire of their missiles, felt that it was the fire of *Pralaya* [the cataclysm] that burns down the world."

Turning to Tibet, we find that the ancient people of this land also talked about atomic aerial warfare. Here's what Noorbergen had to say in this regard: "One of the most amazing literary testimonies to man-made destruction among the ancient advanced civilizations is found in the Tibetan *Stanzas of Dzyan*, translated [in the nineteenth century], [but] the original dating back several millennia. Like the *Mahabharata*, the *Stanzas of Dzyan* depict a holocaust engulfing two warring nations who utilize flying vehicles and fiery weapons.

"'The great King of the Dazzling Face, the chief of all the Yellow-faced, was sad, seeing the evil intentions of the evil Dark-faced. He sent his air vehicles to all his brother chiefs with pious men within, saying, Prepare, arise, men of the good law, and escape while the land has not yet been overwhelmed by the waters.

"'The Lords of the Storm are also approaching. Their war vehicles are nearing the land. One night and two days only shall the Lords of the Dark-faced arrive on this patient land. She is doomed when the waters descend on her. The Lords of the Dark-eyed have prepared their magic Agneyastra [the equivalent of the Hindu agneya weapon--an atomic bomb]....They are also versed in Ashtar [the highest magical knowledge, i.e. technology]. Come, and use yours.

"'Let every Lord of the Dazzling Face ensnare the air vehicle of every Lord of the Dark-faced, lest any of them escape.

"'The great King fell upon his Dazzling Face and wept. When the kings were assembled, the waters of the earth had already been disturbed. The nations crossed the dry lands. They went beyond the water mark. The kings reached then the safe lands in their air vehicles, and arrived in the lands of fire and metal....

"'Stars [atomic bombs?] showered on the lands of the Dark-faced while they slept. The speaking beasts [radios?] remained quiet. The Lords waited for others, but they came not, for their masters slept. The waters rose and covered the valleys....[I]n the high lands there dwelt those who escaped, the men of the yellow faces and of the straight eye'"(9). Is this describing a an ancient atom bombing spree that caused tidal waves?

Elsewhere in this same document, we read: "Separation did not bring peace to these people [apparently the people of Harappa or Mohenjo-Daro] and finally their anger reached a point where the ruler of the original city took with him a small number of his warriors and they rose into the air in a huge shining metal vessel. While they were many leagues from the city of their enemies, they launched a great shining lance that rode on a beam of light. It burst apart in the city of their enemies with a great ball of flame that shot up to the heavens, almost to the stars. All those who were in the city were horribly burned and even those who were not in the city--but nearby--were burned also. Those who looked upon the lance and the ball of fire were blinded forever afterward. Those who entered the city on foot became ill and died. Even the dust of the city was poisoned, as were the rivers that flowed through it. Men dared not go near it, and it gradually crumbled into dust and was forgotten by men. When the leader saw what he had done to his own people he retired to his palace and refused to see anyone. Then he gathered about him those warriors who remained, and their wives and children, and they entered their vessels and rose one by one into the sky and sailed away. Nor did they return"(109).

Sumerian texts describe a battle between two opponents, Ninurta and Zu. Ninurta, we are told, shot "arrows" at Zu, but they were "stopped in the midst" of their flight (interceptor missiles?). As the battle continued, Ea instructed Ninurta to add a *til-lum* to his arsenal, and shoot it into the "pinions," or small cog-wheels, of Zu's wings. Following this advice, Zu's wings were struck and he came spiraling downward in defeat. Interestingly, the *til-lum* that Ninurta was instructed to use against Zu actually translates as "missile." And, when written in its ancient pictoglyphic form, it looks astonishingly like a modern-day missile(171).

The Babylonian Epic of Gilgamesh, dating to about 2000 B.C., recalls a day when "the heavens cried out, the earth bellowed an answer, lightning flashed forth, fire flamed upwards, it rained down death. The brightness vanished, the fire was extinguished. Everyone who was struck by the lightning was turned to ashes"(63).

Though this account doesn't necessarily imply that the devastation described therein was caused by a manmade weapon, the mention of fire flaming upwards and of people turning to ashes are certainly reminiscent of what resulted from the battles described in the Indian and Tibetan records.

But there is another account that dates to the same time, and from the same area, which makes it clear that a weapon capable of inflicting similar widespread devastation was indeed in existence, and was put to use. Ancient cuneiform texts (the "lamentation tablets") describe a terrible destruction unleashed upon Sumer during a battle involving high-tech weaponry: "On the land fell a calamity, one unknown to man: One that had never been seen before, one which could not be withstood." It brought death to all who were in the vicinity--a death "which roam[ed] the street, [was] let loose in the road; it [stood] beside a man--yet none [could] see it; when it enter[ed] a house, its appearance [was] unknown."

There was simply no way of protecting oneself from this "evil which...assailed the land like a ghost....The highest wall, the thickest walls, it pass[ed] as a flood; no door [could] shut it out, no bolt [could] turn it back; through the door like a snake it [glided], through the hinge like a wind it [blew] in....Cough and phlegm weakened the chest, the mouth was filled with spittle and foam...dumbness and daze...[came] upon them, an unwholesome numbness...an evil curse, a headache....The people, terrified, could hardly breathe; the evil wind clutched them, [it did] not grant them another day....The face was made pale by the evil wind." This "evil wind" came as a cloud over Sumer and "covered the land as a cloak" and "spread over it like a sheet"(108).

Another lamentation tablet from the same era talks about "awesome weapons" that were launched from the skies, which "spread awesome rays towards the four points of the earth, scorching everything like fire"(Ibid.).

Around 200 years later, during the reign of King Hammurabi of Babylon, ancient texts told of a powerful weapon that the king used in battle: "With the powerful weapon with which Marduk proclaimed his triumphs, the hero [Hammurabi] overthrew in battle the armies of Eshnuna, Subartu and Gutium....With the Great Power of Marduk he overthrew the armies of Sutium, Turukku, Kamu...with the Mighty Power which Anu and Enlil had given him he defeated all his enemies as far as the country of Subartu"(Ibid.).

In the ninth century B.C., at the time of the reign of Shalmaneser III, the Babylonians were at it again, employing similar weaponry in battle. Texts from this period record King Shalmaneser as saying: "I fought with the Mighty Force which Ashur, my lord, had given me; and with the strong weapons which Nergal, my leader, had presented to me." The weapon of Ashur was said to have a "terrifying brilliance." During one particular war, the enemy, Adini, along with his host of warriors, fled upon seeing "the terrifying brilliance of Ashur; it overwhelmed them"(Ibid.).

With all of this in mind, it needs to be pointed out that there have been vitrified ruins found in the area where the above battles are said to have taken place. Near the site of ancient Babylon, for example, there are remains of a zigguart that show clear signs of vitrification in the remote past. Quoting from researcher Erich von Fange, Noorbergen wrote: "It appeared that fire had struck the tower and split it down to the very foundation....In different parts of the ruins, immense brown and black masses of brickwork had [been] changed to a vitrified state...subjected to some kind of fierce heat, and completely molten. The whole ruin has the appearance of a burnt mountain"(9). This description sounds quite similar to what has been seen in the ruins of Mohenjo-Daro and Harappa, as discussed above.

The ancient Americas had their share of similar tales. The Mayan *Popol Vuh*, for instance, describes the destructive effects of a fire from the sky that put out eyes and decomposed flesh and entrails, which was apparently of human origin. This document also describes how great cities to the north (apparently in what is today the United States) were destroyed at this same time(63). Could this be interpreted to imply that the ancient Native Americans of North America also possessed atomic warfare capabilities?

Possible confirmation of this comes from a legend of a dying totemic cult in northern Canada, near the tundra, which tells of a time "before the cold descended from the north [before the Ice Age?]." It states further that "In the days when great forests and flowering meadows were here, demons came and made slaves of our people and sent the young to die among the rocks and below the ground. But then arrived the thunderbird, and our people were freed. We learned about the marvelous cities of the thunderbird, which were beyond the big lakes and rivers to the south [in the U.S.].

"Many of our people left us and saw these shining cities and witnessed the grand homes and the mystery of men who flew upon the skies. But then the demons returned, and there was terrible destruction. Those of our people who had gone southward returned to declare that all life in the cities was gone--nothing but silence remained"(9).

But do we find any physical evidence in the U.S., or anywhere else in the Americas, of ancient atomic warfare? Apparently so, as we see from this next Noorbergen quote: "The continents of the New World also possess several examples of prehistoric cultures destroyed by a great conflagration. Not far from Cuzco, Peru, near the Pre-Inca forests of Sacsahuaman, an area of 18,000 square yards of mountain rock has been fused and crystallized. Not only the mountainside, but a number of the dressed granite blocks of...[a nearby] fortress...show signs of similar vitrification through extremely high radiated heat.

"In Brazil there is a series of ruins called Sete Cidades, situated south of Teresina between Piripiri and the Rio Longe. The stones of these ruins have been melted by apocalyptic energies, and squashed between the layers of rock protrude bits of rusting metal that leave streaks like the traces of red tears down the crystallized wall surface.

"The most numerous vitrified remains in the New World are located in the western United States. In 1850 the American explorer Captain Ives William Walker was the first to view some of these ruins, situated in Death Valley. He discovered a city about a mile long, with the lines of the streets and the positions of the buildings still visible. At the center he found a huge rock, between 20 and 30 feet high, with the remains of an enormous structure atop it. The southern side of both the rock and the building was melted and vitrified. Walker assumed that a volcano had been responsible for this phenomenon, but there is no volcano in the area. In addition, tectonic heat could not have caused such a liquefaction of the rock surface"(Ibid.).

Vitrified ruins are actually more common than one might think. Iran, Turkey, France, and Scotland are some other areas where such ruins can be found. Speaking of the ones in Scotland, Childress wrote: "There are said to be at least sixty such forts throughout Scotland. Among the most well-known are Tap O'Noth, Dunnideer, Craig Phadrig, Abernathy, Dun Lagaidh, Cromarty, Arka-Unskel, Eilean na Goar, and Bute-Dunagoil....Another well-known vitrified fort is the Cauadale hill fort in Argyll, West Scotland.

"One of the best examples of a vitrified fort is Tap O'Noth, which is near the Village of Rhynie in northeastern Scotland. This massive fort from prehistory is on the summit of a mountain of the same name...

"At first glance it seems that the walls are made of a rubble of stones, but on closer look, it is apparent that they are made not of dry stones, but of melted rocks! What were once individual stones are now black and cindery masses, fused together by heat that must have been so intense that molten rivers of rock once ran down the walls"(15).

There are also many places around the world where huge patches of melted sand can be seen laying about--something known in modern times to be associated with atomic blasts. The surface of the Gobi Desert near Lob Nor Lake is a good example of this. It is covered with vitreous sand which was the result of Red China's atomic tests under Chairman Mao. But this same desert has had such melted, glassy sheets of sand scattered around for thousands of years(61). From whence did these come?

Speaking of such patches of melted sand, Childress wrote: "When the first atomic bomb exploded in New Mexico, the desert sand turned to fused green glass. This fact, according to the magazine *Free World*, has given certain archaeologists a turn. They have been digging in the ancient Euphrates Valley and have uncovered a layer of agrarian culture 8,000 years old, and a layer of herdsman culture much older, and a still older caveman culture. Recently, they reached another layer...of fused green glass. Think it over...

"It is well known that atomic detonations on or above a sandy desert will melt the silicon in the sand and turn the surface of the earth into a sheet of glass. But if sheets of ancient desert glass can be found in various parts of the world, does it mean that ancient atomic wars were fought in the past, or at the very least, that atomic testing occurred in the dim ages of history?

"...Lightning strikes can sometimes fuse sand, meteorologists contend, but this is always in a distinctive, root-like pattern. These strange geological oddities are called fulgurites and manifest as branched, tubular forms, rather than as flat sheets of fused sand. Therefore, lightning is largely ruled out as the cause of such finds by geologists, who prefer to hold onto the theory of a meteor or comet strike as the cause. The problem with this theory is that there is usually no carter associated with these anomalous sheets of glass"(15).

The western Egyptian desert also contains fused sand on a grand-scale(Ibid.). Can we conclude from this that atomic warfare also took place in this region in ancient times? It would appear so. For notice what we find inscribed on the walls of the great temple at Edfu: "So Horus, the Winged Messenger, flew up toward the horizon in the Winged Disk of Ra; it is therefore that he has been called from that day on 'Great God, Lord of the Skies.'...In the heights of the skies, from the Winged Disk, he [Horus] saw the enemies, and came upon them from behind. From his forepart he let loose against them a storm which they could neither see with their eyes, nor hear with their ears. It brought death to all of them in a single moment; not a being remained alive through this"(108).

What appears to be another piece of fascinating evidence of ancient nuclear technology is explained in the proceeding quote from Noorbergen: "It was on September 25, 1972, when Dr. Francis Perin, former chairman of the French High Commission for Atomic Energy, presented a report to the French Academy of Sciences concerning the discovery of the remains of a prehistoric nuclear chain reaction. Perin's first inkling came when workers at the French Uranium Enrichment Center observed that uranium ore from a new mine at Oklo, 40 miles northwest of Franceville in Gabon, West Africa, was markedly depleted of uranium 235. All uranium deposits in the world today contain 0.715 per cent of U 235, but the Oklo mine uranium showed levels as low as 0.621 per cent. The only explanation that could be given for the missing U 235 was that it had been 'burned' in a chain reaction. Evidence in support of this conclusion surfaced when investigators at the French Atomic Center at Cadarache detected four rare elements--neodymium, samarium, europium and cerium--in forms that are typical of the residue from uranium fission! Dr. Perin concluded his report with the opinion that the Oklo uranium had undergone a nuclear chain reaction which had been spontaneously set off by natural causes"(9).

Because very precise conditions must be met for a nuclear chain reaction to occur (such as the need for extremely pure water, not found naturally anywhere on Earth, to act as a moderator to slow down the released neutrons as each uranium atom is split), many scientists have concluded that a natural chain reaction cannot be the correct explanation for this phenomenon. Yet there is no question that a chain reaction did indeed occur at Oklo in ancient times. Could we not therefore conclude that at least the possibility exists that ancient man may have actually been behind this chain reaction?

There's a tantalizing mystery in the long-dead Mexican city of Teotihuacan that has long remained inexplicable. But perhaps it might finally be made sense of in the context of our current discussion about ancient atomic technology. This mystery involves the use of mica in the construction of two buildings in this necropolis.

One is the Pyramid of the Sun. When archaeologists excavated this structure, they found a layer of this unexpected material in its upper tiers. Mica is a fragile mineral that easily flakes off into thin sheets, so it is not a useful substance for building stone edifices. What this mineral IS good for, however, is to provide a protective shielding from radiation and intense heat. Interesting enough, there was discovered directly underneath this pyramid a natural cavern that contained numerous artifacts, indicating that it was once occupied in ancient times. Was this an ancient nuclear fallout shelter?

The other structure in this city that houses a layer of mica is known as the Mica Temple. The sheet of mica found here is quite large (27.5 square meters), and it sits under a rock-slab floor. It turns out that this particular piece of mica originated from a quarry in Brazil, roughly 2,000 miles away. This raises a puzzling question: Why would the builders of this city go to such trouble to transport this large and delicate slab across such a long distance, unless it served an important function? Perhaps there is a yet-undiscovered subterranean chamber beneath this building as well, which also served as an ancient nuclear fallout shelter.

* * * * * * *

If ancient man did indeed possess and utilize atomic warfare capabilities, and if this technology was as widespread as the above information indicates, we could assume that the fighting had gotten so out of hand that it was decided to scrap this technology altogether, hence the reason why we don't see many traces of it today. Or perhaps the technology was simply lost because all of the advanced civilizations that possessed it had wiped each other out. In either case, we can find an important moral lesson here. For if modern man is not careful, he may very well wind up repeating, on a grand scale, a grim scene from his distant past.

In the Mayan *Popol Vuh*, we find an interesting statement that seems to indicate that the loss of the knowledge of the ancients was deemed more a blessing than a curse. Perhaps this was due to the destructive nature of atomic technology. Here's what this old document has to say: "Then the Heart of Heaven blew mist into their [the ancients'] eyes which clouded their sight as when a mirror is breathed on. Their eyes were covered and they could only see what was close, only that was clear to them....In this way the wisdom and all the knowledge of the First Men were destroyed"(208).

The ancients and the "final frontier"

If our distant ancestors were truly capable of airflight and atomic warfare, would it be that much further of a technological stretch for them to have developed the ability to venture outside the bounds of Earth's atmosphere? In other words, were these "primitive" people able to conquer the "final frontier," and travel in outer space? Let us consider some astounding legendary and archaeological pieces of evidence that point us in the direction of this very conclusion.

Looking to an old Babylonian text, the *Epic of Etana*, which dates back between 3000 and 2400 B.C., we uncover a story of a poor shepherd named Etana who happened upon an eagle with injured wings. After being nursed back to health by Etana, the eagle returned the favor by taking him on a flight into the heavens, to show him a view of the world that he had never seen before.

As they first took flight, the bird proclaimed to Etana, "Behold, my friend, the land and how it is! Look upon the sea also. Lo, the land has become like a hill and the sea like a watercourse!"

Rising higher, the eagle pointed out to Etana how the world now looked like a "plantation," with the land appearing as a "hut" surrounded by the "courtyard" of the sea.

Climbing to still greater heights, the land was said to resemble a "grinding stone," and the sea was like a "gardener's canal" (or irrigation ditch).

The story further relates that the two finally ascended so high that no details on the Earth were any longer discernable, and that all became lost in a blue and white haze(9). Soon thereafter, the Earth took on the appearance of a "speck of dust"(15).

Was this an account of a journey into space? Did Etana and his "eagle" trek so far above the Earth that all they could visually resolve was a giant blue-white ball, and then a speck of dust?

This is by no means the only record of such a journey into the "great beyond" in ancient times. In the apocryphal Book of Enoch, mention is made of what also appears to be a trip of this very sort. Here are a few relevant excerpts from this work:

"And they lifted me up into heaven..." (14:9).
"And it was hot as fire and cold as ice..." (14:13).
"I saw the places of the luminaries..." (17:3).
"And I came to a great darkness..." (17:6).
"I saw a deep abyss..." (17:11).

Do these references not sound like allusions to the exact conditions that prevail in outer space? In this realm, without a protective atmosphere, temperatures do indeed range from "hot as fire" in the open sunlight to "cold as ice" in the shade. And the deep recesses of space also fit the description given here of "great darkness" and a "deep abyss."

In the second century of our era, Lucian, a Greek author who had visited Asia Minor, Syria, and Egypt, wrote of a voyage to the Moon in his novel *Vera Historia*. Describing what he saw on his way there, he stated: "Having thus continued on our course through the sky for the space of seven days and as many nights, on the eighth day...[the Earth resembled] a large, shining circular island, spreading a remarkably brilliant light around it."

Chinese historical records mention a certain Hou Yih (or Chih-Chiang Tzu-Yu), an engineer of Emperor Yao (or Yahou), who was acquainted with astronautics. In the year 2309 B.C., as the story goes, he made a trip to the Moon on a "celestial bird." This bird is said to have instructed him regarding the precise timing of the rising, culmination, and setting of the sun. Was this a reference to him getting readings from instruments aboard a spacecraft?

Hou Yih, as the story goes, explored space by "mounting the current of luminous air." Was this an allusion to the flowing exhaust of his craft?

When in space, we are told, Hou Yih "did not perceive the rotary movement of the sun." This makes perfect sense, of course, since the Sun's movement is only detectable on Earth, due to its daily rotation. Out in space, however, the Sun does not "rise" and "set."

Once on the Moon, Hou Yih talked about seeing the "frozen-looking horizon"--a quite apt description of the lunar landscape.

Another description of this world from a close-up perspective was given by Hou Yih's wife, Chang Ngo, who is also alleged to have visited there. She portrayed it as a "luminous sphere...of enormous size and very cold." She also made this astonishingly accurate observation: "the light of the moon has its birth in the sun."

During the lifetime of Hou Yih and his wife, according to an ancient Chinese document known as the *Collection of Old Tales*, many people in China had reported seeing an enormous ship that had appeared on the sea, having brilliant lights that were extinguished during the day. This ship, it is said, could sail to the Moon and the stars, and was thus called "a ship hanging among the stars" and "the boat to the moon." This aerial vessel is reported to have been seen by locals over a period of twelve years. Was this the ship that Hou Yih and his wife used for their lunar excursions? If so, it seems that the general Chinese population was kept in ignorance about it.

And what happened to this ancient spaceflight technology of the Chinese? Why wasn't it transmitted down through the ages? Well, we have already dealt with the problems of technology retention earlier. But a more specific answer, in this case, is provided in the ancient Chinese book *Shi Ching*, which states that when the Divine Emperor saw crime and vice rising in the world, "he commanded Chong and Li to cut off communication between earth and the sky--and since then there has been no more going up or down"(61).

The people of ancient India left us records of space travel as well. The Hindu *Surya Siddhanta*, mentioned earlier, speaks of *Siddhas* and *Vidyaharas* (philosophers and scientists) who were able to orbit the Earth in craft that traveled "below the moon but above the clouds"(Ibid.).

The *Mahabharata*, also cited earlier, talks about "two-storey sky chariots with many windows, ejecting red flame, that race up into the sky until they look like comets...[and they travel] to the regions of both the sun and the stars"(63).

Yet another text from India that we cited earlier is the *Ramayana*. It records this conversation between three passengers on a space-faring vehicle:

Rama: "It seems that the motion of this excellent vehicle has changed."

Vishishara: "Now the vehicle is going away from the closeness of the center of the world."

Sita: "Why is this circle of stars visible...even at daylight?"

Rama: "Queen! It is really a circle of stars, but because of the huge distance we cannot see it at daylight, because our eyes are darkened by the sunlight. But now, with the ascent of the vehicle, it happens no more...[and we can thus see the stars]"(177).

The *Royal Pedigrees of Tibetan Kings* dates to the fourteenth century, but it can be traced as far back as the seventh century of our current era. It talks of the "first seven kings" that reigned about 2,000 years ago, who "walk in the sky"(61). Was this a reference to ancient spacewalks?

Buddhist texts from Tibet and Mongolia speak of "iron serpents which devour space with fire and smoke, reaching as far as the distant stars"(63).

The *India Times* website reported on April 8, 2003 that the Chinese had discovered some ancient Sanskrit documents in Tibet and Lhasa, written in the fourth century B.C. They were sent to Dr. Ruth Reyna of the University of

Chandigarh for translation and analysis. After examining the documents, Reyna concluded that they contained the following information:

1. Directions for building interstellar spaceships with a method of propulsion that is essentially anti-gravitational. It is described as "a centrifugal force strong enough to counteract all gravitational pull."

2. These machines were called "astras."

3. The texts also contain claims that ancient Indians could have sent detachments of men onto any planet in these craft.

4. Also discussed are the secrets of antima (the cover of invisibility) and of garima (a method of making one's body as heavy as "a mountain of lead").

5. Finally, the documents mention a planned trip to the Moon(148).

Coming to the Americas, a Guatemalan Mayan legend mentions "a circular chariot of gold" that was "able to reach the stars"(63).

In addition to the ancient texts that we have just cited, there is also indirect archaeological evidence to support the notion of spaceflight in the ancient world. Though far from conclusive, such pieces of evidence are nevertheless quite tantalizing. We will here cite two prominent examples.

In Japan, clay figurines (known as "dogu") have been unearthed that depict humans wearing what look like spacesuits, complete with helmets that have narrow eye-slits, a breathing apparatus, audio receivers, and antennae(61). Some think that these figures may represent scuba-diving suits, rather than spacesuits. But in any case, they certainly bear the marks of some sort of high-tech gear.

Some amazing petroglyphs exist in Val Camonica, Italy, that portray two men wearing what look like space helmets, both of which have short filaments sticking out of them that resemble an array of antennae. These men are also holding unusual, technologically-advanced-looking triangular devices in their hands.

Section II: Human History: Man in the post-flood world
Part 2: Exploring and settling a new and hostile planet

Post-flood world surveys

Not long after the great flood catastrophe, once the Earth began to stabilize somewhat, we could imagine the survivors, or their early descendants, embarking on missions of world exploration, conducting surveys that would enable them, for instance:

a) to do a damage assessment from the catastrophe, in order to see how badly the Earth had been disrupted;

b) to seek out more hospitable climates wherein to build settlements and grow crops;

c) to locate deposits of natural resources for mining;

d) to establish contacts and trading networks with other flood-surviving settlements (and the descendants thereof) that dispersed around the globe.

There are actually many ancient records that tell of such early explorations of the world. One example of this is found in the Mayan *Popol Vuh*, which talks about how ancient man had "examined the four corners of the horizon, the four points of the firmament and the round circles of the earth"(18).

We also find a couple ancient Chinese literary works that tell of very early world surveys, both of which, by the way, make specific and obvious reference to encounters with the North American continent, long before the age of Columbus. Historian Gunnar Thompson, in his book *American Discovery*, describes one of these Chinese surveys as follows: "Around 2640 B.C., Chinese Emperor Hwang-ti sent out an exploratory expedition to a land called Fu Sang. Details of this journey are so precise that one modern Chinese scholar, Kuan-Mei, decided to trace out the path that this journey took. Kuan's project turned up some surprising results. He discovered that Hwang-ti's crew had crossed the Pacific, landed somewhere in the Northwest United States, and then headed southeast. He believes that they crossed through Yosemite National Park, over the Rockies, and then wound up in the Grand Canyon, or 'Great Luminous Canyon.'"

Interestingly, many Chinese symbols, such as the Taoist Yin-Yang motif, can be found painted and carved on rocks all throughout the Grand Canyon region. Another interesting point to note is that there is a fifteenth century A.D. Chinese map which shows the west coast of the United States, labeling it as "Fu Sang"(78).

Noorbergen provides us with a good description of the other ancient Chinese world survey account, which not only makes reference to an encounter with North America, but with parts of Mexico as well: "One of the oldest Chinese literary works that has survived is called the *Shan Hai King*, or *The Classic of Mountains and Seas*, a treatise on geography. Its authorship is ascribed to 'the great Yu,' who became emperor in 2208 B.C., and the date for the writing of the treatise is approximately 2250 B.C. [This 'great Yu' is said to have] 'measured the earth to its extremities.' For several hundred years after its writing, the *Shan Hai King* [also spelled Shan Hai Ching] was regarded as a scientific work, but during the third century B.C., when many Chinese records were reevaluated and condensed, it was discovered that the geographical knowledge it contained did not correspond to any lands known at that time. Thus, the *Shan Hai King* was reclassified as myth and was relegated to an unimportant position in Chinese literature.

"Within the past few years [as of 1977], however, several portions of the *Shan Hai King* have been reexamined, and the information they contain has altered many previous assumptions concerning the treatise. In the Fourth Book, entitled *The Classic of Eastern Mountains*, are four sections describing mountains located 'beyond the Eastern Sea'--on the other side of the Pacific Ocean. Each section begins by depicting the geographical features of a certain mountain--its height, shape, mineral deposits, surrounding rivers and types of plants and vegetation--then gives the direction and distance to the next mountain, and so on, until the narrative ends. By following these clues and the directions and distances provided, much as one would a road map, investigators have discovered that these sections describe in detail the topography of western and central North America.

"The first section begins on the Sweetwater River and proceeds southeast to Medicine Bow Peak in Wyoming; then to Longs Peak; Grays Peak, Mount Princeton, and Blanca Peak in Colorado; to North Truchas Peak, Mananzo Peak, and Sierra Blanca in New Mexico; then to Guadalupe Peak, Baldy Peak, and finally Chinati Peak, near the Rio Grande in Texas.

"The second section describes an expedition over an even more expansive area. It begins in Manitoba, at Hart Mountain near Lake Winnipeg, and proceeds to Moose Mountain in Saskatchewan; it goes from there to Sioux Pass in Montana; to Wolf Mountain and Medicine Bow Peak in Wyoming; to Longs Peak, Mount Harvard, and Summit Peak in Colorado; then to Chicoma Peak, Baldy Peak, Cooks Peak, and Animas Peak in New Mexico; then on into Mexico, describing the Madero, Pamachic, Culiacan and Triangulo heights, reaching the Pacific Coast near Mazatlan.

"The third section is a tour of the mountains along the Pacific Coast: Mount Fairweather and Mount Burkett in Alaska; Prince Rupert and Mount Waddington in British Colombia; Mount Olympus in Washington; Mount Hood in Oregon; and Mount Shasta, Los Gatos, and Santa Barbara in California.

"The fourth and last section covers several peaks in a small area: Mount Rainier in Washington; Mount Hood, Bachelor Mountain, Gearhart Mountain, Mahogany Peak, and Crane Mountain in Oregon: and Trident Peak and Capitol Peak in Nevada.

"Not only is *The Classic of Eastern Mountains* a geographical survey, but the accounts in each section give the observations and experiences of the surveyors, from picking up black opals and gold nuggets in Nevada to watching the seals sporting on the rocks in San Francisco Bay. They were even amused by a strange animal who avoided its enemies by pretending to be dead: the native American opossum.

"Other portions of the *Shan Hai King*, specifically the Ninth and Fourteenth books, also describe regions in North America. One notable description given in the Fourteenth book is of a 'luminous' or 'great canyon,' a stream flowing in a bottomless ravine,' in the 'place where the sun is born.' Anyone who has witnessed a sunrise in the Grand Canyon will know what the early surveyors had seen. Still other parts of the *Shan Hai King*, currently under investigation, are said to be accounts of explorations further to the east, in the Great Lakes and the Mississippi areas.

"It is very evident from the accuracy of the geographical details and the personal observations in the *Shan Hai King* that an extensive scientific survey of the North American continent was made by the Chinese almost 4,500 years ago"(9).

Perhaps the most fascinating evidence of early post-flood world surveys is the existence of several astonishing old maps. The late Professor Charles Hapgood described these maps in his book *Maps of the Ancient Sea Kings*. He related therein how he had the pleasant fortune, in the course of his research, of coming across a number of medieval maps that were compiled from multiple archaic maps, all now lost to us, which reveal that the ancients were highly advanced mariners, having mapped the entire Earth's surface. Geographically speaking, these maps depict the planet pretty much as it is now (in the post-Earth-expansion period), so the antiquity of the maps that these medieval maps were based upon only went back just so far. However, they obviously did date to an early period of human history--to the time of the post-flood world surveys--for the medieval maps portray parts of the world that weren't even known to exist in medieval times, and, even more significantly, they portray the Earth, as we shall see, under completely different climatological conditions that have not existed since many thousands of years ago, back before the "Ice Age."

One such map that Hapgood came across, which was first examined by naval cartographer Captain Arlington H. Mallery, was compiled in 1513 by a Turk named Piri Reis, an admiral in the Imperial Ottoman Fleet. In his notes contained on the map itself, Reis mentioned how he created this map from 20 charts and world maps that dated back to the Alexandrian Greek era (which were themselves likely based on even older maps). His map sat dormant in the Topkapi Palace Library of the Imperial Palace of Istanbul (Constantinople) until it was discovered around 1930. Drawn on a gazelle skin, only half of this map survives today. But what remains of it is most fascinating. It depicts, for example, the continent of Antarctica, which was not "discovered" until 1819 [Endnote 55]. But as if this isn't enough, this map depicts the Palmer Peninsula and Queen Maud areas of Antarctica being totally free of ice [Endnote 56]. The overall details of the layout of the continents and the proportionate sizes thereof, as well as the latitudinal and longitudinal coordinate markings, have an astounding accuracy that was not matched until the eighteenth and nineteenth centuries. This map also shows the Andes Mountains, also not known to exist in 1513, and the Falkland Islands, not rediscovered until 1592.

Interestingly, the Azores are shown to be much larger than they are today, and there is also depicted a huge adjacent island that no longer exists. However, today, in the exact place of this island, lies the massive, underwater mountain range known as the Mid-Atlantic Ridge. Can it be that the sea level was much lower in ancient times, or that this mountain range sank into the sea since the making of the original maps that Piri Reis copied from? [Endnote 57] Judging from the precise accuracy of the rest of the Piri Reis map, it is not likely that this large, extinct island and the expanded size of the Azores were mistakes. On the contrary, Hapgood determined, after a careful study of this map, that the original map-makers that Reis had copied from were far too competent to make such major slip-ups, having been in possession of a profound understanding of plane trigonometry and spherical trigonometry, as well as having known the precise circumference of the Earth.

Another map discussed by Hapgood in his book is the Oronteus Fineus map of 1531, which was also based on ancient maps that are no longer in existence. This map shows, entirely free of ice, the Ross Sea area of Antarctica and its coastline, which today lies buried under an ice sheet that is over a mile thick. Of course, once again, the fact that Antarctica is depicted at all on a map that was compiled roughly three hundred years before it was "officially" discovered is astounding on its own. But there's more. This map also shows some Antarctic rivers in areas that today are covered under a thick and permanent blanket of ice. However, we know from satellite surveillance that these now-frozen rivers do exist, just as portrayed on the Oronteus Fineus map. So accurate is this map, in fact, that in 1961 a U.S. Air Force cartographic expert, Captain Lorenzo W. Burroughs, made this statement in regards to it: "Beyond a shadow of a doubt, [it] was compiled from accurate source maps of Antarctica"(62).

Still another medieval map that Hapgood discussed is the Hadji Ahmed world map of 1559, showing a broad land-bridge between Siberia and Alaska (Beringia). Though mainstream modern scientists acknowledge that such a land-bridge did exist, they want us to believe that the humans living at the time of its existence were too "primitive" to construct a map that depicted it. Yet there is such a map (or a copy thereof), which stands as a strong refutation to this academic fairytale.

The Zeno map of 1380 is yet another cartographic document that Hapgood refers to in his book. This map, made by the Zeno brothers, Niccolo and Antonio, who voyaged to Greenland and Iceland (and perhaps Nova Scotia) in the fourteenth century A.D., was also copied from ancient documents. It depicts Greenland without any ice cap, showing rivers, detailed mountain ranges, and a central plateau [Endnote 58]. This map, like the others, also exhibits accurate latitude and longitude measurements--something that map-makers, again, were not able to do (or redo) until the last few centuries(79).

While Hapgood went on to describe other maps in his book, the ones mentioned above should suffice to demonstrate the point that the ancients were not, by a long shot, the geographical ignoramuses that we have been led to believe they were. Rather, they were quite advanced in their understanding of the layout of the global landscape--every bit as much as ourselves.

The fact that the above-mentioned maps contain accurate longitudinal coordinate markings is far more of an astonishing feat than what most people might realize at first glance. This is an accomplishment that requires extremely accurate measurements of time as well as distance--something that was not accomplished by man, or so we're told, until the late seventeenth century of our current era.

But it gets even more complicated than this. Making an accurate two-dimensional map on paper of a three-dimensional world requires a knowledge of an extremely sophisticated mathematical and mechanical tool known as map projection. This, too, was allegedly an ability not attained by man until recent centuries.

There is one last map that is worth mentioning--a Chinese world map that surfaced in January of 2006. According to the BBC on January 13 of that year, this map was drawn up in 1763, having been copied from an older map dating to 1418. It was copied by a Mr. Mo Yi Tong during the 16th year of Emperor Yongle. The map depicts Australia, both North and South America, Antarctica, and literally all the rest of the world except the British Isles. The original 1418 map may have itself been a copy of an even older chart. In any case, it clearly reveals that the Chinese had been familiar with lands that were not supposed to have been known, in the case of the Americas, until several decades later, and in the case of Antarctica, until several centuries later.

In light of our earlier discussion about advanced technology in the distant past, including possible airflight and atomic weaponry, the thought of early man having circumnavigated the globe might not seem so surprising. Nevertheless, it is still a most important and fascinating subject to study, which deserves our attention here, as we continue in our endeavor to strip away the blinding biases that have prevented us from developing a clear understanding of ancient history.

It should be mentioned, before continuing, that even though man does appear to have developed airflight technology in remote antiquity, apparently most of the oceanic crossings that took place during the above-described world surveying enterprises were carried out by ocean vessels. And this appears to have remained the case (using ocean vessels to cross the seas) in the centuries and millennia that followed these early post-flood world surveys. For as the globally-dispersed post-flood settlements began to establish trade networks, which were carried on sporadically throughout history, ocean vessels were no doubt found to be best suited for trading, principally because they had much larger cargo capacities, and because airflight technology was probably not widely disseminated at any one given time.

Of all the lands that were explored, settled, and exploited for trade in ancient times, beginning just after the flood catastrophe, there's one region, more than any other, that deserves special attention--a region that was supposed to have been unknown to the rest of the world, until recent centuries.

The ancient "melting pot"

Just as the United States has been called the "melting pot" of the world in modern times, it would appear, as we shall see, that the ancient Americas as a whole, but particularly the regions of the Maya and Olmec cultures, were a virtual melting pot as well, where peoples from around the world had come, mostly for trading purposes, and had mingled their cultural traits with those of the Native peoples.

We have already seen that the Americas were known about, and visited by, the Chinese in very ancient times; not by crossing the Bering Strait on foot, but by trans-oceanic voyages. Although it is not to be doubted that the Bering Strait land bridge was indeed utilized by some people to cross from Siberia to Alaska in the distant past, this was by no means the only, or even the most desirable, method of entering the New World from the Old. As we continue, we will demonstrate, from ancient written and archaeological pieces of evidence, that the Americas were visited by just about every civilization in the ancient world, at one time or another. Some of these contacts were even maintained over extended periods of time, for the primary purpose, once again, of trading goods.

This, of course, is not considered to have been possible, according to contemporary "wisdom," which insists that the ancients were simply not capable of traversing the world's two largest oceans, to reach the Americas. What a pathetic insult this is to the integrity, intellect, and ingenuity of the ancients! But the truth is, as capable as ancient mariners were, there are several natural trans-oceanic currents that can actually carry ships to and from the Americas on their own, without the expenditure of any effort whatsoever on the part of sailors to do so. Elaborating on this matter, Professor Robert Schoch of Boston University wrote: "Columbus himself profited from a current on his trip to the West Indies. He traveled south from Spain to the coast of North Africa, then, near the Canary Islands, he picked up a westbound current that heads directly for the neighborhood of Cuba and Hispaniola. The Spanish discovered another trans-oceanic conveyor belt, called the Urdaneta Route, in the Pacific in 1565. The mariner Andres de Urdaneta was looking for a way from East Asia to Mexico, and he found out something that Asian mariners had likely long known: The northeast-bound Kuroshio Current carries warm water and mild weather from the Philippine Sea across the North Pacific toward the northwestern coast of North America, where the California Current continues the journey south to Mexico. The Spaniards used this route to send their treasure-laden galleons from the Philippines to Mexico"(1).

As it turns out, Pedro Alvares Cabral, the Portuguese sailor who is credited with "discovering" Brazil on April 22, 1500, did so quite by accident. He left Portugal on March 9, bound for India. But as he headed south to sail around the coast of Africa, the powerful current of the North Equatorial Stream carried him and his entourage of 12 other ships off course, all the way to South America. So, as you can see, it is absolutely ridiculous to deny that the ancients were capable of reaching the Americas. In some cases, it must have been completely unavoidable, especially during stormy weather.

To further demonstrate the ease with which the ancients were able to reach the Americas, consider the fact that some people, in modern times, have crossed both the Atlantic and Pacific oceans with mere rowboats, kayaks, and simple rafts(15).

One such person was Thor Heyerdahl. His first and most famous voyage occurred in 1947, when he sailed in an Inca-style raft that he named Kon-Tiki. He journeyed from Callao, Peru, to Papeete, Tahiti--a 4,300-mile trip which demonstrated that the Pacific would not have presented an unsurmountable barrier to such ancient "primitive" vessels that may have been used to cross it(100).

And then, in 1970, assisted by a five-man crew, Thor set sail on a reed ship, the *Ra II* (his first attempt at this, in May of 1969, was unsuccessful), from Morocco to Barbados, a distance of 3,000 miles, taking a total of 57 days. This triumphant journey proved that the Atlantic would also not have been a threat to ancient seafarers who may have sought to cross-navigate it, particularly the Egyptians and Babylonians, who commonly used this same type of vessel(78).

It was due mostly to the superstitious fear of falling off the edge of the "flat Earth" that kept medieval sea voyagers from sailing too far out into uncharted seas. But do understand that much earlier generations of mankind--in possession of advanced knowledge and technology--were not plagued by such foolish inhibitions. On the contrary, they knew that the Earth was round, and thus they trekked across the oceans with the same fearless confidence that today's proud sea captains possess.

The Polynesians are particularly noteworthy as having been more than capable of reaching the Americas (particularly South America) in ancient times, as we find in the following authoritative quote from a Smithsonian publication: "The Polynesians, who were the first human inhabitants of the far-flung mid-Pacific islands, were the most daring deep-sea voyagers and explorers the world has ever known. In the double-hulled ships they fashioned with stone tools, they sailed by stages across the wildest part of the unknown Pacific, from Southeastern Asia, probably all the way to the coast of Peru. They did this centuries before Columbus ventured into the Atlantic....[T]he Polynesians found and populated every inhabitable island in the vast expanse between Hawaii on the north, New

Zealand on the south, Easter Island on the east, and Tonga and Samoa on the west...."(101)[Endnote 59].

So, if the Smithsonian can admit to the Polynesians having made it to the Americas in the distant past, why can't the same be done for other ancient cultures, especially when considering the fact that so many of them had constructed ships that were far more seaworthy than those made by the Polynesians?

Columbus's voyage of 1492 is highly overrated. Far from being a trailblazer, he was, instead, quite a latecomer on the scene. Many, many peoples from the ancient world had been to the Americas long before him, and had left their mark upon various Native American civilizations. This is a fact that numerous ancient Native American legends themselves attest to, telling stories of interactions with people of multiple ethnic backgrounds who ventured here in ships from far away lands. Some Native legends even talk of how their civilizations originated from such trans-oceanic voyagers that came and settled here long ago. For example:

- Early Spanish transcribers of Mayan records reported that "The natives believe their ancestors to have crossed the sea by a passage which was opened for them"(18).

Is this a reference to the natural trans-oceanic currents, discussed earlier, that can carry ships to and from the Americas with little or no effort on the part of sailors?

- A journey by the Mayas to their land of origin across the sea is recorded in the *Popol Vuh*: "The three sons of the Quiche King made a visit to the land situated to the east [across the Atlantic], from which our ancestors came"(Ibid.). Clarifying this point further, this document goes on to stress that the ancestors of the Mayan people came "from the other side of the sea," in the direction "where the sun rises"(1).

- In further regards to the *Popol Vuh*, Schoch informs us that it "describes the first [Mayan] ancestors as 'black people, white people, many were the people's looks, many were the people's languages.' The Quiche Maya may well have understood race very differently from how we moderns do, but clearly they recognized that the people of those early days came not from one background but several. They were a polyglot assemblage of different origins"(Ibid.). The *Popol Vuh* also states that "the white, yellow, and black men lived altogether in harmony..."(5).

- The natives of the Great Lakes area believe that their ancestors came from a land "toward the rising sun [i.e. across the Atlantic]"(18).

- In Delaware, the Leni-Lenapi Indians claim that they originated from "the first land...beyond the great ocean [again, the Atlantic]"(Ibid.).

- The Sioux declare that "The tribes of Indians were formerly one, and all dwelt together on an island...toward the east or sunrise..." They are said to have arrived in the new land by floating in huge canoes "for weeks"(Ibid.).

- In his *Carta Segunda* (1520), Cortes recorded a speech delivered to him by Montezuma, the Aztec emperor, after the Aztecs had anointed the Spaniards with blood from a human sacrifice. Here is what the great Aztec leader had to say: "We have known for a long time, by the writings handed down by our forefathers, that neither I nor any who inhabit this land are natives of it, but foreigners who came here from remote parts. We also know that we were led here by a ruler, whose subjects we all were, who returned to his country, and after a long time came here again and wished to take his people away. But they had married wives and built houses, and they would neither go with him nor recognize him as their king; therefore he went back. We have ever believed that those who were of his lineage would some time come and claim this land as his, and us as his vassals. From the direction whence you come, which is where the sun rises, and from what you tell me of this great lord who sent you, we believe and think it certain that he is our natural ruler, especially since you say that for a long time he has known about us. Therefore you may feel certain that we shall obey you, and shall respect you as holding the place of that great lord, and in all the land I rule, you may give what orders you wish, and they shall be obeyed, and everything we have shall be put at your service. And since you are thus in your own heritage and your own house, take your ease and rest from the fatigue of the journey and the wars you have had on the way"(80).

There's also a wealth of Native American archaeological evidence that points to multicultural integration in the ancient Americas. Here are a few select examples:

- Schoch tells us that "The [Mayan] Chichen Itza structure known as the Temple of the Warriors houses murals that depict three separate races of humans in the same setting. One of the paintings represents a seashore battle scene, in which white-skinned invaders with long blond hair are driven off by dark-skinned warriors, some of them black as Africans, some of them brown like Indians. The whites are shown either naked or wearing short tunics, while the dark-skinned warriors are dressed in the kilt, shield, hand weapons, and elaborately feathered helmet of Mesoamerican fighting men. In one of the panels a white captive is about to be sacrificed by the black warriors, who have stretched him across an altar with his chest upraised to receive the knife that will open his chest for heart plucking"(1).

- Many stone and terra-cotta carvings that have been found, particularly in Mexico, from Olmec and Mayan sites, depict people with obvious foreign facial features. The fact that many of these non-Native faces have beards and/or mustaches make them especially unusual, since Native Americans do not grow facial hair.

- In 1952 in Marcahuasi, about 80 km northeast of Lima, Peru, archaeologist Daniel Ruzo discovered a series of large, ancient megalithic statues of unusual human and animal figures. These statues depict Caucasian, Negro, and Semitic peoples, as well as lions, cows, elephants, camels, and horses, none of which supposedly came to the Americas until after Columbus(61).

- Even finds of human remains confirm the presence of various foreign peoples in the ancient Americas. Schoch tells us that "In a study of skeletons at Tlatlico and other Olmec burial sites, the physical anthropologist Andrezj Wiercinksi concluded that some of the remains revealed African ancestry, a finding that has stirred considerable controversy among other experts in the field. Further south, two necropolises dated to about 300 B.C. on the Paracas Peninsula along the south-central coast of Peru have yielded several hundred mummies, some of which have hair that is wavy, light brown, even reddish. Such hair is more typical of a European than an indigenous American"(1).

Another important quote that informs us of ancient foreign human remains found in the Americas was written by Ivan Van Sertima, in his book *They Came Before Columbus*: "[M]ummies examined in ancient Peru...show that foreign elements, both Negroid and Caucasoid, seemed to have entered the native South American population (400-300 B.C.). Dr. M. Trotter, during a hair analysis on pieces of scalp from Paracas mummies in Peru, reported in 1943 that 'the cross-section form shows so much divergency between the different mummies that they cover all divisions of hair form.'...Also, an examination of skeletons in the area simultaneously conducted by T.D. Stewart, demonstrated the presence of races of greater average height and a different cephalic index (head shape) than the average aboriginal Americans...[which] suggests Negroid elements, mulatto curly-haired Egyptians, and a good deal of [other] racial mixtures"(81).

* * * * * * *

The evidence that we have thus far examined, showing signs of contact between the Old and New World in ancient times, does not even amount to a drop in the bucket. Therefore, to present this case more adequately and forcefully, we will dedicate ourselves, for the duration of this study, to documenting some of the more profound pieces of legendary and physical evidence that point to visits from specific major cultures around the world, who left behind unique traces of their presence all throughout the Americas, and all throughout the centuries and millennia before Columbus.

China and other nations of the Orient

We have already looked at two examples of ancient records that tell of visits by the early Chinese to the Americas: Emperor Yu's *Shan Hai King, The Classic of Mountains and Seas*, and the record of Emperor Hwang-ti's visit to "Fu Sang." But there is much more evidence of a pre-Columbian Chinese presence in the New World. Before looking at such evidence, however, let us first discuss the maritime capabilities of the ancient Chinese. Did they really have adequate vessels and seafaring skills for making such long voyages?

It's evident from Chinese legends, such as those cited above, that the Chinese must have possessed superior sailing vessels in at least the third millennium B.C., at the time that they conducted their world survey. But how about in more recent millennia? Do we have evidence that the people of China continued to construct and sail superior vessels capable of reaching the Americas in the first and second millennia B.C., and even thereafter? Indeed, we do.

For instance, Chinese inscriptions on turtle bones, dating to the second millennium B.C., tell of long distance voyages and the transport of very large armies that could only have been possible with huge, ocean-going vessels.

From the seventh century B.C. to the fifth century A.D., we know with certainty that Hong Kong merchants were sailing in ships, on a routine basis, that were over 100 feet long, to places such as Siberia, the East Indies, and the Persian Gulf(78).

Yes, the evidence is clear that the Chinese, all during their ancient history, were more than capable of crossing the Pacific and reaching American shores. As one notable historical work, *Scribner's Popular History of the United States*, put it: "There is nothing incredible in the supposition that the Chinese may have sailed across the Pacific long before Europeans ventured over the Atlantic Ocean; for they were early navigators; knew in the second century of our era the use of the mariner's compass; and their junks, which have changed little in form since they were first known to Europeans, have been found wrecked upon the west coast of America, at different periods, from the time of the first Spanish voyages in the Pacific"(104).

Elaborating on this further, Gunnar Thompson wrote: "When the first Spanish explorers reached the Pacific, they reported strange sailing ships off the southern California coast. These were foreign merchant ships of the Far East. In 1544, Coronado sighted exotic ships with 'pelican-shaped bows' in the Gulf of California. In 1573, Franciscan friar Juan DeLuco sighted eight 'strange ships' along the west coast of Mexico. During the 17th century, Spanish priests in southern California found Asians working mines and fishing along the coast;

native tribes reported continuous commerce with Asian merchants since the time of their ancestors. Archaeologists later found anchor-stones near Palos Verdes, California. The anchor-stones had been made from local materials by ancient Asian settlers in southern California"(78).

Having established the competency of ancient Chinese mariners, let us now take a look at some fascinating pieces of evidence which show that they truly did reach the Americas, starting off with several more legendary references.

In the fifth century B.C., according to Chinese texts, a voyage was launched to the wondrous "Land of the Blest" across the eastern sea. The main purpose of the journey was to obtain jade and sacred hallucinogenic mushrooms that they called *ling chih*, prized by the Taoists for their ability to bring "enlightenment." This "Land of the Blest" must certainly have been Mexico--a great source of both jade and peyote, and a land where we find many pre-Columbian Chinese cultural influences, as we shall see.

In 458 A.D., a Buddhist missionary by the name of Hui-Shen returned from a trip to Fu Sang (just like Hwang-ti had done earlier, in 2640 B.C.), and astonished Emperor Laing Wu Ti with tales of seeing flying rats (bats?), a sea of black varnish (the La Brea Tar Pits?), big birds with white heads and large talons (American Bald Eagles?), and, finally, merchants who paid no tax (Native Americans?)(Ibid.). Who could argue that this is a description of an ancient Chinese visit to North America?

A couple centuries before this ocean voyage, a description was written in China of a great land to the east that could be nothing but a reference to America. The text, a third-century Chinese poem, states: "East of the Eastern Ocean [the Pacific] lie the shores of the land of Fu Sang. If, after landing there, you travel east for 10,000 li, you will come to another ocean, blue, vast, huge, boundless [clearly a reference to the Atlantic]"(82).

An overseas expedition was sent out in 215 B.C. to the east, across the Pacific to Fu Sang, by the Chinese emperor Shi Huangdi, who sought a special plant, *che* (probably the same as *ling chih*, mentioned above), that was said to bring immortality. Though the people sent on this journey never returned, it was rumored that they did successfully arrive in Fu Sang, and decided to settle there. In fact, the Chinese historical text, the *Shih Chi*, states: "Hsu Fu [the sea captain who led the expedition] found some calm and fertile plain with a broad forest and rich marshes where he made himself a king." Interestingly, this date (215 B.C.) corresponds with the founding of Teotihuacan outside modern-day Mexico City, around 200 B.C.

In addition to this, there are numerous connections between this city and ancient China, such as architectural designs, artistic motifs, pottery styles, and

Chinese writing found on jade beads. But perhaps most intriguing of all is something that comes from China itself--a Chinese text, dating to this same period, which seems to lay out what ended up being the geographical specifications that were followed when selecting the site upon which Teotihuacan was built. This text records the following conversation: "Formerly Duke Huan questioned [his chief minister] Kuan Chung: 'May I ask what should be taken into consideration with regard to topography when establishing a state capital?'" Then the answer was given: "He selects a rich and fertile site with mountains on both sides and bordering a river or marsh. Inside he provides for drainage by an encircling canal which draws its water from a large river." This perfectly describes the layout of the land around Teotihuacan--the Sierra Madre Oriental Mountains and the San Juan River, which was once canalized by its inhabitants(90).

Of all ancient American civilizations, the Olmecs appear more than any other to have been influenced by the Chinese, and may even have been founded by them, as the proceeding quote from Schoch indicates: "There is a Shang tradition that when the last emperor of the dynasty fell at the hands of his rivals circa 1122 B.C., 250,000 people fled east, into the Pacific Ocean. The number appears inflated, but the tradition of migration at a time that fits with the later rise of the Olmecs on the far side of the same sea is intriguing"(1).

La Venta, the first great Olmec settlement, and indeed the largest of all, was founded around 1100 B.C. It was at this very time, or just prior to it (1121 B.C.), that a great political disruption took place in China, spawning what must have been a major exodus from the country, perhaps across the sea to Fu Sang. In that year, the Chou leader, Wu-Wang, attacked Chao, the ruler of the Shang Dynasty, bringing his reign to an end. According to the *Shoo King*, an ancient Chinese historical record, this happened because this leader was debauched, thinking only "of palaces, buildings, terraces, groves, dikes, pools, and extravagant clothes to the neglect and ruin of the people," and because he was "spreading pain and poison over the four seas."

Upon attacking the Shang, Wu-Wang is said to have respected their intelligent men, and thus, according to Choo Hi's *Mirror of History* (another ancient Chinese historical text), he distinguished the families of the clever men of the enemy who had gone away (fled the country). Is it possible that these "clever men" sailed out to sea when they fled the country? Quite possibly so, when we consider that the conquered emperor, as we are told, spread pain and poison "over the four seas." This implies that China, at that time, must have had a widespread, overseas influence, which presupposes an ability to effectively engage in long-distance sea voyages. To further support this argument, it is also stated in the

Mirror of History that, after the conquest, all the thoroughfares were open to the "nine kinds of foreigners" and the "eight tribes of barbarians," no matter from how far or near they brought tribute. Surely all of this interaction with so many foreign countries must have involved ocean crossings by both the foreigners and the Chinese alike.

So if the "clever men" did indeed head out to sea (probably following the wisdom of Confucius, who said, "If the way does not prevail, I will get on a raft and drift out to sea"), it is likely that they encountered stormy weather that blew them clear across the Pacific (something known to have occurred rather frequently, by the way, in the seventeenth and eighteenth centuries A.D.). Stormy weather is here suggested as the mechanism that brought these Shang refugees to the Americas because, according to the same text last cited, there was a storm so savage in 1115 B.C. that it tore up trees and flattened grain.

La Venta has been rightly called the "sanctuary in the swamps," where the weather is typically gray and drizzly. Perhaps this is why the "clever men" chose this site. For we find this statement in the *Book of Feng-Shan*: "Since Heaven is disposed to cloudiness or obscurity, the sacrificial place must be chosen below the high mountains, and atop a small hill, which is commonly referred to as 'Chih,' while earth becomes precious because of sunshine, therefore sacrifices must be made at a circular mound in a swampy land." This is a precise description of La Venta, including the manmade circular mound that is found there. And La Venta was indeed a ceremonial center.

Even the layout of La Venta is in keeping with ancient Chinese specifications. The ancient Chinese work *Chou li* had this to say about planning a city for Prince Ching Wong, in 1108 B.C.: "Now when the kings founded the empire they determined the four cardinal points of the compass and fixed the positions....A space of ground 1,000 li square was styled the king's domain." Without question this applies to La Venta, which was oriented to the four cardinal points, although the alignment today is off by about 12 degrees west of true north. But this is to be expected, due to the Earth's precession. In fact, many other sites in the Americas are offset by this exact same degree and direction--Poverty Point in the Mississippi Valley, for instance. And so is it the case with many ancient Chinese cities--today they are offset by 5 to 12 degrees.

Even the Olmec site of San Lorenzo, not very far from La Venta, shows very clear signs of Chinese influence. This necropolis is surrounded by 20 lagunas, or artificial water holes. Associated with this is a 200-meter-long system of stone drains to control the water level in the complex of ponds. The Chinese version of these were the "dikes" and "pools" that were excessively built by Chao, the debauched Shang emperor mentioned above, which have been excavated at the old Shang capital of Anyang.

Another Chinese similarity at San Lorenzo is the unique group of housing facilities that archaeologists have dug up there. These houses were found to be arranged on two or three sides of family plazas, much like the extended family dwellings of ancient Chinese farmers(91).

There's a great deal more physical evidence to make plausible the hypothesis of contact between the early Chinese and Olmec cultures (along with other Native American cultures). Let us now explore some of this evidence.

A good example to start with is the fact that in both the Chinese and Olmec civilizations, the word for jade also meant "precious"(82). Furthermore, the ancient Chinese buried their dead with small jade celts or pendants in their mouths. This same practice was carried out by the Mayas as well. Additionally, in all three of these ancient civilizations it was a common practice for people to wear disc-shaped mirrors as badges of office(80).

The Chinese have traditionally been credited as the ones who first invented the compass, in the fourth century B.C. However, in the early 1970s, a hematite compass was found in the Olmec ruins of San Lorenzo, Vera Cruz, Mexico, which dates to at least 1000 B.C. It would appear, then, that the Chinese may have gotten this technology from their contacts with Mesoamerica, and brought it back on their return voyage home(112).

A unique style of ancient Chinese pottery was the tripod bowl, with a characteristic inward-bent design that ran all around the vessel, just below the outside rim. This exact same style was also common in several Mexican cultures, such as the Mayas and the Chupicuaros(78).

There have been many paintings and sculpted ceramic and stone figures that have been found in the Americas, particularly at Olmec sites, which depict unmistakable Chinese facial features. Some of these artifacts exhibit goatee beards as well(80). Findings like these beg the question, How could the Native American artists who painted and carved these works of art have depicted Chinese-looking faces unless they had actually encountered people from China?

Many Mesoamerican symbols found on temple facades, monuments, pyramids, murals, altars, ceramics, jade carvings, and various other terra cotta and stone works, dating to between 500 and 100 B.C., are precise mirror images of Chinese symbols that were used in their stone and terra cotta carvings from the same period. Among these symbols are the yin-yang, flight motif, celestial serpent, thunderhead, cosmic eye, and the symbols for the Supreme Being, the life force, and immortality(83).

Schoch points out for us a few more important parallels between ancient China and Mesoamerica, particularly in reference to writing and numerical computations and notations: "Mayan glyphs recall traditional Chinese ideographs in their squareness, the way they are read downward, and their pattern of indentations. And there is the *quipu*, a recording device of strings and knots used as an aid to memory before the spread of writing in ancient times in both South America and China. The Maya wrote numerals with a bar and dot system much like the rod method the Chinese used. And both areas computed with the zero in place-value arithmetic, a mathematical system that was still a relatively recent introduction to Western Europe when Columbus first showed up in the Caribbean. The Maya used the zero before the Chinese did, and it probably was an American contribution to the developing Asian civilization"(1).

Chinese/American contacts resulted in several uniquely-American plants being transported back to China in ancient times. One example is the peanut, or groundnut, which is native to South America, yet has been reported in two archaeological sites in southeastern China, and dated to early in the third millennium B.C.(82) Another example is tobacco. Han Dynasty artworks (200 B.C. to 220 A.D.) show men smoking elbow-pipes. So popular did this habit become, that tobacco farms were set up in ancient China(78).

Speaking of crops, there was also a correlation between the methods employed by the ancient Chinese and Native Americans for fertilizing their planted fields. As Schoch tells us: "Ancient Mexicans fertilized their fields with human waste, a practice unknown in the Mediterranean but commonplace in China. And both areas were adept at irrigation and hydraulic engineering"(1).

Let's now move on to discussing indications of contact between other Oriental nations and the Americas in ancient times.

Patolli is a game of chance that dates way back in the Americas to long before Columbus. It's played with dice, and utilizes beans as counters. It has an exact counterpart in Asia--the ages-old game of *parchesi*(80).

There are striking similarities in the way that bark paper was made and used by the ancient Southeast Asian and Mayan civilizations. Elaborating on this subject, Patrick Huyghe, author of *Columbus Was Last*, wrote: "Unlike the manufacture of true paper, the making of bark paper retains much of the original structure of the bark fibers during manufacture. In the production of bark paper, a layer of bark was widened, thinned, and made flexible by beating and soaking it in water. Two types of bark beaters have been found in Mesoamerica and both resemble those also found in Southeast Asia. One is a longitudinally grooved club. The other is a small, flat stone slab inscribed with parallel lines that was set in a haft or twig racket. [Anthropologist Paul] Tolstoy [of the University of Montreal] found either diagonal grooving or an alternating pattern of deep and shallow grooves on the working faces of racket heads from both sides of the

Pacific.

"To determine of the appearance of bark paper in Mesoamerica was the result of independent invention or of the migration of Asians possessing the technology, Tolstoy set out to compare not only bark paper manufacturing procedures, which involved as many as three hundred variable steps, in two regions, but also how the product was used on both sides of the Pacific. He found the bark-papermaking industries of ancient Southeast Asia and Mesoamerica shared 92 of 121 individual traits. About half of these were actually non-essential to the production of bark paper, while the other half, though essential, were among several known alternatives. But not only were the production techniques similar, so were the product's uses as writing paper, as festoons in rituals, and as payment of taxes or tribute.

"Tolstoy, noting that papermaking technologies in other parts of America were not related to those of ancient Mesoamerica, concluded that Mesoamerican papermaking was distinctly Southeast Asian..."(82).

Various specific techniques of metalworking and alloy production were common between Asia and the Americas. Schoch comments that "Mesoamerican and South American smiths were adept at techniques like various methods of gilding and the diffusion bonding of silver and copper, which were also known in eastern Asia. Two alloys characteristic of Japan, one of gold and copper oxide and the other of copper, silver, and gold, appeared in South America by the end of the first millennium B.C. As early as A.D. 1000, the people of Ecuador and northern Peru were circulating copper money shaped like small axe blades. The Chinese used copper money in the shapes of half-moons, circles, spades, and knives"(1).

Schoch also tells us about the similarities that existed, from way back in time, between the Americas and the Orient in the making of blowguns: "One particular kind of blowgun, the split-and-grooved form, is found in Malaysia, northern Borneo, and western Luzon (in the Philippines). The same weapon was used by the Houma of Louisiana, natives in the upper Amazon, and among the indigenous people near Barranquilla, Colombia. The hemispherical mouthpiece, common in America, also appears in Malaysia. Indonesian and South American blowgun hunters use different trees as the source of their dart poisons, but they tap these trees in the same way and call them by similar names. Both Malaysian and Amazonian hunters cut away the meat surrounding the dart wound before eating the kill, even though the poison has been metabolized and poses no danger to humans. Both Malaysian and Amazonian blowgun users employ salt and lime juice as antidotes to poison despite the lack of evidence, other than superstitious folk belief, that either actually works"(Ibid.).

Pre-Columbian bronze pins for fastening clothing have been found in Asia and South America. The heads of these pins, exhibiting many different stylized designs, very closely parallel each other on either side of the Pacific. It would be hard to account for three or four such design parallels having arisen by chance. But when we realize that there are literally dozens of them, we have no choice but to conclude that we are dealing with evidence of Asian/American trans-oceanic contacts in ancient times(110).

A pottery neck-rest made by the Pre-Columbian Bahia culture was found in Manabi, Ecuador, which has the same basic design as one that was unearthed in New Guinea. Both depict two humanoid figures, one at each end, facing away from each other and suspending the neck supports atop their heads. Understand that finding a neck-rest anywhere in the Americas is very unusual. But when such a find displays close parallels with a similar artifact found across the Pacific, we can be sure that an ocean-crossing contact was responsible for its existence.

There is a Mayan stone carving, dating to about the thirteenth century A.D., which depicts the deity Kukulcan, the "lord of life," wrapped in the coils of a snake that is about to ingest him head first. This figure bears an uncanny resemblance to a twelfth century A.D. Cambodian Mucalinda Buddha statue that shows the sage seated on top of a cobra's coils, with the head of the serpent behind him, looking as though he is about to eat him head first as well(196).

Common to both ancient Mexico and the Far East was the use by priests of lotus scepters. Particularly noteworthy were the Mayan and Javanese utilizations of these items. Both of these cultures often portrayed priests, in their artwork, sitting in the same positions as they held these scepters.

Speaking of the lotus appearing in ancient religious art, another parallel along this line exists between Mexico and the Far East. In both regions, divinities were frequently depicted rising out of lotus flowers, specifically by the Tibetan and Mayan civilizations.

In both Mexico and Bali, certain ancient deities were represented, in religious artworks, in an inverted position, looking as though they're diving. They are shown with their heads and limbs in almost the exact same position, wearing similar belts and loin cloths.

Even in modern times there are similarities that exist between remote areas of Asia and South America, which make us wonder if they represent leftovers from ancient contacts between these regions. Regarding some of these present-day similarities, Schoch wrote: "A number of...similarities link parts of South America, particularly the Amazon Basin, with Southeast Asia. In both areas, tribal warriors hunt human enemies for their heads and preserve them as trophies. People live not in individual family huts but in large communal longhouses built on piles. And dugout canoes are used for river travel in both cultures"(1).

We find a unique architectural parallel between ancient Cambodia and Mexico. On either side of a doorway at the Temple of Banteay Srei, Angkor, Cambodia, seriated columns can be seen set into the wall, simply for decorative purposes. The same style of stonework can also be seen on the outside wall of a Mayan building at Uxmal, in Mexico(176). But this is not the only architectural parallel between these two regions.

Stepped, temple-topped pyramid structures built by the Mayas in Mexico, Guatemala, and Honduras look strikingly like the pyramids in Cambodia(83). Pyramids, of course, are found all over the world--North, Central, and South America, the Canary Islands, Greece, Egypt, China, the Sudan, etc., and all have many key features in common.

For example, aside from the basic shape, all seem to have served an important religious function--a sort of "holy mountain" or "dwelling place of the gods"--where important rituals were performed.

Most were aligned with the cardinal points of the compass, and thus also served as stone calendars, probably helping to track the planting and reaping seasons, as well as the observances of holy days and feasts.

A good many pyramids were utilized as tombs for royalty, often containing mummified corpses of important rulers.

While some pyramids had temples built at their apexes, as described above, many others had temple complexes constructed in association with them, connected via a causeway or tunnel system.

Though most pyramids and ziggurats were stepped, some of them were flat-faced (like the Giza pyramids of Egypt). But even these, in the early phases of their construction, began as stepped pyramids(1).

All of these pyramid similarities clearly indicate a common cultural heritage that dispersed across the globe in ancient times.

Today there are even biological commonalities between the peoples of East Asia and Mexico. These mutually-held physical traits clearly point back to an earlier age when interbreeding was taking place between these populations. The Mayan people, for example, share some unique physical characteristics with East Asians. One of them is the "Mongolian spot"--a bluish spot that appears at the base of the spine in babies, which fades as the child grows. Another one is the Asian epicanthic fold at the corners of the eyes. Finally, the wrinkle lines of the palms of the hands are similar among both of these peoples.

The languages of several Native American cultures had so heavily borrowed from the Chinese language long ago that the speech of these American cultures, even into modern times, has been understood with clarity by Chinese visitors. As Gunnar Thompson explains: "When the 1850's California Gold Rush brought many Asians to America, Chinese immigrants from Kwangtung discovered they spoke the same dialects as the Sioux and Apache tribes. In 1865, two Peruvian scholars reported that natives in Eten, Peru, understood Chinese. In 1874, linguist Stephen Powers reported linguistic evidence of an ancient Chinese colony on the Russian River in California. During the 1980s, a Manchurian visitor in Bolivia was able to communicate with Quechua-speaking natives using her own Manchurian dialect"(78).

Ancient pottery fragments found in Valdivia, Ecuador, very closely match, in every detail, pottery made by the Jomon culture in Japan, during the 4th millennium B.C.(82) Also found in Ecuador were ceramic house models with peaked roofs that closely resemble old Asian reed huts, further clarifying the Japanese origin of the pottery fragments(Ibid.).

Speaking of the Japanese, archaeologists have found an interesting Olmec figure that looks precisely like a Sumo wrestler, both in facial features, hairstyle, and round, protruding belly.

Another correlation exists between the Jomon culture and northern South America, as Schoch tells us: "Possible [further] corroboration of the Jomon-Valdivian connection comes from a weapon found in similar versions in Japan and Korea and in Ecuador, Peru, and Bolivia. Called the star-holed mace, this weapon of war features a star-shaped head hafted onto a handle by means of a round hole. Made in stone, Asiatic star-hole maces date to the Late Jomon Period, in about 1000 B.C. In the Andes similarly shaped maces were made first from stone and later from bronze, and they date to the A.D. 500-1500 period. Since the star-holed mace is not found anywhere in South America outside the cultural area influenced by the Jomon Japanese, it too probably came with them..."(1).

Further evidence of Japanese contact with the Americas comes from certain North American Native tribes. The Haida of the Northwest United States, for example, have facial features that look distinctly Japanese, including the epicanthic eyefold, the Mongolian spot, and facial hair--very rare among Native Americans. There is even a province in Japan, Hida, that sounds very close to this tribe's name. And finally, in the late eighteenth century, Haida tribesmen were observed wearing East Asian-type triangular reed hats. Such attire must have been adopted from Japanese immigrants.

Other North American Native similarities with Japan exist among the Zuni tribe, including oral traditions, dental features, religious practices, linguistic relationships (for instance, they both called their pit dwellings *kivas*), type-B blood frequency, and the use of chrysanthemums in art. Furthermore, between the eighth and thirteenth centuries A.D., there was an overseas exodus of the maritime tribe, the Azumi, from Kyushu, Japan. The very similarity in the name of this tribe with that of the Zuni strongly indicates that this was the specific Japanese tribe that is responsible for all the other Japanese similarities found among the Zunis(78).

Visitors from India

The ancient people of India were just as competent on the open seas, possibly even more so, than the Chinese and other peoples of the Orient. By at least the first millennium B.C., if not before then, the inhabitants of India were aware of a great land far across the Pacific. The name given to this land was *Patala*, meaning "The Opposite Land" (that is, the land on the opposite side of the globe). The record of this far away land is found in an ancient Indian tradition, known as the *Rig Veda Mantra*, which was written down in Sanskrit around 1000 B.C. This written record is most interesting, not only because it seems to make clear reference to the Americas, but because it also talks of how the Earth is round and revolves around the Sun, along with the rest of the planets(78).

Knowledge of, and visits to, the Americas by the ancient inhabitants of India appear to have continued into the early centuries A.D. As Noorbergen wrote: "The sacred Hindu books, the *Puranas*, which date between 600 B.C. and 300 A.D., refer to direct communication between India and distant places around the world. The Indians were well-acquainted with western Europe, which they called Varaha-Dwipa. England was known to them as Sweta Saila, or 'the Island of the White Cliffs'; and Hiranya, or Ireland, as the Irish legends relate, was visited by the Dravidians, a group of men from India. The Irish say that they stayed for only a brief time and had come as surveyors, not invaders. But the Indian books go far beyond western Europe in their recollection. They describe North America, the Arctic Ocean, South and Central America, and other locations"(9).

Ancient Hindu merchants carried on an extensive trade network. In addition to the *Puranas*, mentioned above, the *Jatakas*, which date contemporaneously to the *Puranas*, also describe epic voyages from India to areas abroad, particularly Malaysia and Indonesia. The round trip distance of these voyages was nearly 6,000 miles. These journeys were undertaken on a variety of junks, outriggers, and galleys. Chinese chronicles of this era describe seven-masted Hindu vessels that were the equivalent of 160 feet in length, carried 700 passengers, and were capable of hauling 1,000 metric tons of merchandise(78).

So it appears quite clear that the ancient inhabitants of India did indeed have the capability of reaching the Americas. But is there any physical evidence of such contact?

In Native American burial mounds of the Eastern Woodlands Adena culture (500 B.C. to 200 A.D.), gastropod shells of *Cyprea moneta* have been found in great abundance. This type of shell is native to the Maldive Islands in the Indian Ocean, and was used as currency by ancient Southeast Asian traders(Ibid.). How do you suppose these shells got to North America in ancient times, and in such large amounts?

A common practice among ancient Mayan priests involved the piercing of male genitals, as a sort of "penance" to appease the gods. This same practice was also common with the ancient Hindus of the same era(Ibid.).

Figurines of seated individuals in the meditation position, looking strikingly like Buddhas, are common in Mayan art. Other "Buddha" figures have been found in Ecuador, but they appear to have been influenced more by Southeast Asia. One Chinese Buddha from the sixth century is depicted making specific gestures with each hand (called mudras), precisely matching the gestures made by an eighth-century Mayan maize god. The combined signs made by both hands signify that a gift of courage was being offered.

The name of Buddha's mother, incidentally, was Maya (or Mai). Also, "Maya" was the name of the Danava, an "anti-god" in India. Furthermore, the hrim-mantra is known as the mayabija. Some believe that the Mayan Native Americans derived their name from one or more of these sources.

The lotus symbol, common in ancient India, China, Egypt, Babylon, and many other cultures, which represented fertility and rebirth, can also be found all throughout ancient Mexico (the Mayas used it mostly), and throughout Nazcan art as well(Ibid.).

The elephant has always been highly regarded by the Hindus. There's even an elephant deity, Ganesh, revered in India since ancient times, who is the god of good luck. No surprise, then, that elephant motifs have turned up on occasion in Native American art. This is most unusual, since the elephant was not supposed to have been known in America until after Columbus. In an Indian burial mound in Louisa County, Iowa, for instance, an elephant-adorned pipe was found in 1879.

Of course, elephant motifs may not necessarily imply contact with ancient India exclusively--they may also imply an African contact, which we will discuss later. But another ancient American elephant motif--this one a Mayan carving from Honduras--contains a feature that seems to settle the case for contact with India specifically. Shown in this carving are two men riding on what must be interpreted as an elephant, with one of the riders wearing an Indian-style turban(Ibid.).

In both India and Nepal, Buddha and various deity figures were often portrayed wearing a serpent belt around their waists, where the snake either had two heads or else there were two snakes coiled around each other. Likewise, various cultures in Mesoamerica also pictured some of their deities wearing double-headed serpent belts.

Many tri-faced deities are found in the religious iconography of both India and Mesoamerica. These gods are shown to have one head with three faces, one looking forward, another looking to the left, and the other to the right(196).

There was a two-headed deity worshiped in ancient India, known as Agni. At the same time, there was a very similar, two-headed god that was revered in Mexico.

One ancient Indian work of art portrays a winged deity emerging from the center of a coiled snake. This exact same imagery has been found along the northern coast of Peru.

Another Hindu deity pops up in ancient Mexico, in slightly modified form. It's a Huastec sculpture of an adolescent with a skyward-gazing head. His body is divided vertically, and only one side of him is covered with incised carvings. Similar to this are the seventh-century statues of Harihara, half Vishnu and half Shiva. These idols, from Cambodia, are also divided vertically down the middle, and are regarded as two-in-one(90).

Still another Hindu god that found its way to ancient American shores, apparently originating on the island of Sumatra, was one that wore a mushroom-looking hat on his head. This exact same hat is found on a deity that was once venerated in Guatemala. And perhaps not so surprisingly, both deities, on either side of the Pacific, are situated in the same seated position.

Hindu symbols are frequently found at ancient sites in the Americas. One good example of this is a stylized "S" symbol, where both ends are bent around in a partial coil. It appears in pre-Columbian Mexico, such as in the Aztec document *Codex Borbonicus*.

Another example is from Mayapan, Mexico, where carvings of footprints with swastikas on them have been excavated. This exact same motif was often used by followers of Guatama Buddha in India during the early centuries A.D. Swastika symbols have also been observed on pre-Columbian pottery found in Peru.

It's no wonder we find so many Indian Buddhist and Hindu influences in ancient America, since missionary work was, along with trading, a powerful motivating factor for ocean-crossing voyages among the Hindus and Buddhists of ancient India. One dignitary, in particular, who was known for inaugurating such ventures was the great Mauryan emperor Ashoka in the third century B.C. He began as a Hindu but later converted to Buddhism, and went to great lengths, as many other Indian emperors had done, to spread his faith (or faiths) abroad. It is acknowledged by mainstream academia that his endeavors, specifically, had stretched as far as Rome and Egypt. So why should we doubt that the Americas were out of his reach?

Here's another astonishing archaeological find that shows contact between ancient India and the Americas: In Tiahuanaco, archaeologists excavated a figure wearing a turban. Both the turban and the facial features look like they are of Asian Indian origin.

If people from ancient India did indeed journey to Peru in ancient times, perhaps Easter Island--2,200 miles off the South American coast--served as a stop-off point for them, since the rongorongo script of that island is, for the most part, a mirror-image copy of the picture-writing employed by the ancient peoples of Harappa and Mohenjo Daro, located in modern-day India and Pakistan. But the original settlers of Easter Island were apparently not from India. Credit for that seems to belong to the ancient Peruvians. For an old legend of Easter Island claims that a white chief-god named Tiki was the founder of their race. Meanwhile, the Incas speak of a white chief-god named Kon-Tiki, whom their forefathers had driven out of Peru, across the sea.

It's worthy of note, in this context, that the first Europeans to visit Easter Island in the eighteenth century found it to be inhabited by white natives. Were these the descendants of Tiki, or Kon-Tiki? If so, this would explain the massive stone heads on the island, which required expert stone masons. These stone idols stand, on average, about 30 feet high, and are topped by colossal red stone "hats." They weigh in at between 5 and 25 tons each, are nearly eight hundred in number, and are scattered all over the island. Such creations like these were not the work of primitive, isolated islanders. But they would have been no challenge for the builders of the great stone cities in Peru and Bolivia, such as Tiahuanaco. Perhaps visitors from India to Peru were first introduced to Easter Island by the early Peruvians(199). The reader should be aware, too, that the stonework of the walls of Ahu Tahira at Vinapu on Easter Island are the mirror-image of the masonry of Inca and pre-Inca peoples, where the stones fit tightly together like a jigsaw puzzle, with each piece being of a different shape and size.

Coming back to artifact similarities, animal figures on wheels were common toys in ancient India, China, and Mesopotamia. But there have also been found very similar toys in Tres Zapotes, Vera Cruz, Mexico(62).

There are even some curious linguistic similarities between ancient cultures in the Americas and India, as well as other Asian countries. Schoch elaborates on this thusly: "In India, the melanotic chicken is called *kharcha*; in the language of the Arawak of northern South America it is *karaka*. The Chinese names for chicken are *ke*, *ki*, or *kai*, depending on dialect. The names in the various Mayan tongues are *ke*, *ki*, *ek*, and *ik*. To the Japanese a hen is *mendori* and a cock *ondori*; among the Tarahumara of northwestern Mexico they are *otori* and *totori*"(1).

We also find an indication of American influence in twelfth- and thirteenth-century India. In at least three pre-Columbian Hoysala stone block temples near Mysore, Karnataka, there are depictions of deities holding ears of corn, some still in their husks with the curled silk strands sticking out of the tip. Corn, of course, grew only in the Americas before Columbus. Yet here we find that it showed up in India hundreds of years before the time of Columbus. Obviously, then, we have here yet another clear sign of pre-Columbian trans-oceanic contact between the Old World and the New. Visitors from ancient India to the Americas must have brought corn back with them to their native land.

Sumerians/Babylonians/Assyrians

Like the ancient people of China and India, those of Sumer also wrote about apparent visits to the Americas in the distant past. As Gunnar Thompson tells us: "Sumerian tablets of the third to fourth millennium B.C. tell of voyages beyond the western sea (the Atlantic) and establishment of colonies in the distant land (America)"(78).

Sumerian tablets also make frequent reference to the "Lake of Manu," sometimes called "cloud lake" (lake in the clouds?), which was said to be in the "Mountains of the Sunset" in the "Sunset Land." Could this be describing anything but Lake Titicaca high up in the Andes, which, to the Sumerians, truly was in the "Sunset Land"--a land far to the west?

The Peruvians have a legend that tells of how their distant ancestors were immigrants that came from "the sunrise across the water"(110). Was this a reference to the Sumerians or the Babylonians? In any case, here's the real question to ask: Did the Sumerians/Babylonians actually have sea-worthy ships that were capable of ocean crossings?

The Babylonians conducted an extensive overseas trade network with East Africans and Hindus in the third millennium B.C., confirmation of which has been made from cuneiform inscriptions and ceramic trade goods found at various archaeological sites throughout these regions. The Babylonian sea vessels used in these ancient trans-oceanic trading excursions were made from reed bundles wrapped with hemp ropes and covered in a tar-like substance known as bitumen. They were large enough to carry up to 28 tons of trade goods and reached lengths of over 100 feet. These boats were often reinforced with bamboo poles and trunks of palm trees. Certainly ships of this sort were capable of journeying across the Atlantic to the New World(78). But what do we have for evidence of such journeys?

At a burial site in Waywaka, Peru, gold foil and lapis lazuli beads were uncovered that date back to the sixteenth century B.C.--the exact same time that the Babylonians were engaged in ocean-bound trade with these very items.

Other connections between Peru and Babylon from this same time include divination from entrails, conical clay pegs placed in the hollow spaces between foundation stones of buildings, and a unique braided beard pattern on royal personages (don't forget that beards on Native Americans are highly unusual to begin with).

A goddess was featured in Babylonian art that held a vase with flowing streams of liquid coming out its left and right sides. This exact same scene has been found in Zapotec art, in Mexico.

It was a common custom in both Assyria and Mexico for dignitaries to be shaded from the Sun by attendants holding umbrellas over their heads. This scene was often portrayed in ancient artworks on both sides of the Atlantic(200).

Patterns painted on Mesopotamian pottery have sometimes been found to closely match those painted on ancient American pottery. For example, one pottery piece from Elam (in modern-day Iran) has a pattern that perfectly matches one found on a pottery item from Northwestern Argentina. The design involves intermittent squares formed by two black and two white triangles with their apexes pointing in toward the center, with multiple vertical lines separating each square.

Another example of pottery patterns held in common between Mesopotamia and the Americas is the repeating stepped pyramid motif, such as what is found on a pot from Asur (an ancient Assyrian city) and Salvador. In both these examples we find that even the number of steps on the pyramids (3) is the same.

The patee cross was a recurring symbol in Babylonian art. It was used as one of many representations of the Sun. It also appeared in ancient Mexico as a symbol for the yearly cycle.

There have been numerous Sumerian cuneiform inscriptions discovered all over the Americas, but probably the most famous ones have been found in Bolivia. A set of three such etchings, known as the Pokotia Inscriptions, appear on a stone statue, and have been dated to the second millennium B.C.(Ibid.) When translated, they read as follows:

Distribute the opening of the Oracle to mankind.
Proclaim [that Putaki's] offspring [are to] witness esteem.
Act justly [now], to send forth the oracle to nourish knowledge.
Appreciate the cult.
[All to] witness the divine decree.
Send forth the soothsayer to capture the speech [from the oracle] to make clear the ideal norm [for living, as a guide for mankind].
[Citizens] witness in favor of this human being to create wisdom [for all mankind], and send forth [an example of good] character [Indeed]!

Good Putaki, a wise man and progenitor of [many] people.

Take an oath to witness character and wisdom.
Witness the deity's power [to make for you] a righteous soul(118).

Also found in this area was the Fuente Magna Bowl bearing Sumerian inscriptions as well, two of which translate thusly:

Girls take an oath to act justly [in this] place.
[This is] a favorable oracle of the people.
Send forth a just divine decree.
The charm [meaning the Fuente Magna Bowl itself] [is] full of Good.
The [Goddess] Nia is pure.
Take an oath [to her].
The Diviner.
The divine decree of Nia [is], to surround the people with Goodness/Gladness.
Value the people's oracle.
The soul [is to] appear as a witness to the [Good that comes from faith in the Goddess Nia before] all mankind.

Make a libation [in this] place for water and seek virtue.
[This is] a great amulet/charm, [this] place of the people is a phenomenal area of the deity [Nia's] power. The soul [or breath of life].
Much incense to justly make the pure libation.
Capture the pure libation [or, Appear as a witness to the pure libation].
Divine good in this phenomenal proximity of the deity's power(117).

If trade was carried out between Mesopotamia and the Americas, as this study proposes, we would expect there to be similarities in trading customs between these two regions in ancient times. And we do indeed find such similarities.

For instance, Indo-Sumerians marked trade goods with personalized cylinder seals for identification purposes, as did the ancient Peruvians and Central Americans (particularly the Olmecs and the Mayas). Such seals were elaborately carved and worn on a necklace on both sides of the Atlantic(78).

A symbol that was used quite heavily in Sumerian and Babylonian art was the birth (or life) symbol. It was an upside-down U with outward-bending curls at the ends. This same symbol has also turned up in numerous places in Pre-Columbian art.

Assuming the Olmecs were indeed visited by Mesopotamians, it would have almost been inevitable for the Olmecs to have incorporated at least some of their deities into their own pantheon. We see this having been done, most definitively, with a goddess whose sculpted image was found in Chalcatzingo, Mexico. This deity is found on a cliff face overlooking agricultural lands, seated in the mouth of a shallow cave. In her lap she holds a large tablet, while in front of her are spirals of smoke. Outside the cave can be seen great clumps of corn growing as big drops of rain fall thereon. And it happens that all the elements from this scene are common to the Mesopotamian grain goddess, Nisaba. She too carried a tablet, which cuneiform texts describe as "the tablet of the favorable star of heaven." Moreover, the Sumerian hymn to Nisaba that follows contains further parallels to the goddess of Chalcatzingo:

O Lady colored like the stars of heaven
Holding [on her lap] the lapis lazuli tablet,
Nisaba born in the great sheep fold by the divine Earth...
Mouth opened by the seven flutes...
Dragon, emerging brightly on the festival,
Mother goddess of the nation...
Pacifying the habitat with cold water [rain],
Providing the foreign mountain-land with plenty...
In order to make grain and vegetable [or corn, in the case of Mexico] grow in the furrow,
So that the excellent [grain] can be marveled at...
In the *Abzu* of Eridu sanctuaries are appointed,
Nisaba, woman born in the mountain [cave](91).

Unavoidably, as Mesopotamian trading ships journeyed to the Americas in ancient times, their design would have exerted an influence upon ship-building styles of Native Americans. And this is just what we find. In fact, commenting on this very issue, Schoch wrote: "To this day the people who live on the shores of Lake Titicaca preserve a singularly striking artifact that points to a connection with the ancient Middle East. Mesopotamian and Egyptian paintings and inscriptions from the second and third millennia B.C. show the gods sailing through the heavens in high-bowed, high-sterned boats made of reeds tied in tight bundles. Boats of that design are still in use in the marshes of the Euphrates delta in modern Iraq, on Lake Chad in the southern Sahara, and on the Mediterranean island of Sardinia. The same kind of vessel is likewise found on Lake Titicaca, along the northern coast of Peru, and on Easter Island, 2,350 miles off the coast of Chile. Reed boats were used in Mexico as well, until the middle of the twentieth century. Set any of these vessels against the ancient Mesopotamian inscriptions, and even a nautical expert would have trouble saying which is Old World, which New"(1).

In the course of trading, we would also expect terminologies and beliefs to be exchanged between Mesopotamia and the Americas. And our expectations here are not met with disappointment either.

As a case in point, both Babylonian rulers and Peruvian chiefs were referred to as "Lords of the Four Quarters." In addition, the Incan empire was known as *Ta-huantin-suyu*, or "The Four Quarters of the World"(78).

Both Sumerian and Peruvian cultures had a goddess called "Mama," and believed in the existence of a sacred mountain with twin peaks, where the Sun rested at night(84). The Sumerians called their twin peaks Mashu, while the Incas had their Machu Picchu(78).

Turning to the practice of writing, we see at least one parallel here as well. Both Mesoamericans and Middle Easterners made paper in ancient times, and both coated the writing surface with lime(1).

In the area of mathematics there was one specific similarity that was common to both these regions in ancient times--the concept of the number zero. Each side of the Atlantic had assigned this number its own symbol, and made frequent use of it in calculations. This number was totally unknown to all other cultures of the ancient world.

These two parts of the world were also fascinated--even obsessed--with astronomy. They built sophisticated observatories, for example, and maintained meticulous records of eclipses(Ibid.).

Public construction projects utilized the same type of vassal labor system in these two vastly-separated regions. Schoch explains: "Pyramids and other major public works projects in Mesoamerica and the Middle East were constructed with corvee labor--unpaid labor due from a vassal to his lord or rendered to a lord as part of annual taxes"(Ibid.).

Even in the realm of medical procedures we find the parallels between Mesopotamia and the Americas to have been so precise that they couldn't possibly have arisen independently. As Schoch put it: "Both Peruvians and ancient Middle Easterners practiced trepanning, surgically opening the skull for medical or ritual reasons. Trepanning is known from various locations around the world, and alone its existence in different places is itself not remarkable. Prehistoric Mexicans performed the operation too, drilling a round hole in the skull to expose its contents. The curious similarity between the Middle East and Peru is that the procedure was done in the same way. Rather than drilling a hole, the surgeon or priest made four overlapping knife or saw cuts in the skull, the same pattern as the start of a game of tic-tac-toe, then removed the square of bone in the middle"(Ibid.).

Ancient Mesopotamians benefited from their American contacts in many ways, particularly in the realm of agriculture. A good example of this is the pineapple, which was originally native to South America alone. An undeniable representation of this fruit has been found on a ninth-century B.C. Assyrian stone mural. This pineapple shows all the expected features endemic to it, clarifying that it could be no other plant. It has a stem with buds, crown with pointed leaves, and the classic diamond-shaped pattern on its skin. Ceramic pineapples have also been found in some ancient Egyptian tombs(78). And it is to this corner of the ancient world that we now focus our attention.

Egypt and its kindred nations

Superior in the disciplines of astronomy, mathematics, navigation, time measurement, and mapping, the ancient Egyptians had all the necessary skills to transverse the Atlantic Ocean and reach American shores. But in spite of such skills, the Egyptians weren't exactly the best ship builders in the ancient world. The biggest problem with their ships was that they lacked keels. However, tension cables suspended above the decks seemed to be adequate enough to keep the ends of their boats from sagging(78). Though this was a rather clumsy form of boat construction, these ships apparently were sufficient enough to carry Egyptians to the Americas, as Thor Heyerdahl has shown. Or perhaps the Egyptians may have employed, on occasion, the services of the Phoenicians to take them across the ocean, just as Pharaoh Necho II employed their services around 600 B.C. to circumnavigate the coast of Africa, in search of better trade routes [Endnote 60]. In any case, there is an abundance of evidence to show that there was indeed an Egyptian presence in the Americas, long before the time of Columbus, and that this presence exerted an extensive influence on Native American cultures, particularly the Mayas and the Olmecs.

There are some fascinating ancient records that talk about Egyptian knowledge of, and apparent journeys to, the Americas in ages past.

For instance, Gunnar Thompson informs us that "A 5th-century Greek philosopher, Proculus, reviewed Egyptian records during his visit to Alexandria in 445 A.D. He reported documents confirming Egyptian knowledge of a distant continent in the western sea"(Ibid.).

In the third millennium B.C., Pharaoh Manetho sent a fleet into the Atlantic to explore "the great continent beyond."

In this same millennium (around 2180 B.C.), the funeral text of Pepi II described a great boat journey to see "the inhabitants from beyond the western horizon" (across the Atlantic)(Ibid.).

Queen Hatshepsut was the only woman in the entire history of Egypt who ruled as a pharaoh. In a papyrus scroll dating back to 1500 B.C., we find that she, by sending out oceanic expeditions, had defined "the boundary West to the land of the going down of the sun"(91). Could this be a reference to anything other than the Americas?

But most interesting is the twelfth century B.C. account of Pharaoh Rameses III, who ventured to a land in "the inverted waters." When one realizes that the Egyptians were well aware of the roundness of the Earth, this term, "the inverted waters," takes on a very special significance, because, from the vantage point of Egypt, the western Atlantic lies on the other side of the globe. In other words, this part of the world was/is upside-down, or "inverted," to the Egyptians.

Interestingly, Rameses IV, son of Rameses III, wrote of his father, saying that he had in his grasp "the Ocean and the Great Circuit...to the ends of the support of the sky"(78).

A common motif in Egyptian art, drawing on the "land of the inverted waters" theme, is an upside-down boat, such as on the ceiling of the tomb of Seti I and Rameses VI, where the "solar boat" is first shown right-side-up as it approaches the Underworld, and then upside-down as it enters therein. The Mexican version of this is an Olmec carving that depicts a seated deity (Tlaloc) with an inverted reed boat over his head, looking remarkably like an Egyptian vessel. This seated god is encircled by a snake, similar to a scene that pops up in ancient Egyptian art, where the Nile god (or goddess) is shown in a seated position encircled by a snake as well. Sometimes the Egyptian solar boat is pictured with solar symbols (crosses within circles) above it. And, no surprise, the Olmec solar boat has crosses, or Xs, on it as well. Also, Tlaloc is shown holding a satchel, probably containing incense. Egyptian and Assyrian deities were also often pictured holding satchels that looked just like Tlaloc's, although the Egyptian version was used more for holding seeds, rather than incense(197).

In the tomb of Rameses III, there is one room that is inscribed with a text called "Litany of the Sun," which has on its ceiling a depiction of a fleet of seven ships among the stars. Compare this with the following passage from the writings of Bernardino de Sahaguun, a sixteenth-century Catholic priest who recorded many oral traditions related to him by the elder natives of Mexico: "Concerning the origin of this people the account which the old people give is that they came by sea...and it is certain that they came in some vessels of wood, but it is not known how they were built; but it is conjectured by one report which there is among all these natives, that they came out of seven caves, and that these seven caves are <u>seven ships</u> or galleys in which the first settlers of this land came." He then went on to say that, according to what he was told, these first settlers who "crossed the water" (the Atlantic) had come looking for "a terrestrial paradise," and that they settled near the highest mountains they found. Compare this further with the mention in the *Popol Vuh* of the early ancestors (from across the sea) who decided to set out on an ocean journey in search of a paradisal city called "seven caves [ships?], seven canyons."

Could this be a reference to the legendary paradisal city of Iaru, mentioned in ancient Egyptian writings? And could the highest mountains that de Sahaguun mentioned, which the early Mexican ancestors settled near, be an allusion to the sacred and mythical twin-peaked Manu of the ancient Egyptians, believed by some scholars to be St. Martin--a twin-peaked mountain near the Mexican Gulf

Coast? In his inscriptions, Rameses III actually stated that he reached this very twin-peaked mountain in the far west of the world(91). Further along this line, Gunnar Thompson wrote: "The Mayan *Popol Vuh*...[tells] the story of an ancient fleet of [seven] ships coming from the east. Aboard the ships were explorers seeking an earthly paradise and the resting place of the sun. This was a Native American perspective of the great Egyptian quest for *The Abode of the Sun*"(78).

The Egyptians wanted to follow the Sun into the "Underworld" (or the land in "the inverted waters"). They sought to reach this "Underworld" so that they could be united with the Sun at its rising. Pertinent to this point is a reference in the *Popol Vuh* to the Mayan ancestors from across the ocean, who "just wore their hearts out there in expectation of the sun." They were hoping to witness the birth of the Sun, says the document, but they didn't find what they were looking for, and were forced to conclude that "this is not our home here." Consequently, some decided to leave and continue their search elsewhere(91). Perhaps they moved on to Teotihuacan (in Mexico City) or Tiahuanaco (in Bolivia), both of which also have many striking Egyptian parallels, particularly in regards to the Sun being a central focus in art and religion.

A possible further connection between Rameses III and Mexico lies in the fact that his name appears to be preserved in the Mayan *Popol Vuh*, which states that descendants of the ancestors who had emigrated to the world of the Maya decided one day to return to their homeland. They crossed over the sea to the sunrise (the Atlantic), and met Naxcit, the ruler of the trans-Atlantic kingdom. Naxcit is very close to Nakht, which was part of Rameses III's Horus name--one of five names for this pharaoh(89).

Rameses III is sometimes depicted in ancient Egyptian art as a mighty conqueror over his enemies. One scene shows him standing over a submissive enemy, with a club in one hand and a clump of the victim's hair in the other, as he prepares to strike him. The vanquished foe has his hands raised in a posture of surrender, with his pointer finger extended, and the pharaoh's legs are spread apart, as though he is stepping toward his human target. Amazingly, in a pre-Columbian picture book from Mexico, we find a scene of a king and a vanquished enemy that matches in every particular the Rameses scene described above(Ibid.).

Egyptian and Mexican parallels seem to be endless in number. Another one involves the Egyptian deity Anubis, a jackal-headed god. Regarding the Mexican equivalent thereof, we are informed by the respected chronicler Sahagan that the first people who came to Mexico brought with them a god called Coyotlinauatl--a coyote-like god (the coyote being the Mexican version of the Egyptian jackal, of course). As to why the Mexican rendition of this deity had a different name than its Egyptian original, the Mayan *Popol Vuh* has the answer. It tells us that the "first people" who came "across the sea from the sunrise" had brought their scriptures with them, and that the names of their gods were changed after they had arrived(Ibid.).

Returning to the Underworld for a moment, there are almost endless parallels between ancient Egyptian and Mexican beliefs regarding this "realm of the dead." Graham Hancock tells us about a few more: "Is it a coincidence that the peoples of ancient Central America preserved a parallel vision [with ancient Egypt] of the perils of the afterlife? There it was widely believed that the underworld consisted of nine strata through which the deceased would journey for four years, overcoming obstacles and dangers on the way. The strata had self-explanatory names like 'place where the mountains crash together,' 'place where the arrows are fired,' 'mountain of knives,' and so on. In both ancient Central America and ancient Egypt, it was believed that the deceased's voyage through the underworld was made in a boat, accompanied by 'paddler gods' who ferried him from stage to stage. The tomb of "Double Comb,' an eighth-century ruler of the Mayan city of Tikal, was found to contain a representation of this scene. Similar images appear throughout the Valley of the Kings in Upper Egypt, notably in the tomb of Thutmose III, an Eighteenth Dynasty pharaoh. Is it a coincidence that the passengers in the barque of the dead pharaoh, and in the canoe in which Double Comb makes his final journey, include (in both cases) a dog or dog-headed deity, a bird or bird-headed deity, and an ape [monkey] or ape-headed [monkey-headed] deity?

"The seventh stratum of the ancient Mexican underworld was called Teocoyolcualloya: 'place where beasts devour hearts.'

"Is it a coincidence that one of the stages of the ancient Egyptian underworld, 'the Hall of Judgment,' involved an almost identical series of symbols? At this crucial juncture the deceased's heart was weighed against a feather. If the heart was heavy with sin it would tip the balance. The god Thoth would note the judgment on his palette and the heart would immediately be devoured by a fearsome beast, part crocodile, part hippopotamus, part lion, that was called "Eater of the Dead'"(20).

We also find unique parallels between ancient Egyptian and Mexican architecture, namely with pyramid structures. Although nearly all pyramids in the Americas more closely resemble those in Cambodia, which themselves resemble Babylonian ziggurats, there are a couple that appear to have been influenced by the flat-faced Egyptian-style constructions. As Gunnar Thompson wrote: "Only two New World pyramids evoke images of Egyptian pyramids. These are the earthen pyramids at Panache, Colombia, and La Venta, Mexico. They have smooth sides without steps, and there is no room at the top for either a temple or an astronomical observatory. These two pointed structures stand in stark contrast to truncated temple pyramids found from Peru to the Upper Great Lakes. All of these conformed to the Mesopotamian pyramid tradition which had been introduced between 5000 and 3000 B.C."(78).

Another architectural parallel exists between Egypt and the New World. Earlier we talked about the pre-Incan ruins of Sacsahuaman, located just outside Cuzco, Peru. The unique way that these stones are cut, so as to fit together like puzzle pieces, is a feature held in common with some stone ruins in Gizeh, Egypt.

The corbeled arch was employed by both the Egyptians and the Mayas. It can be seen, for example, in the Red Pyramid at Dashur, Egypt, and at some of the Mayan ruins of Honduras.

Still another architectural similarity exists between Egypt and the New World--this time in Bolivia (as well as Cambodia). In these three places, a very unique method was employed to provide stability to structures that were built with giant stone blocks. The blocks were etched with deep grooves that had metal clamps wedged in them, to hold the stones together more securely. These clamps, by the way, were all roughly the same shape(89).

There is a striking parallel between the quarrying methods employed by the Egyptians and the pre-Inca peoples of Peru. Both groups drove wooden wedges into regularly-spaced holes that were made in rock surfaces. Next, water was poured on the wedges, causing them to expand and thus crack the rock along the desired course where the wedges were placed(81). For such a method to have been independently developed, on both sides of the Atlantic, is highly unlikely.

Moving on to mummification, it happens that close correlations exist here as well, between ancient Egypt and Peru. Schoch writes: "The New World mummification practice most like the ancient Egyptian method is found in Peru, where the high, dry climate of the Andes favored the preservation of the dead better than the sweltering humidity of Mesoamerica. As in Egypt, the Peruvian cadaver was eviscerated by insertion of a hooked instrument through the anus with which the insides were pulled out, then was preserved by being rubbed with various oils and resins. The Peruvians, like the Egyptians, embalmed the internal organs in special containers known as canopic jars. And like the Egyptians, the Peruvians mummified dogs and buried them with the dead"(1).

The practice of infant head deformation among royal families was quite common in both Peru and Egypt. The high forehead gave the ruling class a distinction from the "common folk." This practice also occurred in Mesopotamia and other places around the ancient world(78). It was done by fastening boards around the heads of infants, which would cause the soft, undeveloped bones of the skull to deform in an elongated fashion, as they grew.

Both the ancient Peruvians and Egyptians used balances for weighing goods, and utilized the same materials to make them. The stems were made of either wood or bone, and the trays were constructed out of copper, suspended by strings(197).

The close connections between ancient Peruvian and Egyptian tools are almost innumerable. Their bronze axe-heads were made in precisely the same unique shape, as were their tweezers and fishing hooks. Their bows and arrows, hair combs, milling stones, metates for grinding grain by hand, spindle whorls, and their weaving looms were all made in the same fashion, usually from the same materials, and very often it's difficult to tell which of these is Old World and which is New, when viewed side-by-side. They even used the same material for making textiles--cotton, employing the same type of weaving technique that produced a very fine quality of fabric, rivaling what we manufacture today.

Egyptian and Olmec sailors were often portrayed in the exact same seated position, with the right leg folded under the buttocks and the left leg folded such that the foot was planted flat on the floor and the knee pointing straight up in the air. Another precise similarity is the fact that one or more of the sailors' arms were occasionally deliberately omitted from the carved or drawn representations, with round sockets standing in the place where the arms should be. It is obvious that such mirror-image parallels, from either side of the Atlantic, cannot be so easily brushed off as mere coincidences.

Interesting parallels existed in burial customs between ancient Egypt and Mexico.

For example, both the Olmecs, Mayas, and Egyptians buried their kings with death masks covering the face.

Another example is from the Ramessid period of Egypt, when the pharaohs were mummified with their arms crossed over their chests. The Mexican version of this practice is represented by an Olmecoid stone carving found at Oaxaca, which features a deceased man (signified by his exposed rib cage) with his arms crossed over his chest, in typical Ramessid fashion(89).

It was believed by the ancient Egyptians that when a king died, he was assisted by the four sons of Horus who provided him with a rope-ladder that he would use to climb to the "great beyond" in the sky. As Pyramid Text 2078 puts it: "These four gods, Friends of the King--Imsety, Hapy, Duamutef, and Qebehsenuef, the children of Horus of Khem--they tie the rope-ladder for this King, they make firm the wooden ladder for this King, they cause the King to mount up to Khepera when he comes into being in the eastern side of the sky." And now for the Mexican parallel: In *Codex Vindobonensis*, the Aztec god Qetzacoatl is pictured climbing on a rope-ladder that dangles from the sky.

Many ancient Mexican artifacts have been found that depict people with unmistakable Egyptian features. One touts a conical headdress with "side flaps," as well as having strongly-accentuated eyes. Collectively, these features are exclusively Egyptian and must represent an early trans-Atlantic visitor to Mesoamerica from the "Land of the Pharaohs," long before Columbus.

Sitting in the Jalapa Museum is an Olmecoid figure with arms crossed over his chest, wearing an Egyptian-looking wig. Though badly weathered, the Egyptian resemblance is unmistakable. It would appear that this carving is an ancient Mexican copy of an Egyptian ushabti (a "servant of the dead" statuette), many of which were placed in tombs so that they could serve the dead in the afterlife(91).

In Mexico, at the Cerro de las Mesas Olmec site, there lies a stone carving (Stele 5) that looks remarkably Egyptian. It portrays a man with a squared-off artificial beard dangling below his chin, as commonly found on many Egyptian monuments(86). False beards such as this have also been observed at later Mayan sites.

At Palenque, a Mayan ceremonial center in the Mexican state of Chiapas, a tomb was discovered deep inside a pyramid that belonged to a king by the name of Pacal Votan. On the lid of the sarcophagus found therein, the king is pictured lying at the base of a tree. This is in keeping with the ancient Mayan belief that the dead were to rest forever under the cool shade of the Yaxche tree. In Egypt, the deceased were said to lie under the shade of the sycamore tree.

There's another underworld tree in Egyptian mythology--the "tree of life," called the Ished tree--which some pharaohs are pictured lying underneath (such as Rameses III). It's important to note that, of all the cultures in the ancient world, only the Mayas and the Egyptians are known to have embraced this concept of the dead lying in the shade of a tree(91).

Some Olmec kings were portrayed, in carvings and paintings, with their heads and feet in profile, while the chests were shown in full frontal view--a typical Egyptian art style(115).

Many close Egyptian/Mexican relationships existed in regards to royal vestments. We will now review several examples.

One Olmec stele, at Alvarado, depicts a bearded ruler who wears a double crown made up of a falcon that is situated directly beneath a serpent. This is quite similar to the Egyptian pharaoh's double crown, with the vulture and serpent arranged in the same relative positions, one beneath the other.

Numerous pharaohs wore an animal tail as part of their royal attire, as was sometimes done by Olmec dignitaries. And the tail, in both cultures, was attached in the same manner--above the belt.

Among the ancient Egyptians and Mayas, both priests and kings wore spotted feline skins, complete with the heads, tails, and claws still attached. While in this garb, the king or priest was usually shown holding a libation or purification vessel.

In Egypt, the sign of a queen or princess was a royal headdress with double horns pointing straight up on the top. We find the same type of double-horned crown on a Mayan royal personage carved on a stele from Izapa, Mexico.

Yet another commonality between Egypt and the New World was in the methods used for casting metal figurines. Again we quote from Schoch: "[W]hen either an Egyptian smith or a Mesoamerican smith wanted to create a metal sculpture, he used the same method of lost-wax casting, a complicated and highly skilled process taught by master to apprentice. It is not the sort of thing two groups come up with independently"(1).

Several symbols endemic to Egypt have turned up at New World archaeological sites, the most famous of which is the ankh. On a Mayan terra-cotta vase from central Mexico, an unmistakable Egyptian ankh symbol is clearly portrayed(89). There is also a conspicuous ankh knot found on a headdress of the goddess of subsistence and fertility of the ancient Mexican Totonac people. A similar headdress with an ankh knot is found on a famous bust of Queen Nefertiti of Egypt. But it needs to be pointed out that both of these artifacts come from vastly different time-periods. The Mexican piece dates to between 300 and 800 A.D., whereas the Egyptian counterpart had its origin in the fourteenth century B.C. Nevertheless, the similarities are too striking to rule out cultural diffusion. It may be that the Totonac headdress and its ankh knot design were handed down over many generations, having been initially adapted from a much earlier Egyptian contact.

A common Egyptian hieroglyph is a triangle with a vertical line that runs from the center of the base partly up toward the apex, looking like a tepee. This same symbol pops up on Olmec cylinder seals from Chiapas, Mexico.

There are also many religious interconnections between the two regions of the ancient world now under consideration, some of which we will now turn our attention to.

A popular religious belief in ancient Egypt was that the Sun god "opens up the double doors of the Underworld." Thus Egyptian art often pictured the hands of the Sun god parting the Earth, or spreading apart two hills, from whence a beetle (or "solar insect") emerged. Nearly this exact same scene is depicted in the Aztecan *Codex Bodley*. Present in this work are almost all the elements in the Egyptian version, particularly the hands of a deity dividing the Earth, with a beetle-looking bug waiting to emerge from beneath the spreading gap.

The next religious particular we will scrutinize is incense and its ritualistic uses. According to Schoch: "In the ancient Middle East, incense played a key role in worship. Two of the gifts the Magi brought to the newborn Jesus were frankincense and myrth, particularly costly and highly prized types of incense. The ancient Egyptians formed incense into small balls that were burned in a sticklike censer with a bowl at one end and the head of an animal at the other. At Medinet Habu, Rameses III is shown tossing incense balls into such a censer. The *Codex Selden* depicts a Mexican priest likewise tossing balls of incense into a censer, and Mexican censers commonly had a bowl at one end and a carved animal head at the other. There is even a distant similarity in the names for incense. Plutarch, a Greek historian of the first century A.D., said that the Egyptians called incense *kephi*. The ancient Mexicans named it *copal*, a phonetically similar word still used today to refer to certain aromatic tree resins"(1). Another commonality with censers, on both sides of the Atlantic, is the fact that the bowls rested on single carved human hands that held them in place.

R.A. Jairazbhoy, in his book *Ancient Egyptians and Chinese in America*, tells us of another religious parallel: "In the Parque La Venta at Villahermosa is the Olmec sculpture of a short kneeling man. He holds an object in his hands which appears to be a stone trough. This posture is typically Egyptian in which the object held is either an offering table, such as Rameses II holds on a sculpture from Abydos, or libation basins as this Olmec sculpture seems to represent.

"Fortunately another libation ceremony survives in Mexico which proves conclusively its Egyptian ancestry....The manner in which the Underworld gods Mictlantecuhtli and his consort pour crossed libation streams from vases over the head of another god in *Codex Borgia*, is identical to a pair of Egyptian gods pouring crossed libation streams from vases over the Pharaoh's head"(91).

Rameses III is sometimes depicted holding a ball-shaped incense jar in each hand, as part of a religious ceremony. Likewise, incense burners found in Mayapan depict religious devotees holding round incense jars, or incense balls, in the same manner as Rameses III, with arms outstretched and hands cupped underneath.

In addition, there are also many similarities in the deities worshiped by both of these regions of the ancient world. To illustrate this point, let's consider the Egyptian god Min. This deity was always pictured with his left arm uplifted and a flail behind his back. In a cave in Oxtotitlan, Mexico, an Olmec god is pictured in the exact same position--left arm uplifted and a flail behind his back(89).

The Egyptian god Osiris was sometimes depicted in an acrobatic stance, with his feet bent around behind him, resting on top of his head. A precise parallel to this was found in Tlatilco, Mexico. It's a clay figure of a man (or god?) situated in the exact same position(86).

Egypt's ancient sky goddess, Nut, was often pictured holding up the sky symbol above her head. Likewise, an Olmec carving from Portrero Nuevo shows two dwarf figures (gods?) that are holding up four Egyptian sky symbols over their heads.

Both Egyptian and Olmec priests were commonly portrayed holding artificial phalli in one hand and oars in the other(Ibid.). Could these practices, on either side of the Atlantic, really have both arisen on their own, independent of each other? This is certainly what the "experts" want us to believe.

A Mayan poem recorded in the *Popol Vuh* tells of a ritual in which shooting arrows toward the four quarters of the heavens makes a man a god. Amazingly, during the Egyptian Sed Festival, a pharaoh had to shoot arrows at the four quarters of the heavens before he, too, could become a god(Ibid.).

Another ritual held in common between ancient Egypt and Mexico, the nature of which is not fully understood, was the act of striking a ball with a stick. All we know about it is that, according to an Egyptian text, the pharaoh, in performing this act, was "making a given life" (whatever that means). In Mexico, priests were depicted, in figurine form, as holding a ball in one hand and a stick in the other, just like their counterparts in Egypt. But as if this isn't "coincidental" enough, the performers of this ritual, on either side of the Atlantic, are depicted wearing a twin-feathered headdress(Ibid.).

Still another ritualistic practice anciently shared between these two widely-separated localities was focused around the goal of restoring speech to the dead, for the afterlife. Artistic depictions of this ritual being performed are found in both the Egyptian *Book of the Dead* and in an Olmec painting deep in a cave at Juxtlahuaca. Each of these two scenes depict a snake-like implement being applied to the mouths of the dead persons, both of which are situated in kneeling positions. In each instance, the officiating priests are wearing leopard or jaguar skin robes, with the tails of the animals dangling between their legs.

In Egypt, the restoration of speech was also achieved by touching the mouth of the dead, or the mouth on his statue, with the bleeding foreleg of a bull. Similarly, the Mayas, in their *Popol Vuh*, state that their ancestors who came across the sea from the sunrise could make a statue speak by touching its mouth with the blood of deer and birds(Ibid.).

As we can see, Egypt had left its mark on Olmec and Mayan religion in many different ways, but some of them were quite unpleasant indeed. Schoch elaborates: "To the modern mind, one of the most disturbing aspects of

Mesoamerican civilization was the heart sacrifice. Practiced by the Maya and the Toltecs and taken to a level of bloody overindulgence by the Aztecs, the heart sacrifice ritually offered a still-beating human heart to the sun god as a propitiation. The Egyptians did practice human sacrifice,...in the burial of retainers in the provincial Middle Kingdom, and in the subsequent Nubian renaissance, but there is no record of heart sacrifice in northern Africa or neighboring regions. Still, a fascinating parallel arises in Egyptian mythology.

"The Egyptian *Book of Caverns* shows the enemies of the Sun with their arms pointed behind them and their hearts at their feet, blood fountaining from open chests. The Egyptian *Book of the Dead* tells how the goddess Amemit ate the hearts of condemned sinners after they confessed their sins to her. The Mexican goddess Thazolteotl performed much the same task by consuming sins confessed during one's lifetime. In Egyptian belief, the heart of the dead one was weighed by the jackal god Anubis to determine the worthiness of that individual to enter paradise. The Mexican equivalent was Coyotlinauatl, the god whose name survives in English as 'coyote.' The notion of sin, the centrality of the sun, the devouring goddess, and the form of a divinity drawn from a canid animal run parallel in both Egypt and Mesoamerica"(1).

There were actually several methods of human sacrifice that were common to ancient Egypt and Mexico. Another one was the practice of burning victims alive. Still another method was to drown the victims. The Mayas, to ensure a good upcoming crop-planting season, would drown a virgin by degrees while asking her to intercede with the gods and implore them to send rain. Not surprisingly, the Cenote of Sacrifice at Chichen Itza was found to contain many skeletons, all of them females. Old World parallels of this practice occurred in both Egypt and China. In the Egyptian version, an attractively-dressed virgin was thrown into the Nile, while in the Chinese version, virgins were floated down a river in beds and thrown overboard as brides to the Lord of the River. A still further sacrificial parallel was the custom of killing servants at the time of a ruler's death, who would attend to his needs in the afterlife, being buried with him in his tomb. This was a common practice among the Egyptians, Mesopotamians, Chinese, Mayas, and Aztecs, to name but a few(90).

The next set of parallels is even more gruesome than those described above. So gruesome, in fact, that I would not even mention them except for the fact that they present yet another strong case for ancient Egyptian contact across the Atlantic. Schoch gives us this elucidation: "The ancient Egyptians had their own proclivities to bloodthirstiness, commonly practicing extraordinary cruelties on defeated enemies and prisoners of war. When Rameses III defeated the invasion of Libyans--the so-called sea peoples--in the twelfth century B.C., he ordered the penises of all the captured Libyan soldiers cut off. Scribes counted the number of the defeated by tallying the piles of severed organs, an occurrence recorded in the Medinet Habu reliefs. An inscription reports that one heap numbered 12,555, another 12,860. Rameses III was not only torturing, and in some cases killing, his enemies with this butchery, but also ensuring that there would be no offspring to mount a further attack against him.

"The event had a mythological parallel. When Horus battled Seth in revenge for the death of Osiris, Seth lost his testicles in the fight. He was mocked for his infertility; in the words of the ancient text, 'his seed will be destroyed.'

"A relief sculpture dating from the earliest stages of Mexico's Monte Alban culture, in the Valley of Oaxaca, circa 500 B.C., shows a large number of figures conveniently referred to as Los Danzantes, 'the Dancers.' Three of the dancers, all males, display oddly twisted limbs--possibly the result of agonizing pain--and they grasp their crotches as if they were wounded. Near the three stands a figure holding some kind of instrument, possibly a surgical device for amputating the penis. Yet another similar Danzantes relief shows a miniature figure holding such a device, and red paint that has survived the long passage of time since 500 B.C. stains the thighs of the dancer. In earlier Olmec artwork, jaguar warriors are depicted brandishing the severed penises of vanquished foes, just as North American Indian fighters displayed the scalps they took in battle.

"To the Egyptians the severed phallus was not just a tally for counting prisoners of war. It was also revered as the source of virile fertility, the fountain of seed that grew crops. Herodotus reported that Egyptian women carried models of greatly oversized phalluses around their villages and pulled strings to make the models move like the real thing. The same Medinet Habu reliefs from the reign of Rameses III that depict the piles of severed penises also detail the Egyptian religious cult of the phallus. A priest carries a life-sized statue of the phallic god Min, whose awesome male endowment greatly exceeded human reality, in a company of standard-bearers. Another relief, in Luxor, depicts the phallic god lying on his back, much like the reassembled corpse of Osiris experiencing the magical erection that led to the birth of Horus.

"Many of the same details are found in ancient Mexico's phallic cult. To this day, the descendants of the Maya perform a ritual dance with string-manipulated mock phalluses. A drawing in the *Codex Borbonicus*, one of the few surviving ancient Mexican writings, depicts similar artificial phalluses. An Olmec relief at Chalcatzingo shows a procession of standard-bearers approaching a phallic figure who lies on his back with his erection pointing up like Osiris'. A statue of Min shows this Egyptian god displaying his immense phallus while he holds his right arm up and bent at the elbow, a posture very like the physical attitude of an Olmec painting of a phallic figure at Oxtotitlan"(1).

The Egyptians also appear to have exerted an influence on the tools used by scribes in the Americas. Gunnar Thompson explains: "Numerous double-chambered inkwells have been found at Mesoamerican sites. Although of local manufacture, they are the same size and shape as Egyptian inkwells. Two chambers were necessary [on both sides of the Atlantic] for keeping separate the two primary colors, red and black. Inkwells have been found at Olmec sites dating from 500 B.C. and Mayan ruins of the seventh century [A.D.]"(78).

There are further Egyptian influences in the realm of the written and spoken languages of ancient Mexico. Quoting again from Schoch: "The Egyptians, as is well known, wrote in hieroglyphics. So did the Mayas and Aztecs. Writing among the Olmecs is a point of scholarly controversy....Still their art includes a figure of a kilted individual who looks very much like an Egyptian scribe, holding what could be a scroll. Mesoamerican and Egyptian hieroglyphics share a similar conception of the relationship of the image to the word and follow similar principles, which are quite different from the ideas and processes that underlie an alphabet. An occasional claim has been made that Egyptian glyphs appear in Mesoamerica, but so far the idea cannot be sustained. The forms of writing in Mesoamerica and the Middle East look similar, but the content differs.

"There may, however, be a close connection between the spoken languages. Mary LeCron Foster has presented a telling body of technical linguistic evidence that underscores a possible relationship between the ancient Egyptian and the Mixe-Zoque languages of southern Mexico, which include the Mayan tongues and are thought to have derived from the still-unknown Olmec language. Quechua, an indigenous language spoken to this day by the Indians of the Andes, is similar to the Mixe-Zoque languages and, according to Foster, contains additional words with Semitic roots, most likely from Arabic"(1).

We have already looked at many parallels between Egyptian, Mayan, and Olmec religious artworks, beliefs, and practices. But there were many other American cultures that shared such traits with ancient Egypt. We will now look at some samplings of this (along with a non-religious ceremonial parallel and some burial custom similarities) from several South American civilizations.

Turning to Lake Guatavita in Colombia, we find Egyptian fingerprints there in the form of the legend of Eldorado. Many believe that this legend had to do with a land rich in gold. But this was not the case. Instead, it involved a coronation ceremony that was acted out whenever a new king (a "golden man") took the throne.

The ritual began by the heir to the throne (the son of the recently-deceased king) secluding himself in a cave. He then journeyed to Lake Guatavita where he boarded a raft. On the raft were four incense burners situated around the would-be king as he sat there, covered from head to toe in resin and gold flakes, so that he "gleamed like a ray of the sun."

Upon reaching the center of the lake, according to some seventeenth-century chronicles, he "plunged into the water like a great golden fish. As he emerged, the multitude broke out into cries of joy."

We can rightly interpret this ritual as follows: The entering into the cave represented a journey into the Underworld, where the former dead king had departed. As his young son emerged from the cave with his body painted in gold, this denoted his complete union with his father, the Sun god. Next, by sailing to the center of the lake, diving in, and then emerging with the gold paint washed off, he became the earthly incarnation of his father, and thus the rightful, divinely-appointed new king.

In the Egyptian version, the body of the dead king was painted in gold. From there, according to the *Book of the Dead*, the deceased royal personage ventured to various pools and other places in the Underworld. At the lake of Tchefet, he was arrayed in the apparel of Ra (gold). In the lake of Tchesert, after a ride on the solar boat, he was plunged to wash himself of all impurities. Instead of four incense burners, this boat had four blazing flames for the ka (soul). Both vessels, in Egypt and Colombia, contained sunshades or fans--one placed before the king and the other behind him. Both of these shades or fans had radiating sunbeams on top of them. Interestingly, one of Egypt's sacred lakes associated with this ceremony was called Uakh, whereas the ancient Colombians had a sacred lake called Ubaque.

There were numerous close resemblances between Egypt and pre-Columbian Colombia in the area of burial rites. In 1841 two Muiscan Indian mummies were found in caves. They were in pristine condition, and it could clearly be determined that their organs had been removed through a left-flank incision, which was the exact same manner in which Egyptian mummies had their organs removed. The Muisca bodies had been covered with a resin, just like their Egyptian counterparts were. And, exactly like in Egypt, it was only royalty that were mummified.

Ancient Colombians, just like ancient Egyptians, stored the preserved viscera in canopic jars. This similarity is remarkable enough on its own. But the styles of the jars were also in close correlation, where the lids, in the shape of a human head, were formed in the likeness of the deceased. Some of the heads on the Colombian lids were even adorned with Egyptian-looking wigs.

Obelisks were popular religious monuments among the ancient Egyptians, representing their Sun god, Ra. These same monuments also turned up in ancient

Peru and Bolivia.

For instance, at Chavin de Huantar, high up in the Andes, stood an obelisk right in the center of the complex. There was also said to have been one outside the temple at Tiahuanaco, high up in the Andes as well, and another one allegedly stood in front of the Pyramid of the Moon at Puma Puncu.

As an interesting side note, the obelisk at Chavin de Huantar was in the form of an anthropomorphic knife. By comparison, the god Ra was called a "knife" (since he killed his nocturnal enemy every day at dawn). Likewise, in Mexico the Sun deity was also referred to as a knife, and was represented in stone in the shape of this very implement.

We know from sixteenth-century Spanish chronicles that as late as the era of the Incas, in Cuzco, there was an obelisk erected near a solar temple that was covered with gold leaf. So also was this the case in the eighteenth dynasty of Egypt and onward--the tops (and sometimes the shafts as well) of obelisks dedicated to the Sun were plated with gold.

Just as the Egyptians spoke of a triple aspect of the sun god ("Khepra in the morning, Ra at noon, and Temu in the evening"), the Incas spoke of a triple aspect of their own solar deity (Apu, Churi, and Guauqui).

A common pre-Columbian practice in both Peru and Mexico was to bore a hole into stone deities that ran from behind the statue to the mouth of the idol, so that a hidden human agent could give the impression that the image was actually talking to the religious devotees. The later Incan version of this "talking deity" was used for providing advice to those who made inquiries to the statue. The Egyptian equivalent of this was a portable god in the Ramessid period that would nod its head in response to questions. And it must have been this Egyptian practice that served as the inspiration for the American versions, since the sixteenth-century missionary, Sahagun, informs us that the ancestors who came from across the sea had carried their god on their backs, and "the god went advising them."

The head of Osiris, the Egyptian god of the dead, was sometimes superimposed on the back of a crocodile in religious art. An exact counterpart to this existed in ancient Peru (on a Nazcan pot), as well as in Mexico, among the Mayas and Olmecs(91).

The Americas also appear to have left their mark in ancient Egyptian society. This mark was made in the form of high concentrations of cocaine and nicotine that were found in several Egyptian mummies (coca and tobacco, the sources of cocaine and nicotine, are native to the Americas). This discovery was first made while examining the mummy of a female named Hemut Taui. In 1992, Svetla Balabanova, a German scientist, examined a total of nine mummies that showed such concentrations, revealing that use of these narcotic substances was not an isolated occurrence. The obvious implication of these findings is that there must have been an international "drug trade" in the ancient world that involved the Americas(119).

Leaving Egypt, we now proceed to examine other kindred nations of the region, to see if they too had reached the Americas in ancient times.

The first and foremost nation to consider is Phoenicia, which was located along the coastal plain of what is now Lebanon and Syria. The Phoenicians were expert mariners who operated an extensive trading empire that extended throughout the Mediterranean, across all the coasts of Africa, and up and down the western coastline of Europe, including the British Isles. They flourished from the thirteenth century B.C. on, until they began to decline in the sixth century B.C., finally disintegrating in 332 when Alexander the Great stormed the Phoenician city of Tyre.

There is good reason to believe, based on ancient textual and archaeological evidence, that the vast trading network of the Phoenicians may have included the Americas. We'll now look at some of the more outstanding pieces of this evidence, starting off with a quote from a Reader's Digest book, *Mysteries of the Ancient Americas: The New World Before Columbus*: "In 1872 slaves working on a Brazilian plantation near the Paraiba River were reported to have uncovered four pieces of stone tablet inscribed with unknown characters. A copy of the inscription but not the stone itself reached Dr. Ladislau Netto, Director of the Museu Nacional in Rio de Janeiro. After diligent study, Netto announced that the inscription recorded a visit to Brazil by Phoenician mariners centuries before the time of Christ. Unfortunately, neither the tablet nor the sender was ever located and Netto's translation was ridiculed by the academic establishment of his day.

"A century later, however, an American scholar, Cyrus H. Gordon, took a new look at the Paraiba inscription--and came up with some startling conclusions. The text, he claimed, contained expressions and grammatical forms unknown to linguistic students of the 19th century. How, then, could forgers have produced such a text? Gordon's translation reads, in part: 'We are sons of Canaan from Sidon, from the city where a merchant (prince) has been made king. He dispatched us to this distant island....We sailed from Ezion-geber into the Red Sea and voyaged with ten ships. We were at sea together for two years around Africa. Then we got separated by the hand of Baal and we were no longer with our companions.' He dates the accidental landfall in Brazil to the 6th-century B.C. reign of Hiram III of Tyre..."(85).

Could this Paraiba inscription have resulted from the following sea voyage, described by Diodorus of Sicily in the first century B.C.? Reflecting back in history, he wrote: "[I]n the deep off Africa is an island of considerable size...fruitful, much of it mountainous....Through it flow navigable rivers....The Phoenicians had discovered it by accident after having planted many colonies throughout Africa"(125)[Endnote 61].

This is not the only text from the Old World that talks of an apparent Phoenician journey to the Americas. Another account of this nature was written by the first century B.C. Greek geographer Strabo, who had this to say: "[F]ar famed are the voyages of the Phoenicians, who, a short time after the Trojan War [circa 1200 B.C.], explored the regions beyond the Pillars of Hercules, founding cities there and in the central Libyan seaboard. Once, while exploring the coast along the shore of Libya, they were driven by strong winds for a great distance out into the ocean. After being tossed for many days, they were carried ashore on an island of considerable size, situated at a great distance to the west of Libya"(87). (As you may recall, this is how Cabral discovered Brazil.)

A second-century Roman educator, Claudius Aelianus, also talked about there being a huge island on the far side of the Atlantic that was known to the Phoenicians. So isn't it obvious that they were on to something here?

According to Diodorus, the Phoenicians kept their discovery of this western land a closely-guarded secret to prevent competition in the trading business, and to preserve this region as a safe refuge in case their homeland was ever invaded. He even mentioned that the Phoenicians were aware of the fact that the Americas were divided into two major continents. They called the southern continent Colchis, and the northern continent Asqua Samal, meaning "Great North Land"(78).

If the Phoenicians had come to the Americas in remote history, perhaps conducting an extensive trans-Atlantic trading network over many centuries, we could imagine that their ships would sink on occasion during a storm, somewhere off an American coastline, and that one of those sunken vessels, or at least some of the remains thereof, would have survived to modern times. But have such remains ever been found on the western side of the Atlantic? Yes indeed. A Phoenician shipwreck that had carried a cargo of amphorae (large clay vessels usually used for carrying and storing wine) was found off the coast of Honduras in 1972(86).

Other forms of archeological evidence exist to bolster the argument that the Phoenicians conducted trade across the Atlantic. Patrick Huyghe wrote: "In a cave in Paraguay...an inscription records what appears to be an extensive visit to the area by ancient Phoenician mariners. The inscription, which is located a thousand miles inland at the upper reaches of the Paraguay River, is given in two alphabets, Ogham and Iberian, but one language, Phoenician. According to [the late epigrapher Barry] Fell, it reads: 'Inscription cut by mariners from Cadiz exploring'"(82).

The Phoenicians had apparently traded as far up as the modern state of Maine. For just ten miles off the Maine coast, on Monhegan Island, an Ogham inscription was found, given in the old Gaelic tongue of the Celts, which reads, "Ships from Phoenicia, Cargo Platform." Epigrapher Barry Fell, who examined this platform, believed that it was a loading and unloading dock for Phoenician trading ships. The posted inscription was probably designed to notify the Celts of New England where to trade their goods with the Phoenicians (more will be said on the New England Celts later).

Further south, on the shore of Mount Hope Bay, near Bristol, Rhode Island, a similar find was made. What was discovered here was a high-sterned watercraft carved on a rock, with an inscription below it claiming the spot for Phoenician mariners(88).

There have also been Native artworks found in Mesoamerica that give strong hints of a Phoenician presence there in ancient times.

A prime example of this is a very interesting Mayan incense burner that was dug up in Iximche, Guatemala. It features a distinctly Phoenician-looking face with a beard(87).

The Carthaginians inherited the maritime trading empire of the Phoenicians in the sixth century B.C., shortly after the Phoenicians were overwhelmed by the Persians (under Cyrus) in the year 538 B.C. Like the Phoenicians themselves, the Carthaginians were excellent mariners who appear to have also reached the Americas. One of the reasons for making this assertion is that Aristotle, in his book *On Marvelous Things Heard*, gave an astonishing account of a Carthaginian discovery of a land in the far western Atlantic, which has got to be a reference to the Americas. Here's what he had to say: "In the sea outside the Pillars of Hercules they say that an island was found by the Carthaginians, a wilderness having wood of all kinds and navigable rivers, remarkable for various kinds of fruits, and many days' sailing distance away....[T]he Carthaginians...observed that many traders and other men, attracted by the fertility of the soil and the pleasant climate, frequented it because of its richness, and some resided there..."(99).

Patrick Huyghe tells us about another ancient record of Carthaginians reaching the Americas: "[Barry] Fell finds additional support for Carthaginian contact with the New World in the work of Plutarch, a Greek biographer who lived during the first and second centuries. In A.D. 75 he composed a dialog, *On the Face in the Moon*, in which Sextius Sylla, a Carhaginian antiquarian, recounts a story he had heard from another Carthaginian regarding pilgrims setting out on religious migrations to a western land across the Atlantic.

"The story states that by sailing northwest from Britain you pass three island groups where the sun sets in midsummer. These sound like the Orkneys, Shetlands, and Faeroes. After five days at sea, you then encounter at an equal distance another island called Ogygia. This appears to be Iceland. Another five days of sailing, the story continues, brings you to 'the great continent that rims the western ocean.' Greenland fits the direction and distance indicated. Then by sailing south along the coast, you pass a frozen sea, apparently a reference to the Davis Strait, and finally come to a land inhabited by Greeks. These Greeks live, said Plutarch, around...the same latitude as the north end of the Caspian Sea. Since the bay referred to appears to be the Gulf of St. Lawrence, the land inhabited by the Greeks should be either Newfoundland or Nova Scotia.

"Fell has found a large number of Greek roots in the vocabulary of the Algonquin Indians of the American Northeast and believes that the Greeks Plutarch says settled there were actually North African or Libyan Greeks. He thinks they were brought to the New World by Libyan mariners recruited from such cities as Cyrene and Ptolemais during the fourth and third centuries B.C. This was a period, he points out, when Carthaginian interests in North Africa were at their peak"(82).

There is also a fair amount of physical evidence to back up the claim that Carthaginians reached the Americas.

Carthaginian coins, found especially in North America, are one good piece of this evidence. Barry Fell pointed out that such coins have been plowed up by farmers in Kansas, Connecticut, Arkansas, and Alabama, and that all of these finds were discovered at sites in the vicinity of navigable rivers, or near natural harbors on the coast. Additionally, all of these coins dated to the fourth and early third centuries B.C.(88)

While on the subject of Carhaginian coins, Schoch brought out a very significant point: "A Mount Holyoke College geologist named Mark McMenamin has found what he thinks are cryptic world maps on Carthaginian coins dating to 350-320 B.C. In McMenamin's interpretation, the coins offer a schematic view of the Mediterranean and a landmass to the west that can only be the Americas"(1).

It's not too difficult to imagine that, if Egyptians, Phoenicians, and Carthaginians made it to the Americas in the remote past, Libyans, too, right next door to these peoples, might possibly have made it across the Atlantic as well. In fact, the Libyans were highly competent mariners, and had built ships that were far more sea-worthy than those of the Egyptians. But do we have evidence for

Libyan contact with ancient America?

Ancient Libyan journeys across the Atlantic left their traces almost exclusively in the form of rock inscriptions. These inscriptions have been found primarily in North America. Patrick Huyghe cites some striking examples: "A rock-cut inscription found by an archaeologist near the present-day Zuni reservation in northwestern New Mexico describes a marriage or fertility sacrament. It is written in a Libyan alphabet used prior to the Moslem conquest of about A.D. 700. [Epigrapher Barry] Fell has...identified what appear to be Libyan letters in Zuni art. More Libyan inscriptions have been found under a rock overhang of the Rio Grande in Texas. One--in Ogham and Libyan, or Numidian, letters--reads: 'a crew of Shishonq the King took shelter in this place of concealment.' Both Libya and Egypt have had several kings by this name.

"Some Libyan visitors appear to have entered the Mississippi from the Gulf of Mexico and penetrated inland to Iowa and the Dakotas and westward along the tributaries of the Mississippi. It seems they left behind numerous records of their presence along the cliffsides of the Arkansas and Cimarron rivers in Oklahoma. Some of these inscriptions are bilingual, in ancient Libyan and ancient Egyptian. These bilingual inscriptions are not uncommon in America, and are really not surprising. Then, as today, ships' crews were probably multi-ethnic in origin.

"One of the most notorious of these bilingual inscriptions turned up in Davenport, Iowa, and suggests that perhaps a colony of Mediterranean people settled in America in about 800 or 700 B.C. The story of its discovery begins in 1874, when the Reverend M. Gass opened a small Indian burial mound in Davenport. After locating several skeletons, Gass and two student assistants came across a tablet, or stele, engraved on one side with various animal figures, a tree, and a few other marks, and on the other side with a ceremonial scene and a series of strange inscriptions.

"While some have claimed this find a hoax, the inscriptions, which are in Libyan and Iberian scripts, had not been deciphered at the time that the stone was found. Yet both of these scripts, translated by Fell, yield intelligible and mutually consistent readings regarding how to regulate the calendar.

"Still more Libyan inscriptions have been found in the Northeast--in Quebec, New Hampshire, New York, and Pennsylvania. The Libyan inscription found in a shell midden in 1888 at Eagle Neck, near Orient, Long Island, certainly could not be a forgery, as the ancient Libyan language was not deciphered until 1973..."(82).

A fascinating piece of evidence that points to an Arab presence (possibly Libyan) in the ancient Americas is carved on Monument 13 at La Venta, in

Mexico. It's a portrayal of a man wearing a distinctly Arab-looking turban. To his right are three symbols, one of which looks unmistakably like an elephant.

Further evidence of an early Arab presence in the Americas is the existence of two medieval Arab maps, both of which appear to have been compiled from much older charts. One was made by al-Idrisi in 1154, and the other by Ibn Said in 1250. These two maps depict a very large island across the Atlantic, which was given the name *Ansharus*, meaning "Far Land"(179). Was this a denotation of the American continents? If not, what else could it possibly have represented?

At the same time these maps were being compiled, and perhaps with their aid, Muslim Arab traders and explorers appear to have made it to the Americas. Among them were the Brothers Al Magrurim, whose voyage is known to us by the writings of the twelfth-century geographer Al Idrisi (*Nuzhet al Mushtaq*, 1154). He told of how they sailed from Lisbon on a quest for new islands in the western Atlantic that might prove profitable for trade. He described how the fishing was very rich in the North Atlantic--a clear reference to the famous Grand Banks of Newfoundland. He also made mention of the whalebone huts of the Inuit in Labrador. He further talked about the native inhabitants of Saun, across the Atlantic, being beardless and having breath of "wood smoke" (tobacco smokers)--Native Americans, no doubt.

With the ancient Egyptians, Phoenicians, Carthaginians, and Libyans having embarked on America-bound ocean voyages, we could easily picture Negroes, especially Nubians, accompanying them on such trips, since there was much cultural interaction between all these populations. Some ancient Negro peoples may have also undertaken their own independent voyages to the Americas. In fact, we can be certain of this, according to the testimony of several ancient documents.

For example, a West African tradition, recorded by Arabian scholars in the fourteenth century, tells of how a Ghana chief sailed across the Atlantic in the ninth century B.C.

In addition to this, the Mayan *Popol Vuh* mentions "blacks" who came from "The Land of the Sunrise"(78).

But what do we have for physical evidence of blacks in the ancient Americas?

Some of the most obvious signs of a Negro presence in pre-Columbian America are the huge carved stone heads that have been unearthed at several Olmec sites, which have unmistakable Negroid features. Interestingly, these Negroid heads are depicted wearing helmets that are very similar to the type worn by Ghana chiefs who reigned contemporaneously with the Olmec civilization

[Endnote 62].

There have also been found other, smaller-scale Olmec sculptures that depict people with Negroid features. Some of them even contain representations of scar tattoos, which were rare in the Americas but quite common in Central and West Africa, as is still the case today. Additional Olmec artworks show Negroid figures that have “cornrow” hairdos, providing yet further proof of an African presence in early Mexico.

Other cultures that long post-dated the Olmecs, both in Mexico and Peru, appear to have had contact with black Africans as well, as reflected in their art.

Such pieces of evidence of a black presence in the ancient Americas are further supplemented by the discovery of Negroid skeletal remains at archaeological sites throughout Mexico. According to Gunnar Thompson, “Several anthropologists have identified Negroid bones in ancient American burials. A Polish forensic specialist, Andrzej Wiercinski, identified Negroid bones at three Mexican archaeological sites: Tlatilco, Cerro de Las Mesas, and Monte Alban”(78).

Apparently, at least some of the blacks who came to the Americas in pre-Columbian times had established permanent settlements, since their descendants were found still living here when the Spanish first arrived. Commenting on this phenomenon, Huyghe wrote: “Shortly after Columbus’s trips to the New World, the Spanish encountered a number of large black settlements in South and Central America. Balboa, while crossing the Isthmus of Panama in 1513, not only discovered the Pacific Ocean, but also a village in which the Indians were holding black prisoners. He asked the Indians many questions about these people but learned only that the Indians were at war with a nearby settlement of tall black men. ‘These were the first Negroes that had been seen in the Indies,’ according to Lopez de Gomara, who wrote a history of Mexico in 1554. Later in the sixteenth century, another Spanish party found a large settlement of blacks waging war with the Indians on an island off the coast of Cartagena, Colombia”(82).

Pre-Columbian Negro maritime traders benefited in many ways from their contacts with American cultures. For instance, maize and cassava were found to be in use in Nigeria before the Europeans arrived.

Greco-Roman visitors

The Romans were expert ship builders and sailors. Merchant ships called *pontos* were quite common throughout the empire, and reached lengths of over 100 feet. But the largest merchant vessels were giant grain freighters known as *frumentariae*, which were elongated versions of the *pontos*. These ships were 200 feet or more in length, and were capable of hauling 1200 tons of grain. They had about ten times the storage capacity of Columbus' flagship, the *Santa Maria*. First-century passenger vessels were equally impressive, capable of carrying 600 travelers from port to port on the Mediterranean(78).

As far as the Greeks go, their ships weren't substandard either. The two main purposes that their vessels were employed for were warfare and trade. But it's obviously the Greek trading ship that we are more concerned with here, since this is the type of boat that would have been used to reach the Americas, if in fact the Greeks did make it here in ancient times. The earliest Greek trading ship was powered by sails instead of oars, and could achieve speeds of up to 5 knots per hour. It averaged about 150 tons by 400 B.C. But later, by 240 B.C., this type of boat weighed in at between 350 and 500 tons, had two to three masts, and was about 60 feet long. This shows that the Greeks at least had the capability to reach the Americas long ago.

In addition to being in possession of superior sailing vessels, the Greeks and Romans also had sophisticated navigational tools that enabled them to engage in ocean travel with relative ease. Gunnar Thompson offers us some insight on these innovations: "Several navigational tools helped sailors find their way to foreign ports. Greeks and Romans used magnetic compasses with dials. Although historians usually credit the origin of European compasses to 13th-century Arabs, a Greek legend attributes the compass to a 12th-century B.C. hero named Hercules. In the Greek legend, the Sun God Helios gave Hercules a golden cup with a floating magnet. Greeks called the magnet *Lapis Hercles*. In the 12th century B.C., the Greek poet Homer gave another account of the mariner's compass. In a mythical adventure called the *Odyssey*, Homer related that Greek sailors used a device that enabled them to sail through fog and darkness without fear of shipwreck"(Ibid.). If the Greeks did indeed have the compass this early in history, then they obviously beat both the Chinese and the Olmecs with the invention of this navigation tool.

Recall, also, the geared Greek bronze "computer" (the Antikythera device) found among the debris of a sunken first century B.C. Greek ship, which we discussed earlier. This mechanism, which gauged the movements of heavenly bodies, must certainly have served to guide sailors in their long-distance sea voyages.

The ancient Greeks had produced many fascinating maps that indicated an awareness of a great body of land beyond the Atlantic Ocean. Although none of these maps have survived to our time, tantalizing written records and copies of these maps reveal enough details about some of them to paint us an illuminating picture of what the originals must have looked like.

For example, Strabo of Amesaia wrote *Geography* around 25 B.C., which was based on charts and travel accounts that were at the Alexandrian Library in Egypt. In it he gave a description of the layout of land masses from Spain to Serica (China). He also talked about distant lands called *Epeiros Occidentalis*, or "Western Continents," beyond the Atlantic(Ibid.).

Another good example is a map made by Claudius Ptolemy in 140 A.D. He had put together a world map based on compilations of then-available nautical charts and mariner's reports. But Ptolemy can't legitimately lay claim to being the genius behind the creation of this map. He himself attributed its layout to a colleague of his--Marinus of Tyre. Today, all that remains of it are some copies--perhaps not very good ones--found in Muslim archives. But what we do find on them is quite intriguing.

The original map apparently showed 180 degrees of the Earth's circumference, rendering an accurate depiction of the relative geographical positions of Europe, Africa, India, and Asia. The map did seem to have some drawbacks, however (assuming that these copies are accurately drawn): both Europe and Africa are shown with exaggerated sizes. But the most striking feature of this map is that it shows a strip of land at the far eastern tip of Asia. This piece of land, which is not named on the map, extends twenty degrees south of the equator and then connects with *Terra Incognita*, which stretches east from Africa. Though many have criticized Ptolemy for this inaccuracy, the point is that he (or his colleague, Marinus) was aware of a large, distant body of land across the Pacific.

Interestingly, although this body of land is not named, there is a city that is listed there called "Cattigara," ten degrees south of the equator. According to Marinus, Cattigara was located eighty degrees east of Asia, which would place it in South America. We are told in the diary of a Greek merchant, Alehandro, that he reached Cattigara in the early second century A.D. He left the island of Borneo in the western Pacific and headed due east on a journey that lasted "innumerable days"(Ibid.). Did this man reach South America?

Homer's *Odyssey* mentions a mythical land in the western sea called Elysium, which means "Isle of the Blest." As you may recall, this is very close to a fifth century B.C. title that the Chinese ascribed to a land across the Pacific--the

"Land of the Blest."

Another Greek myth relates the story of the immortal Hercules, who journeyed to the "Other World" to find the sacred apples of Hesperides. Interestingly enough, when Columbus reached the Caribbean island of Hispaniola, he claimed to have found this "Other World" of ancient legend.

The very word Hesperides itself means "Land of the Golden Apples." Some have argued that this legend cannot be a reference to the Americas, since apples were not supposed to have been known in this land until the arrival of Columbus. However, the word "apples" was applied by Europeans to both apples and tomatoes, and tomatoes were originally native to the Americas. It's important to understand that the first tomatoes found by Europeans in the Americas were of a golden variety, and the Spanish referred to them as pomos de oro, or "golden apples."

A fifth century Latin attorney, Martianus Capella, stated that Hesperides was located "in the most secret recesses of the sea." In other words, way out in the deep Atlantic. This is, of course, exactly where the Americas are located.

The book *The Quest for America* tells us about another ancient Greek reference to the Americas: "*Timaeus* contains one very odd, gratuitous detail. When speaking of the [Atlantis] islanders and their empire, Plato says that their rule extended not only into Europe and Africa but also to a number of lesser Atlantic islands; and that there was a route from Atlantis, by way of these islands beyond it, to the 'opposite continent surrounding the ocean.' There, also, the Atlanteans had colonies. In other words, there was--and still is--a continental land-mass on the far side"(80).

Still another ancient Greek saga hints at a possible knowledge of a great land-mass across the Atlantic. The author was Theopompus, one of Plato's younger contemporaries. *The Quest for America* makes the following comments about this saga: "The original is lost, but a Roman writer quotes part of it. This extract purports to be a legend about Silenus, the scandalous old comrade of the god Dionysus and the Satyrs. It relates how he visited King Midas (of the golden touch)....Silenus, according to this, told Midas that beyond the known world and across the Ocean there is a continent of indefinite extent"(Ibid.).

Pausanias, another Greek, writing around 150 A.D., made this thought-provoking remark: "West of the Atlantic are a group of islands whose inhabitants are red-skinned and whose hair is like a horse"(105). Can you think of a better description of the Native inhabitants of America, or perhaps of the islands in the Caribbean?

Claudius Aelianus, in his *Varia Historia*, citing Theopompus as his source, wrote: "Europe, Asia and Libya [Africa] are islands, around which the ocean flows, and the only continent is the one surrounding the outside of this world. He [Theopompus] explained how infinitely big it is, that it supports other large animals and men twice the size of those who live here [recall the large size of skeletons exhumed from many burial mounds in the United States, as mentioned in Appendix C]. Their lives are not the same length as ours, but in fact twice as long. There are many large cities, with many styles of life, and laws in force among them are different from those customary among us....[Some of the inhabitants] live in peace and with great wealth; they obtain the fruits of the earth without the plough and oxen, and they have no need to farm and cultivate.

"The inhabitants are not less than twenty million. Sometimes they die of illness, but this is rare, since for the most part they lose their lives in battle, wounded by stones or wooden clubs (they cannot be harmed by iron). They have an abundance of gold and silver, so that to them gold is of less value than iron is to us"(209).

Here again we have an obvious allusion to the Americas and their Native inhabitants. Perhaps the only point that needs elaboration from this last quote is the mention of how these inhabitants could not be harmed by iron. This must clearly refer to the fact that Native Americans did not use iron as a weapon of war, but instead employed stones (arrowheads) and clubs (tomahawks), as this same quote indicates.

A Roman geographer, Pomponius Mela, wrote about Epeiros Occidentalis, or the "Opposite Continents." He called one of these fabled lands Alter Orbis, the "fourth part" of the world, and said that it was inhabited by a people called the Antichones.

Plutarch, in 75 A.D., thusly described the sea route to the "Western Lands" (the Americas, of course) that were customarily sailed by ancient seamen: "West from Britain lies the Isle of Ogygia [Greenland], and from there--equidistant--are three other islands to the west in the general direction of the setting sun in summertime. The natives have a story that Cronus [the god Saturn] is confined by Zeus to one of these islands which lies along the edge of the ocean. It is about 5,000 stades from Ogygia. Some areas of the sea are slow of passage--others are frozen." The inhabitants of these "Western Lands" are referred to in Greek legends as Hyperboreans, or "people dwelling beyond the north wind." This only makes sense, since travelers sailing west from England had to endure the Arctic winds near Iceland and Greenland before reaching America(78).

We could go on and on with citing such ancient accounts of a knowledge of the Americas in the ancient Greco-Roman world. But what about physical evidence of contact?

Evidence for ancient Greek contact with the Americas is rather scanty.

Nevertheless, it does exist. What may have happened is that Greek influence on American cultures probably came about more indirectly, through the Romans, who brought parts of Greek culture over with them. The Greeks may have made some Atlantic-crossing voyages themselves, but perhaps not as many as the Romans.

One apparently Greco-Roman influence that was exerted on the Americas in pre-Columbian times was the adoption in South America of a particular musical instrument that was most commonly used in ancient Greece, but perhaps brought here by the Romans. Researcher Urana Clarke informs us that "The countries at the eastern end of the Mediterranean Sea were exchanging musical ideas and instruments with Greece at least 5,000 years ago. The desire for trade carried men by land and by sea from that part of the world to places as far west as England and Ireland, as far east as India and China. In their travels they spread customs and habits so widely that it is difficult now to decide which country was the first to have them. The aulos, a pipe with double reeds like an oboe and with the piercing sound of a bagpipe, was a favorite Greek instrument. It is, however, almost exactly like pipes found in the Orient, on the island of Java, and in the South American countries of Bolivia and Peru. Its shrill, disturbing music was sometimes used for war, sometimes for wild dances and celebrations. The Scottish and Irish bagpipes were forms of the aulos"(93).

A very unique, square-shaped labyrinth design found on coins struck in ancient Crete has turned up on cliff walls in Arizona. This design is so precisely similar on both sides of the Atlantic that we are forced to conclude that the American version could not have been developed independently of the Greek version. While it is not apparent what this symbol represented, it clearly does point to a trans-Atlantic connection in the latter centuries preceding our current era.

Roman shipwrecks have been discovered off the shores of the Azores, which may have been a stopping point for the Romans, en route to the Americas(86). This is no idle speculation. For it so happens that Roman shipwrecks, or at least pieces of cargo from Roman ships, have been found off the shores of the Americas as well.

In 1971, for example, a scuba diver found two Roman-style amphorae at a depth of forty feet in Castine Bay, off the coast of Maine. Scholars identified the amphorae as Iberic Roman from the first century A.D. Further investigations found a third amphora off the coast of Jonesboro, Maine, a short time later(78).

Another find of this nature occurred in June of 1976, when a Brazilian diver, Jose Roberto Teixeira, spear fishing near the llha do Governador in Guanabara Bay, fifteen miles off the coast of Rio de Janeiro, found three large

Roman amphorae. Elizabeth Will of the department of classics at the University of Massachusetts (Amherst) identified the amphorae as having been manufactured in the Moroccan port of Zilis in the third century A.D.(82)

Roman coins have also frequently been found throughout the Americas. In Beverly, Massachusetts, for instance, eight coins were found by treasure hunters with a metal detector, back in 1977. These coins all dated between 337 and 383 A.D. In addition, whole hoards of Roman coins have been found in Tennessee, North Carolina, Georgia, Ohio, Oklahoma, and Venezuela [Endnote 63], all dating to within the first few centuries A.D.(78)

Distinctly Roman ceramic oil lamps have turned up in Alabama, Connecticut, and Peru. There have also been found bronze swords near Merida, Mexico and Bedford County, Tennessee; a Roman fibula, or clasp, in Guanabara Bay, Brazil; a bronze chalice near Roanoke River in Virginia; and a bronze Athenian medallion near Red River in Oklahoma(Ibid.).

The Roman arch has turned up in at least one location in the Americas--Cuenca, Ecuador.

Both the Greeks and Romans were known for lining the fronts of their temples and market places with sectioned columns. This same architectural style was employed by the Mayas, especially in the Mercado, or the Market of One Thousand Columns, at Chichen Itza, Mexico. On both sides of the Atlantic, the individual column sections were designed to interlock via a knob and socket mechanism in the center of each piece(Ibid.).

A large variety of distinctly Roman stone mason symbols have turned up at Comalcalco, a Mayan site near the coast of Tabasco in southeastern Mexico. Here, there is an interesting stepped pyramid made entirely of mud bricks that were sun-dried, then fire-baked to produce a durable ceramic. In the late 1970s, Neil Steede, with permission from Mexican authorities, conducted a survey of more than 4,600 of these bricks. Over 10 years later, based on this survey, Barry Fell determined that a third of them contained inscribed mason's marks that, for the most part, precisely resembled those found on bricks produced in Roman brickyards during the first half of the first millennium A.D. A total of about 50 different marks appear on the Comalcalco bricks. One of these marks is the Greek Cross, which was evidently the work of Christians. Other marks include concentric arcs, double axes, stars, crossed squares, ladders, Calvary Crosses, and Iberian letters. Could all this be written off as coincidence?

Roman soldier helmets, with their rooster-like crests on top, have been found in Mesoamerican works of art, depicted on the top of the heads of what are apparently soldiers.

In 1933, an archaeologist named Jose Garcia Payon uncovered a small terra cotta head in a burial at Calixtlahuaca, in the Toluca Valley just west of Mexico City. The burial was located underneath two undisturbed cement floors, which dated to a pre-Hispanic period sometime between 1476 and 1510 (Cortes didn't reach this area until 1519). The head, however, was clearly of Roman origin (judging from the facial features and hairstyle), dating to the time of the Severin emperors, between 193 and 235 A.D.

But what was this older artifact doing in a much younger burial? It's actually not unusual to find ancient artifacts in Aztec burials from more recent centuries. Another example of this was the discovery of a 3,000-year-old Olmec greenstone mask in a 500-year-old Aztec tomb inside the Great Temple of Mexico--Tenochtitlan.

So there's no reason to doubt that this Roman head is quite old, probably left behind by early Roman visitors and later found by the Aztecs who included it among their grave goods. Perhaps they attached some spiritual significance to it, viewing it as a creation of their early ancestors.

Burial mounds made by the Native American tribes of what is now the Eastern United States were almost exact copies of funerary earthworks made in ancient Greece and surrounding regions. Homer described the construction of a symmetrical mound over Archilles. Alexander the Great is also said to have built a great mound over his friend Hephaestion. And not very far from Greece, the Scythians did the same. Herodotus wrote thusly about the burial of a Scythian king: "[T]hey set to work to raise a vast mound above the grave, all of them vying with each other, and seeking to make it as tall as possible"(47).

It would appear that the Americas (particularly South America) also left a trace of influence upon the ancient Romans, as a result of their visits west across the Atlantic. For during excavations in the volcanic ash-buried city of Pompeii, a painted picture of what is unquestionably a pineapple was uncovered. The pineapple, as stated earlier, was originally native to South America, and was supposedly not known to the Old World until after Columbus(55).

"For at a distant date this ancient world
Will westwards stretch its bounds and then disclose
Beyond the main a vast new continent
With realms of wealth and might."
- Seneca (3 B.C. - 65 A.D.), from his play, *The Medea.*

Hebrew encounters

In several Hebrew scriptures, mention is made of "islands" on the other side of the sea. Along with these, there are several other scriptures which contain similar, yet somewhat more subtle, statements that nevertheless appear to be allusions to the Americas. Here are some typical examples:

- "...and the kings of the isles which are beyond the sea." - Jeremiah 25:22.

- "...declare it in the isles afar off..." - Jeremiah 31:10.

- "...the isles afar off..." - Isaiah 66:19.

- "...ships of the western coastlands..." - Daniel 11:30.

- "Sing unto the Lord a new song, and his praise from the end of the earth, ye that go down to the sea, and all that is therein; the isles, and the inhabitants thereof." - Isaiah 42:10.

Going back to the days of King Solomon, the nation of Israel was, at least at that time, a first class trading power. This was most likely attributable to the fact that Solomon had achieved close relations with the Phoenicians, who may very well have shared with him their knowledge of the Americas, as well as their maritime capabilities that enabled them to successfully sail such long distances as the Americas [Endnote 64]. The ancient Hebrew scriptures tell us of this close relationship, which would have allowed for the sharing of trade secrets. In the biblical book of I Kings, chapter 9, verses 26 through 28, we read: "And king Solomon made a navy of ships in Ezion-geber, which is beside Eloth, on the shore of the Red sea, in the land of Edom. And Hiram [probably Hiram III, king of Tyre] sent in the navy his servants, shipmen that had knowledge of the sea, with the servants of Solomon. And they came to Ophir, and fetched from thence gold, four hundred and twenty talents, and brought it to king Solomon." A similar statement is made in II Chronicles 8, verses 17 and 18: "Then went Solomon to Ezion-geber, and to Eloth, at the sea side in the land of Edom. And Huram sent him by the hands of his servants ships, and servants that had knowledge of the sea; and they went with the servants of Solomon to Ophir, and took thence four hundred and fifty talents of gold, and brought them to king Solomon." So prominent of a figure was Solomon in world trade at this time that II Chron. 9:22 further says of him that he "passed all the kings of the earth in riches and wisdom."

The location of the city of Ophir, mentioned above, has been a great mystery to historians and archaeologists alike. Where was this elusive city that was the source of so much gold that Solomon had imported to Israel? Some think that it may have been in Peru, an area that was very rich in gold in ancient times.

Further reasons to believe that Peru may have been the location of Ophir come to us by two map-makers who lived centuries after King Solomon.

The first was Ptolemy. He deduced that Tarshish, a stopping point on the way to Ophir, must have been situated on an island somewhere east of India, which he called Taprobana, meaning "on the seaway," or "on the main route." It is believed to be either Ceylon, Java, Sumatra, or Australia. But wherever Taprobana was, could we not conclude that its meaning, "on the seaway," or "on the main route," implied that it was a stop-off point en route to Peru?

The second map-maker, Pomponius Mele from the first century A.D., wrote an accompanying text on his map that referred to Taprobana as "the principal entry point to the Other World." Was this "Other World" a reference to South America--the source of Solomon's gold? Interestingly, Amerigo Vespucci later used this same term--"the Other World"--to refer to South America(179).

The fact that the round trip to Ophir is said to have taken three years (see I Kings 10:22 and II Chronicles 9:21) provides us with yet more evidence that it was most likely in far-away Peru.

But there is even archaeological evidence that Peru was the location of Ophir.

In December of 1989, an American explorer in the highland jungles of Peru had found evidence indicating that King Solomon's legendary gold mines may have been in that area. The explorer, Gene Savoy, said that he found three stone tablets which contained the earliest known writing discovered in the Andes region. The inscriptions, he reported, were very similar to Phoenician and Hebrew hieroglyphs. This discovery was reported by the December 7, 1989 *San Francisco Chronicle*, which stated: "The hieroglyphs on the tablets are similar to those used in King Solomon's time and include one identical to the symbol that always appeared on the ships he sent to the legendary land of Ophir, which the Bible described as the source of his gold, Savoy said"(127).

As we move forward several centuries from the time of Solomon, we see strong indications of a series of return Hebrew visits to the Americas, occurring during Roman times. It would appear that some Jews, seeking refuge from Roman invasions of Israel during times of revolt, had fled to far-away lands like the Americas in the early centuries A.D.

We will now consider some of the better pieces of evidence for such voyages to the Americas during this period.

About 35 miles south of Albuquerque, New Mexico, lies a 500-foot-high hill called Hidden Mountain. The whole area of this large hill is believed by some to have once been the location of an ancient Hebrew settlement, dating to the late second century B.C. It contains the remains of over 100 small dwellings, a large rectangle-shaped enclosure roughly 100 by 150 feet, several ancient Hebrew inscriptions, an equinox observation site, a star chart petroglyph, and, most famous of all, the Los Lunas Decalogue Stone--a 100-ton basalt stone at the foot of the hill that contains, as its name implies, the Decalogue, or the Ten Commandments, which are written in a form of ancient Hebrew that dates to about 2,000 years ago. The rectangular enclosure, built on the summit of the hill, is quite similar to Middle Eastern Bedouin enclosures that were used in ceremonies to protect tent sites and animals. The smaller dwellings, which measure 3 by 7 feet, are also of Middle Eastern design and were constructed to house only one or two people each. Besides all of this, there is another Hebrew inscription carved into a rock on the summit of a nearby hill overlooking the Decalogue Stone, which states, "Yaweh is our mighty one"(100).

Newark, Ohio, is home to a famous Hopewell Indian ceremonial center that dates back about 2,000 years. In 1860, David Wyrick, a surveyor for Licking County, Ohio, dug up a strange artifact from one of the mounds in the Newark complex. Found 14 inches below the surface, it was 4 ½ inches long and earned the name "Keystone" because of its shape. The inscriptions on all 4 sides of this stone turned out to be supplications to God, written in ancient Hebrew. These 4 inscriptions read as follows:

Qedosh Qedoshim, "Holy of Holies"
Melek Eretz, "King of the Earth"
Torath YHWH, "The Law of God"
Devor YHWH, "The Word of God"

Just 5 months after this discovery, Wyrick made another one, of a similar nature. Only ten miles south of the Newark site, he dug up a small rectangular sandstone box from another Hopewell burial mound. Inside the box was a small tablet about 5 inches long, which depicted Moses on one of its faces, with an abbreviated version of the Ten Commandments on its elongated sides. The inscriptions were in the same ancient Hebrew script as Wyrick's previous find. Both of these discoveries, as you might guess, were labeled as forgeries until seven years later, in 1867, when David M. Johnson, a local Newark banker, unearthed, in the same area as the "Moses Stone," another artifact (the Johnson-

Bradner Stone) that contained the exact same type of Hebrew script(Ibid.). Thus it looks as though all of these finds dated to the same period--about 2,000 years ago.

There have also been Hebrew coins found in North America, dating to roughly the same time-period as the above-mentioned Ohio artifacts. Referring to these coin finds, Patrick Huyghe wrote: "In Kentucky,...a number of inscribed Hebrew coins have been found dating to Bar Kokhba's rebellion against Rome, which took place between 132 and 135 A.D. The coins were dug up in 1932 in Louisville, in 1952 in Clay City, and in 1967 in Hopkinsville..."(82). Did certain Hebrews flee to North America to escape Roman oppression during this rebellion, carrying these coins with them?

Speaking of Hebrew flights from Roman oppression, there's a fascinating written record that tells of just such a flight, apparently to America, from the Roman armies that invaded Jerusalem in 70 A.D., as a result of the rebellion that began in 66 A.D. This record is found in the writings of Flavius Josephus, which reads as follows: "The Hebrews fled across the sea [the Atlantic] to a land unknown to them before"(78).

Still another indication of a Hebrew presence in North America during this same era was discovered in 1889. In that year, in Bat Creek, Tennessee, a 5-by-2-inch stone was found that was inscribed with eight Hebrew characters. This stone was unearthed by John Emmert, a field assistant who was employed at the time by the Smithsonian Institution. He discovered the stone aside two brass bracelets and what looked to be polished wooden earspools, all beneath the skull of one of nine skeletons in the burial mound he was assigned to excavate. At the time of the finding, the Hebrew origin of the inscription was not recognized. It wasn't until over fifty years later, when it was realized that the artifact was being displayed upside-down at the Smithsonian, that the Hebrew letters "LYHWD" were discerned. And then, it took until 1972 before Cyrus Gordon, a Hebrew scholar at Brandeis University, realized that these letters belonged to the Hebrew style of the Roman period. This enabled him to translate the inscription as "a comet for the Jews," which was actually a standard phrase from the period of the Bar Kokhba rebellion (cited above). This phrase was associated with a then-popular prophecy regarding a comet that would herald victory for the Jewish people(82).

The famous "Star of David" symbol has been observed at two Mayan sites, one in Uxmal, Mexico, and the other at Copan, Honduras. It needs to be pointed out, however, that this might not necessarily be indicative of Hebrew contact with the Mayan people, since the origin of this symbol can be traced back to ancient India, where it represented the tantric union of Kali and Shiva. Yet, when we consider the fact that there are several Hebrew loan-words that have been discovered in the ancient Mayan language from the LaVenta area of Mexico, the chances of this symbol in the Mayan world being evidence for Hebrew-Mayan contact in pre-Columbian times becomes all the more likely. The discovery of these loan words was made in 1871 by Jose Melgar, a Hebrew scholar.

Circumcision was practiced from the Yucatan down to Colombia, and it was usually performed on babies by a priest. However, this might not necessarily indicate a Hebrew diffusion, since the Egyptians allegedly practiced this ritual long before them(197).

Another Mexican/Hebrew parallel is the four-horned altar. Both cultures, on either side of the Atlantic, carved four "horns," one on each corner of the square-shaped top surface, and these horns were oriented in the same outward-curving direction.

Still another Mexican/Hebrew parallel is the use of phylacteries--leather pouches containing small pieces of parchment inscribed with selected quotes from sacred texts that are attached via leather straps to the wrists and forehead, used during morning prayers. In Vera Cruz, Mexico, a Mayan stele was found that portrays a bearded, rabbi-looking man with a cord wound around his forearm and palm, and then fastened around his thumb and fingers. There is also something dangling from his elbow, hidden from view, which must obviously be a phylactery box. There is even something hanging from a strand around the forehead, which may yet be another phylactery box. The whole scene is unmistakably Semitic(172).

In the seventh century B.C., it was customary among the Hebrews of Jerusalem to bury their dead with a U-shaped stone around the head. The stone was placed flat with the two ends of the U touching the shoulders. A parallel to this existed among the Mayas of Mexico, where stone yokes have been unearthed in graves, situated around the heads of the deceased in the same fashion as in the Jerusalem burials(202).

Coming back to North America, we see a whole host of cultural traits among the pre-Columbian Native tribes in this region that closely paralleled those that existed among the ancient Hebrews. Here is a small sampling of these traits:

- Branding certain animals as "unclean."
- Abstaining from eating blood.
- Belief in a single deity, or "Great Spirit."
- A religion with "high priests."
- Puberty rites.
- Holy days in the spring and fall (corresponding with Passover and Succoth).

There was also a two-day fasting period that coincided with the Day of Atonement, or Yom Kippur.
- Lunar calendar.
- Divisions of the population into "tribes"(Ibid.).

In addition to these commonalities, the Yuchi tribe, which was originally from Florida and Georgia but was later forced to migrate to Oklahoma, had many religious practices, unique among Native Americans, that date back to pre-Columbian times and positively had to originate with the ancient Hebrews. For instance, according to the Hope of Israel website, "[e]very year on the fifteenth day of the sacred month of harvest, in the fall, they [the Yuchi people] make a pilgrimage. For eight days they live in 'booths' with roofs open to the sky, covered with branches and leaves and foliage. During this festival, they dance around the sacred fire, and call upon the name of God.

"The ancient Israelites had a virtually identical custom, in many respects. In the harvest season in the fall, on the 15th day of the sacred month of harvest (the seventh month), they celebrated the 'festival of booths' for eight days. During this time they lived in temporary booths, covered with branches, leaves and fronds. This festival dated back to the time of Moses and the Exodus from ancient Egypt (Leviticus 23)"(173).

The Cherokees, from the same region as the Yuchis, talked in their old legends about a great king, or "Great One," named Yehowa. He was a man, yet a spirit--a great and glorious being. His name was never to be mentioned in common talk, and he required of his followers that they rest every seventh day.

Yehowa is said to have created the world in seven days at Nu-ta-te-qua, the first new moon of autumn, with the fruits all ripe. He made the first man from red clay, and the first woman was made from one of this first man's ribs.

The Cherokees also taught that the world was once nearly destroyed by Yehowa in a flood, and that it would again be destroyed one day by fire. They further believed in a future judgment day, sometime after death, when the good would be rewarded and the wicked would be punished.

Furthermore, Mariano Edward Rivero and John James von Tschudi, in their 1857 book *Peruvian Antiquities*, pointed out an ancient Native American/Hebrew parallel of even greater significance, which was apparently still extant in their day. They wrote: "But that which most tends to fortify the opinion as to the Hebrew origin of the American tribes is a species of ark, seemingly like that of the Old Testament [the "ark of the covenant"]; this the Indians take with them to war; it is never permitted to touch the ground, but rests upon stones or pieces of wood, it being deemed sacrilegious and unlawful to open it or look into it. The American priests scrupulously guard their sanctuary, and the High Priest carries on his breast a white shell adorned with precious stones, which recalls the Urim of the Jewish High Priest..."(Ibid.).

All of these parallels smack of contact between Hebrews and Native Americans in the remote past. There's simply no other way of explaining them.

The Celts

In 55 B.C., a Roman fleet of warships crushed the Celtic fleet off the coast of Brittany. The subsequent Roman invasion of England drove Celtic survivors to Ireland and across the Atlantic to seek out a new homeland. Later, in the fifth century A.D., Saxon invasions drove out more Celts from their native land. And in the ninth century, Norse invaders drove out still more(88). During these three periods of mass-exodus, doesn't it seem reasonable to assume, realizing what superb seamen the Celts were, that they, or at least some of them, would have been able to find their way to the friendly shores of America, to find solace there?

Along with superior seafaring skills, the Celts were also in possession of ocean vessels that were more than capable of making trans-Atlantic voyages. After successfully defeating the Celts in 55 B.C., Julius Caesar described Celtic naval vessels in his *Commentaries on the Gallic Wars*. He talked about how their single-masted ships were constructed in the shape of huge swans, and were even taller than Roman galleys. Their hulls were solidly built, able to withstand strikes by Roman battering rams, and their oxhide sails seemed immune to Roman fire arrows. As Caesar himself admitted, the only reason the Romans were able to defeat the Celts in sea battles was because they had them outnumbered, and were a bit more cunning in battle(78).

The ancient Celts are known for their folklore of great sea adventures, which collectively are called *Imrama*. These stories are many in number. But there is one such tale that seems to stand out from the rest in that it gives detailed descriptions of known places, time-distance relations of the journey, natural phenomena observed, native flora and fauna indigenous to specific far-away lands, and sensible details of navigational strategies that were undertaken on the voyage. This tale is known as the *Navigatio Sancti Brendani Abbatis*, or *The Voyage of St. Brendan the Abbot*. A best-seller during Europe's Middle Ages, this book records, as its title implies, the epic voyage of St. Brendan, along with an accompanying party of devoted monks who sailed to an interesting, unknown land on the other side of the Atlantic, in the middle of the sixth century(82).

Known more for his dedication as a faithful abbot to roughly three thousand monks, Brendan ranks as the third most popular of Irish saints, trailing just behind St. Patrick and St. Columba. But, historically speaking, he should be more well known for his sea-faring accomplishments, which, unfortunately, are afforded far too little attention by modern historians.

Paul Chapman, a navigator who ferried airplanes across the Atlantic during World War II, became fascinated with the account of Brendan's sea voyage. So fascinated, in fact, that he set out to plot the course that Brendan followed nearly fifteen centuries earlier. Drawing on the detailed navigational maneuvers mentioned in this classic work, and the descriptions of land masses encountered, Chapman believes that the highlights of Brendan's voyage included first the Faeroes, then the Azores, then on to Barbados, and finally, "a Great Land." The description given of this "Great Land" (which was also referred to as the "Promised Land"), and the duration of time that it took to reach it, make it difficult to believe that Brendan could have arrived at any other location than the eastern coast of North America(94).

==========================

A closer look at the St. Brendan voyage

According to Chapman, Brendan and his crew set sail from Ireland toward the Northwest in January, probably in the year 564 A.D. After 15 days with favorable winds, they hit a period of strenuous rowing. Then, after a duration of "forty days" (a term used frequently in the legend that probably referred to a long period of time), they spotted an island that was "exceedingly rocky and high." As they closed in on this island, they observed a "very high bank like a wall and various rivulets descending from the top." This description appears to fit Vagar, an island in the Faeroes. Not long after this, they encountered a port, and when they went ashore, they came across a dog that led them to a village.

After resting for three days in the local village, Brendan and his crew were once again sea-bound, heading west. It was the end of February when they reached an island full of large white sheep. This chunk of land was probably Sando, another island in the Faeroes. After gathering up provisions, they set sail again and soon encountered a treeless black island. As the crew began to kindle a pot to prepare a hot meal, the island suddenly began to move. It was quickly discovered, as the story goes, that what they had landed on was a whale. This part of the story would appear, quite obviously, to be a bit of fancy folklore added to give the tale a bit more flavor.

Not long afterward, this same crew came upon another island where a little river flowed into the sea. This was probably the island of Stromo in Saxon Harbor. On this island they saw the "whitest birds," most likely ptarmigans--birds which have snow-white feathers during the winter months.

By early June, Brendan and crew again set sail, seeing nothing but the sea and the sky for three months straight, which means that they must have been heading south, since they would have struck land during that time if they were heading in any other direction.

When they finally came across another island, it took them "forty days" (a long time) to find a port. This island was probably in the eastern Atlantic--most likely one of the Azores, judging by the later comments about sailing west into the Saragasso Sea. Specifically, it was probably Flores, an isolated land-mass that lacks a natural harbor. Here they met up with a group of two dozen Irish monks whose monastery was founded by St. Ailbe eighty years earlier. Brendan and his party remained there until the Christmas season had passed.

At the beginning of the new year, 565, Brendan and his company again headed westward, but they were "carried through various places." Because the Azores are in the midst of the prevailing westerlies, their vessel was blown back to the east. It wasn't until several weeks later that they spotted another island that had a natural harbor. Once ashore, the crew gathered fish from a nearby stream and water from a "lucid pool." But after drinking the water, they fell sick for a few days. This island appears to have been Sao Miguel, which harbors a pool containing mineral water that is unfit for human consumption.

Upon recovering, the crew sailed off once again by the end of February. After three days, the wind died down, which made it possible for them to finally head in the desired westerly direction. For twenty days they drifted, after which time another westerly wind carried them back toward the east. Then, after another "forty days" (again, a long time), they encountered an immense beast ejecting spume from its nostrils. Suddenly, another creature appeared to challenge the first. What they actually saw was probably a killer whale preying upon a whale of another species.

"On another day," Brendan and crew spotted a spacious and heavily-wooded island off in the distance. The only reasonable candidate here would be an island in the Caribbean, probably Barbados. One of the reasons for this assumption is the fact that the legend makes no mention of inhabitants on this island, which was also the case when the Spaniards first arrived there roughly ten centuries later. Another reason for this assumption is the fact that Martin Behaims's globe of the world, completed in 1492, pictures Barbados with the name "Isle of St. Brendan." Behaims wrote on his globe that Brendan had reached this island in 565. Also, the chart that Columbus himself used showed "Brendan's Isle" some 57 degrees west of Spain, which is fairly accurate (Barbados is actually 54 degrees west of Spain). So it would appear that brave old Christopher Columbus wasn't "sailing blind" after all.

Brendan and company remained on this island for three months, due to a storm that brought strong winds, rain, and hail. While a three-month storm seems highly unlikely, it is important to point out that Columbus, during his fourth trip to the New World in 1502, encountered a fierce storm of similar length. Speaking of this tempest, he wrote: "The storm in the heavens gave me no rest," and he said that it lasted an astonishing "88 days."

After the storm had finally subsided, Brendan and his crew set sail "toward the northern zone," a significant shift in direction. Later, "on a certain day," they saw an island off in the distance with a flat surface and no trees--a clear reference to Barbuda, the only island in the area that matches this description.

Soon they set sail once again. Not long into this leg of their journey, "a very large bird" dropped on their ship a branch with red grapes. Later, when they came ashore on another nearby island, they discovered that it was covered with the same type of grapes. The bird was probably a flamingo, and the grapes were clearly sea grapes, which grow in clusters in sandy soil, having a purplish-red color. This island appears to have been Great Inagua, the most southerly of the Bahama Islands. Brendan and his companions remained on this island for "forty days," after which time they departed with as many grapes as their ship could hold.

Next, guided by the winds and currents, this brave band of seafarers headed northwest along the Bahamas. And after celebrating the Feast of St. Peter, on June 29, Brendan discovered "a clear sea" through which he was able to see a multitude of large beautiful fish (a likely reference to a coral reef, such as the Great Bahama Bank, extending for more than 340 miles and containing a wealth of large tropical fish). This would have meant that Brendan's ship had gotten caught in the Gulf Stream, which flows north along the eastern seaboard of the United States. From there it flows northeast to Newfoundland, and finally to northern Europe, from whence the Brendan crew began their journey.

One version of the Brendan legend tells of how, after leaving the Bahamas and heading north up the Gulf Stream, Brendan and company disembarked in the "Promised Land," or "Great Land," which was sunny and warm and abounding in fruit. After 40 days of exploration, they reached an uncrossable river. And because the land seemed to stretch indefinitely beyond this river, they gave up attempts to find its limits (did they reach the Mississippi River while cutting across this "Promised Land," or "Great Land"?).

From this point, Brendan's party sailed to a location north of Newfoundland and south of Greenland. The record states that "on a certain day" they encountered "a column in the sea," which was probably an iceberg. Brendan said that, upon approaching it, he could not see its summit because it "was covered with a strange curtain." This was most likely an allusion to a mist that had formed around it, due to a warm air current that moved in. Brendan said of this icy column (iceberg) that it was "harder than marble" and that it was like "a very bright crystal."

Next, the crew sailed north-northeast for eight days. Brendan talked of how they soon encountered an island and, in fancied form, he described what appears to have been a volcanic eruption that they had witnessed. This would mean that they probably came upon the submarine ridge off the Reykianes Peninsula, on Iceland's southeastern corner.

Heading further north, Brendan spotted a high mountain in the ocean that was "very smoky at its top." Perhaps this was Oraefajokull, the largest of three volcanoes on the south coast of Iceland, which is known to have been active in the sixth century. Tragically, the legend states that one of the monks accompanying Brendan had gone ashore to explore the black sands at the base of the volcano, and was consumed by a hot bed of volcanic ash.

As they continued sailing on, a strong wind carried them southward. Ten days later they happened upon an island that was "small and exceedingly round," seemingly a rock sticking out of the sea. This description fits Rockall--a large stone in the middle of the ocean about five hundred miles southeast of Iceland. In a cave at the top of this giant rock, Brendan found an Irish monk named Paul who told him that he sailed there in just seven days. Based on this information, Brendan was able to find his way back to Ireland, reaching there at the end of August, in the year 565(82).

========================

The legendary evidence for pre-Columbian Celtic contact with America seems conclusive enough. But what about the archaeological evidence?

Throughout New England, the Ohio Valley, Virginia, and parts of upstate New York, numerous stone monuments have been found that date to long before the time of Columbus. These structures very closely resemble the dolmens built by the Celts in Ireland and elsewhere in Western Europe, over one thousand years ago. Probably used as some type of burial marker, these lithic creations were made of huge, crudely-chiseled slabs or boulders suspended slightly above the ground by three to four smaller stone pedestals.

Other structures, found mostly in New England and New York, resemble the above-ground stone chambers built by the Celts, primarily in Ireland, Scotland, and Portugal. Some of the American versions have often produced, upon excavation, stones containing clearly-delineated Ogham script(87)[Endnote 65]. Many of these chamber constructions are also astronomically aligned, just like their European counterparts. They are built of dry stone walls (not held together by mortar), capped with flat stone slabs for their roofs, and then covered over with dirt so that the whole edifice looks like a small hill, having an opening on one side. All of these features are held in common with the European versions. And as far as their function goes, they seem to have been used as either ceremonial centers or burial chambers, or possibly both.

Stone circles, like Stonehenge in England, are not unique to western Europe. They have been found all over the Americas as well. Those found in the northeastern United States are particularly reminiscent of the European variety, although they're not built on such a large scale. They consist of elongated stones that are erected vertically and arranged, of course, in a circle. They are also, like those in Europe, aligned with the cardinal points, and were most likely used, as those in Europe were used, as calendars, to help determine the change of the seasons for the precision timing of crop-planting and annual religious observances(78).

Cairns are another type of lithic construction found in the Eastern United States and Western Europe. They are orderly-stacked piles of rocks that served, like the above-ground stone chambers described a few moments ago, as either burial markers or some type of ceremonial construction, or maybe both.

Other stone creations held in common between North America (particularly in New England) and Western Europe are the fertility cult monuments.

One example of this is the phallic menhir. These stones, whether in Europe or the U.S., are usually found alone, in a standing position, and probably had no astronomical significance whatsoever. They most likely served only as objects of religious veneration.

Another example is the Men-a-tol, or female stone. These are somewhat flat, circular stones that stand erect, bearing a large hole in their centers. As with the phallic stones, they appear to have only served as religious icons.

Many other signs of a Celtic presence in North America have been unearthed over the years--mirrors, swords, metal helmets, dishes, steel bows, mummies with auburn hair, a fourteenth-century glass bottle, an optical lens, and even a statue of a Catholic Madonna with Child. Such artifacts have been found from the Midwest to the Carolinas, as well as in northern New England(Ibid.).

Human remains have also been found, in various places throughout North America, that are from people who appear to be of Celtic descent. One good example of this is a series of finds from Holliston Mills, Tennessee. In an ancient burial site dating to the first millennium B.C., skulls were uncovered that were identified as being from definite Western European (as well as Asian) stock.

At a nearby site (Snapp's Bridge), Celtic and Basque writings have also been found, which help to clinch the case that the Holliston Mills skulls are indeed of Celtic origin.

Another group of skulls of obvious Celtic lineage was found in a mound outside of Boston. However, this find was dated way back to several thousand years before the Tennessee remains, and probably represents early ancestors of the Celts(30).

A popular motif in Celtic art was the spiral. Often numerous spirals were carved on the same stone, closely tiled together. We find this same exact design in the Americas, especially among the Anasazi Indians.

Viking voyagers

One would have to be entirely ignorant of the history of the Norse to not believe that they were capable of reaching the Americas long before Columbus. The Vikings sailed in streamlined boats modeled after old Phoenician galleys. Known as *skeilds*, or "longboats," they had pointed ends, shallow keels, and square-shaped sails. These single-masted ships were made of oak, with overlapping planks either riveted together or fastened with iron clinch nails. Furthermore, these vessels were an average of 80 feet long and carried anywhere from 40 to 60 men. The longest of them on record, reported in the *sagas*, was King Canute's battleship, which was over 260 feet from stem to stern. But as seaworthy as these boats were, they were not the vessel of choice for long-distance voyages. The Vikings most favored a ship called the *knorr* for such trips. This boat was more sturdy and had high freeboards for easier transport of large cargoes and farm animals on the high seas. It was common for vessels of this type to reach lengths of 50 to 100 feet(78).

From the eighth century up to the twelfth, the Norse were the uncontested masters of the Atlantic. With help from bands of seaborne raiders from Denmark and Sweden, the Norse were able to conquer Ireland and the Faeroe Islands before the end of the eighth century. From there they established extensive settlements in Iceland, and had invaded England and France in the ninth century. By the end of the tenth century, they had colonized Greenland and ventured as far as the Caspian Sea. Not long after this, they crossed the desert to Baghdad, and then, in the eleventh century, they rounded Spain, fought in the Mediterranean, and completely swarmed Sicily(82).

The Norse were not only brave (and often ruthless) warriors, but bold and adventurous explorers as well. They were constantly in search of new lands to expand their territory, establishing colonies as they went along. But not all Norse seafarers had such self-seeking (and often barbaric) ambitions in mind. Some went in search of new lands simply for missionary purposes, hoping to win over converts to Catholicism. Whatever their motive, however, the Norse during these centuries (the eighth to the twelfth) had spread out all across the Atlantic, and thanks to findings such as those in L'Anse aux Meadows, Newfoundland (dated to about 1000 A.D.), it is now impossible for even the most staunch skeptic to deny any longer the pre-Columbian presence of Norsemen in the Americas. But are these Newfoundland findings the only proof of a pre-Columbian Viking presence in the western Atlantic?

As they traveled, the Norse made crude geographic sketches of the lands they encountered. Centuries later, several Europeans made maps of the North

Atlantic that were based on these centuries-old Norse charts. One such map-maker was a Dutch historian, Christian Friseo, who copied an old document found in an Iceland archive. In a captivating notation on the bottom of this map, Friseo wrote: "Greenland and nearby regions toward the North and West from an ancient map drawn in a crude manner many hundred years ago in Iceland in which they were then known lands." This map shows nine main land masses:

England
Norway
Rifland (Russia)
Iceland
Frisland (an unknown island, perhaps no longer existent)
Greenland
Baffin Island (Helleland)
Labrador (Markland)
Newfoundland (Promontorium Winlandia)

So we have here yet more confirmation that the Norse unquestionably reached America long before Columbus(78).

Who hasn't heard of the famous story of Leif Erikson (son of Eric the Red) and his travels to "Vinland"--the Land of the Grape Vine--in the latter part of the tenth century? Interestingly, the layout of the land described in Erikson's story corresponds remarkably well with the coastline of northern Maine, which, incidentally, is an area rich in wild grapes. Erikson, in fact, is believed by many to have been the one who established the L'Anse aux Meadows settlement in Newfoundland, discussed above(1).

The Norse sagas relate how, in the year 982, an Icelandic Christian missionary monk named Ari Marson, while sailing the Atlantic, was blown off course by a tempest and wound up in an unknown land. Many see in this story a correlation with the Mayan legend of a bearded white visitor, Kukulcan, also known to the Aztecs as Qetzacoatl, who came to Mexico from beyond the sea of the rising sun (the Atlantic) around the year 1000, and who was later deified by both the Mayan and Aztec cultures.

This visitor is said by the Aztecs to have taught them how to use metals, and that he had spoken out against human sacrifice--things that we would expect from a Norse missionary.

It is also said of Qetzacoatl that he visited a nearby area known as Tula, which is very close to the old name for Scandinavia, Tule. Is this where Tula got its name?

Another interesting parallel here is that the Norse had decorated their ships with feathered dragons, or feathered serpents. And perhaps not so coincidentally, the name Qetzacoatl means "feathered serpent."

Also, Qetzacoatl was later represented by the Aztecs as a white-robed man with crosses on his sleeves. This sounds just like a Catholic missionary.

Furthermore, he was occasionally depicted as a bearded white man with a hood over his head, much like a monk.

Additionally, he was sometimes portrayed as having red hair, which would even further confirm that we are dealing with a Norse presence in pre-Columbian Mexico(82).

The ancient Peruvians had their equivalent of Qetzacoatl/Kukulcan--Viracocha--who was likewise depicted as a white man with a beard [Endnote 66]. Another representation of him shows him wearing large earrings with crosses on them. Is this the same Norse Christian missionary that visited Mexico?

========================

Christianity in pre-Columbian America

There were many fascinating and precise parallels that existed between the beliefs, symbols, and practices of Christians (more specifically, Catholics) and those found among the Native peoples of the Americas, particularly in Mexico. Indeed, there were so many of them that we must rule out independent development. Although there were some modifications in the New World versions, their Old World origin, via the Norse, can still plainly be discerned.

Let's now turn our attention to some of these parallels.

The Cross

When the Spanish first arrived in Mexico, they found that there was a tremendous reverence for the symbol of the cross, and that many of the beliefs surrounding it were amazingly reminiscent of Christianity.

Following Columbus' expedition was one led by Francisco Hernandez de Cordova, who focused on exploring Mexico. After sailing around the Yucatan to Campeche, he and his crew met a friendly band of Natives who led them to their town. Writing of this incident, de Cordova stated: "There were other idols there that carried the sign of the cross, all painted so that we admired them as something never before seen or heard about."

A year later, another Spaniard, Grijalva, led an expedition to an island called Ulua, not far from Cozumel. Juan Diaz, the chaplain of this mission, wrote: "They [the inhabitants] worship a large white marble cross with a gold crown on top, and they say the one who died on it is more magnificent and resplendent than the sun."

One of Cortes' captains, Andres de Tapia, wrote this about Cozumel: "We found in front of [an] idol, at the foot of [a] tower, a cross made of lime that was an estado-and-a-half high [9 feet]....In this city they had a principal god who at one time had been a man. They called him Qetzacoatl....[He] is supposed to have worn a white vesture like a monk's tunic, and over it a mantle covered with red crosses."

Crosses, in fact, were found everywhere the Spaniards went, all the way down to Peru. And they were made, not only of stone, but often of wood as well, and were held in very high religious esteem by the Natives. Also commonly found were Maltese cross symbols in various locations around Mexico.

Baptism

Sahagun, an early Spanish priest who worked in Mexico in the sixteenth century, said that he "perceived inexplicable resemblances between [Catholicism] and the Aztec practices of baptism by water." Babies were anointed with water shortly after birth, in the presence of invited friends and family, after which time a meal was presented by the parents of the child for all the guests.

As the baby was anointed, the officiator would address a female deity, referring to her as "Our Lady," asking that the child be cleansed from all filthiness that he/she inherited from his/her parents.

Next, the forehead of the infant was touched with water, and a new name was given to it--a baptismal name, just as is done in the Catholic faith.

Another prayer was offered up at this time, asking that sin, "which was with us before the beginning of the world, might not visit the child."

In the Yucatan, the word for baptism was caput-sihil, which means "to be reborn," or "born anew." This, of course, harkens back to the Christian concept of being "born again."

Any adults who were thus anointed, not having been anointed as babies, were to live a new life from that point on, clean from their former evil ways--the equivalent of repentance in the Christian religion.

Baptism was also practiced, sometimes by full immersion, both in Peru and North America (among the Hopi, particularly). Baptismal fonts were even constructed in ancient Peru that look just like those made in Asia Minor in the early centuries A.D.

Communion

Near the Yucatan, the Natives made amaranth-seed dough in the shape of bones. When these bone-shaped rolls were eaten, it was said that the devotee was eating the flesh and bones of their chief deity, Huitzilopochtli. These dough bones were blessed, of course, before they were eaten, and the ceremony was considered a "sacrifice."

The Peruvians had their version of this sacrament as well, mixing the dough with the blood of a freshly-sacrificed sheep.

Lent

Prior to their feast of the god of war, the pre-Columbian Mexicans were accustomed to observing a 40-day fast. Usually these fasts involved abstaining from meat. Interestingly enough, some cultures in Mexico fasted every Friday as well.

The Peruvians also observed a seasonal fast involving abstaining from meats, except that their version was observed during the fall equinox, instead of the spring.

As an interesting side note, during this fast, blood of young children was placed above the lintel of every door of each person's home that partook in the fast. This appears to have originated with the biblical story of the Israelites putting blood above the doorposts in Egypt, during the tenth plague (the death of the firstborn).

Parallels with prayer

While in prayer, the pre-Columbian Mexicans often addressed God as "Creator," "Lord of all Lords," and "King of all the Universe." They also frequently summoned him as "Father."

Confession

Confessions of sins were made to priests, or to the chief deity, at which time the penitent sinners were usually told to make restitution to the ones offended by their misdeeds. If something was stolen, for example, the one confessing was instructed to return the item to its rightful owner.

There was also a policy of confidentiality on the part of the priest--that he would not reveal anything he heard during a confession.

It was believed that nothing could be hid from God (or the gods)--that even one's thoughts were open to his (or their) scrutiny, and thus even evil thoughts needed to be confessed to obtain full pardon. Forgiveness was considered important in order to escape future punishment by the chief deity.

It was also believed that each person, endowed with the freedom of choice, was responsible before God (or the gods) for his or her own evil actions, and that when forgiveness was granted, the slate was wiped clean, as though the evil was never committed.

Confession frequently involved a scapegoat, usually an elderly woman. After sins were confessed in her presence, she was killed by the priest, and with her death came the death of the confessed sins, along with the atonement for them.

Among the acts deemed to be sinful, and thus warranting a confession, were: murder, abortion, adultery, fornication, sodomy, drunkenness, and theft, some of which called for the death penalty, usually in the form of stoning.

Penance

Often a quite painful penance was assigned after confession. Sometimes the penitent sinner was told to pierce his earlobe or his tongue, and to pass a twig there-through, to "make up" for his sins. Other times the penance was much more humane and even beneficial, involving helping the poor or the sick.

Divine law

The pre-Columbian Mexicans (namely the Aztecs) had their own version of the ten commandments (or in this case, eight). After performing a sacrifice, the priest would recite the following list of commandments, as a reminder to the people of how they were supposed to live:

- You shall fear, honor, and love the gods.
- You shall not use the names of the gods on your tongue or in your talk, at any time.
- You shall honor the feast days.
- You shall honor your father and mother, your kinsmen, priest, and elders.
- You shall not kill.
- You shall not commit adultery.
- You shall not steal.

- You shall not bear false witness.

Blood sacrifice

Although blood sacrifice is not a Christian practice, the specifics we will look at are so closely aligned with what is found in the Old Testament, that it is likely that they were adopted from exposure to this text through contact with Christian missionaries, like so many other religious beliefs and practices we've been looking at. Of course, the same may have resulted from Hebrew contacts, which we discussed earlier. So in this case it's hard to tell for sure which of these two origins we're dealing with. But either way, trans-oceanic contact is clearly implied.

When the ancient people of Israel strayed from Jehovah and adopted the practices of the Canaanites, one of the things they did was to sacrifice their own sons and daughters to the god Moloch. The ancient Mexicans sacrificed their children as well.

Animals, of course, were also sacrificed in Mexico, as in Israel. And there were many striking parallels in how these sacrifices were carried out in both these regions.

Very often, the Mexican sacrificial victim was chosen because it was "without blemish." It had to be a spotless sacrifice, just as in ancient Israel.

The Mexican sacrificial rituals also involved the use of incense, which was likewise common among the Hebrews, as prescribed in the Old Testament.

In ancient Mexico, as in ancient Israel, quail were brought to the priest who would twist the head, breaking the neck. From there, the corpse was tossed at the foot of the altar.

The pre-Columbian Mexicans also sacrificed lambs, lamas, rams, ewes, guinea pigs, pigeons, and various other birds. These, of course, were the very types of animals offered in ancient Israel.

Also offered in Mexico were the first fruits of the field--again, just like in Old Testament Israel.

The Trinity

The Mexicans worshiped a trinity of deities, known as Topa, Topiltzin, and Yomoletl, translated as Father, Son, and Heart of Both. They were each worshiped separately, yet recognized as a single unit.

Interestingly, the name of the deity that represents the Son in the Mexican trinity means "he who was flayed and ill-treated."

The pre-Columbian people of Peru worshiped a deity known as Tangatanga, which they described as being one in three and three in one.

<u>The virgin birth</u>

What appears to be an echo of the Christian concept of the virgin birth of Christ is found with the Mexican deity Uitzilopochtli, of whom it was said that "no one [no man] appeared as his father."

<u>Life after death</u>

The early Mexicans believed in an existence beyond the grave--a good place where the upright went, and a bad place, known as Metnal, where the evil doers went. Here dwelt a malevolent deity (the devil), who was the chief of a whole host of lesser evil deities (demons).

The Peruvians also believed in an afterlife. To them, man was comprised both of a body and a soul. The soul, they said, outlived the body, and would go to either a good or bad place, depending on how the person lived. Yet they also believed in a future resurrection of the body, which is why they practiced mummification. In addition, they believed in a future judgment, when all would have to stand before the chief deity to give an account of their deeds performed in the earthly life.

<u>Adam and Eve</u>

In Cholula, Mexico, the goddess Cioacoatl is said to have bequeathed the sufferings of childbirth to all women. She is also reckoned to have been the one through whom sin came into the world. Additionally, she was often represented with a serpent by her side, and her name signified "the serpent woman."

As far as Adam goes, he is said, in the Bible, to have been formed from the dust of the ground. Compare this with the following Incan prayer for reception of light and knowledge of deity:

Oh if I might know!
Oh if it could be revealed!
Thou who made me out of earth,
And of clay formed me.
Oh look upon me,
Who art thou, Oh Creator?

Coming back to ancient American Eve parallels, in the Mayan *Popol Vuh*; the first woman was created as the first man slept, and this same woman is mentioned as having reflected on the possibility of death should she eat a wondrous fruit.

Parting of the Red Sea

The Spaniards recorded an old Mexican tradition of a great man who, after suffering many hardships on behalf of his countrymen, gathered the multitude of his followers and persuaded them to flee to a land where they could live in peace.

Having made himself leader of the people, he went to the seashore and moved the water with a rod that he carried in his hand. Then the sea opened, and he and his followers went through.

The enemies, seeing this opening, followed after them. But the waters returned to their place, and the pursuers were never heard from again.

The Mayan *Popol Vuh* states that the ancestors "passed through as though there were no sea...the water divided itself...and they were able to walk across."

The Flood

We have already seen that nearly every ancient culture around the world, including Native American civilizations, had their own versions of a global flood legend. But the details of the Aztec version very closely match those found in the Old Testament.

The Aztecs claim that the flood came as a result of mankind's bad behavior, as a punishment from God, or the gods. Only one man, Tapi, along with his wife, had survived this disaster. They survived because Tapi built a boat, and he stocked it with two of every kind of animal.

After the flood rains had stopped, a vulture was sent out in search of land, but it did not return because it fed on the bodies of the dead giants floating in the waters.

Next, a humming bird was sent off, which returned with a twig in its mouth, signifying that land was nearby.

In Peru, the flood legend is accompanied by a belief that the rainbow serves as an assurance that the world will never again be destroyed by a flood(201).

Babel

We already took this issue up earlier--how that the concept of man having once spoken a single language was found, not just among the Hebrews, but many cultures around the world, including Native American peoples. We will now quote again from the Native American sources cited earlier.

The Mayan *Popol Vuh* states: "Those who gazed at the rising of the sun [the ancients]...had but one language....This occurred after they had arrived at Tulan, before going West. Here the language of the tribes was changed. Their speech became different. All that they had heard and understood when departing from Tulan had become incomprehensible to them....For the tongue[s]...had already become different....Alas, alas, we have abandoned our speech! Why did we do this?...Our language was one when we departed from Tulan, one in the country where we were born"(18).

The Navajo tribe says that in long ages past, all men "spoke one tongue," but that soon after the great flood there "came many languages"(17).

And now we will look at another relevant quote from a source not previously cited. This quote comes from a Native Mexican prince by the name of Ixtililxochitl, from Texcoco, who wrote a history of his people just prior to the arrival of the Spaniards. Once they arrived, they translated this work and found that it contained this amazing recollection of what sounds like nothing else but the story of the Tower of Babel: "The men made a very high and strong...tower, to protect themselves in it when the second world would be destroyed. At the best time their languages were confounded, and not understanding one another, they went away to different parts of the world"(201).

========================

In further reference to Qetzacoatl/Kukulcan/Viracocha, we actually have a reliable time-frame for the arrival and departure of this pre-Columbian visitor. According to Mayan records, he arrived in Mexico at the end of Katun 4-ahau (somewhere between March 12, 968 and November 25, 987 A.D.). He must have left at a ripe old age, because the time of his departure, the Aztec date of 1-acatl, was 1051 A.D., give or take a few years. This was the very period that Leif Erikson and his companions were sailing their trans-Atlantic exploration and missionary voyages.

The departure of Qetzacoatl/Kukulcan/Viracocha occurred simultaneously with a major celestial event that has been definitively dated to the same time mentioned above. This event was the appearance of a wondrous spectacle in the heavens. Miguel de Quetzalmazatzin, recording the history of the Mexican people in 1670, wrote that, when Qetzacoatl departed, “a star was seen smoking against the sky.” Another Spanish chronicler, Domingo de San Anton Munon Chimalpahin Quauhtlehuanitzin, stated that “one star smoked above...which astonished the Tultecas.”

This “smoking star” was not a comet, for comets only appear in the sky, visible to the naked eye, for around six months maximum. Yet this smoking star remained in the sky for a much longer duration than that. Quetzalmazatzin said that this object was visible for eleven years. He wrote: “eleven years above smoked one star in the sky.”

So what was this object? It had to be an exploded star--a supernova. And it just so happens that there was a supernova--a major one--that exploded in the year 1054, the remnant of which is believed to be the Crab Nebula.

In both China and Japan, this same spectacle was witnessed and recorded. To-to, in his *Sung Shi* (History of the Sung Dynasty), wrote: “In the first year of the period *Chih-ho* (1054), the fifth moon, the day *Chi-chou* [July 4] [a guest-star] appeared approximately several [inches] south-east of Tien-quan [Zeta Tauri]. After more than a year it gradually became invisible.

“On the day *Shin-wai* [of the third moon of the first year of the period *Chia-yu*, or April 17, 1056] the Chief of the Astronomical Bureau reported that from the fifth moon of the first year of the period *Chih-ho* [June 9 to July 8, 1054] a guest-star had appeared in the morning in the eastern heavens, remaining in the Tien-quan [Zeta Tauri], which has only now become invisible.”

Another Chinese document, the *Sung-hui*, proclaims: “On the twenty-second day of the seventh moon of the first year of the period *Chih-ho* [August 27, 1054] Yang Wai te said: ‘...I have observed the appearance of the guest-star; on the star there was slightly an iridescent yellow color....

“‘Originally, this star had become visible in the fifth moon of the first year of the period *Chih-ho* [June 9 to July 8, 1054] in the eastern heavens in Tien-quan [Zeta Tauri]; it was visible by day, like Venus; pointed rays shot out brilliant on all sides; the color was reddish-white.’”

We now turn to the Japanese account, as recorded in the *Mai Getsuki*: “In the middle ten-day period of the fourth moon of the second year of the period Ten-Ki [May 20-30, 1054] and thereafter, between [one and three a.m.], a guest-star appeared in the orbit of Orion; it was visible in the eastern heavens. It shone like a comet in Tien-quan [again, Zeta Tauri] and was as large as Jupiter”(205).

There is an obvious time-length discrepancy here--the Mexican account of this event said that it lasted eleven years, whereas the Chinese chronicler, To-to, said the time was just over a year or so. This may be due to an error on the part of the Mexican record, since the spectacle that was being recalled had occurred centuries earlier. But nevertheless, both of these accounts doubtlessly point back to the selfsame event, which clearly remained visible for far too long to have been a comet. It was, instead, unquestionably the supernova explosion ("guest star") of 1054. Thus we can see that Qetzacoatl/Kukulcan/Viracocha was most likely Leif Erikson, or one of his companions, who came to Mexico in the late tenth-century and left in the mid-eleventh, at the very time of the appearance of this "guest star."

There are actually what are believed to be two petroglyphic representations of this supernova explosion of 1054 A.D., both of which are located in Chaco Canyon. One shows the star as it appeared when it first exploded, exhibiting radial lines jetting out from the center. The other one must have been drawn about a year or so later, as the ejected material, still visible to the naked eye, began to form wispy filaments that look like crab legs (hence the name Crab Nebula). This particular glyph, in fact, looks just like modern photographs of this interstellar cloud of dust and gas.

Let us continue on with our discussion of physical, archaeological evidence for a Norse presence in the ancient Americas.

In the state of Maine, bronze axe-heads and other tools of an obvious Norse origin (some of which contain runes that give the owners' names) have been found at several locations(78).

At Rocky Neck, near Gloucester, Massachusetts, a bronze battle-axe was found that now resides in the Goodwin Collection, Wadsworth Atheneum, Hartford, Connecticut. It bears a Tifinagh inscription which reveals that it was a royal award given to battle veterans and widows of warriors who died in combat(30).

Other, similar bronze axe-heads have been found all over the Americas, from Nova Scotia to Cozumel, Mexico. While we must not rule out the possibility of forgeries here, there have been far too many reports of findings like these, in many cases by highly-credible researchers, to not take them seriously, or at least most of them.

Traces in North America of a Norse presence before Columbus are actually more common than one might think. Gunnar Thompson tells us that "Between 1946 and 1950, civil engineer Arlington Mallery and his associates excavated numerous Nordic sites from Virginia to Newfoundland. Mallery found remains of 20 Nordic iron-smelting furnaces along the Ohio Valley, and he located scores of Nordic habitation sites along the St. Lawrence river. He identified 14 Norse sites on Newfoundland and many others along the shores of adjacent Sop's Island. An associate, James Howe, found remains of 16 Norse iron smelting furnaces in Virginia's Roanoke Valley and collected 400 pounds of miscellaneous pieces of worked iron. Mallery identified Nordic iron furnaces on hilltop fortifications and furnace mounds built in accordance with northern European designs. The mounds provided a draft for furnaces by raising them above surrounding trees. Evidence of ancient furnaces included stone draft channels, slag from peat-bog iron, and pieces of solid iron 'blooms' which formed inside the ancient smelters. Mallery noted the variety of artifacts at Nordic sites: 'Easily identified among the items were many Viking-type tools, spikes and rivets, scribers for marking wood, caulking tools used in building Viking ships, chisels and axes, boat spikes and boat rivets. The chisels and axes were formed by welding together thin sheets of iron by cladding. The rivets were duplicates of rivets found in a Viking ship, the *Oseberg*, which was discovered in 1903 under a mound on the shores of Oslofjord, Norway'"(78).

If the Norse had truly come to North America in pre-Columbian times, we should be able to find some Norse cultural traits present in ancient Native societies of this region. And indeed, we do find such things. One clear example is in the realm of housing design and construction of the Iroquois people, as William McNeil, author of *Visitors to Ancient America*, explains: "[T]he Iroquois...built longhouses similar to those of the Vikings....Both types of longhouses had a single entrance on one side, windows set high in the walls along the long side of the building, and oval roofs to shed the rain and snow. The Indian villages...were even arranged in an orderly fashion like European villages. Houses were built side by side along wide avenues. Gardens were arranged around the outskirts of the village and a town square was located at one end of the village. The similarity between the structures of the Indians and the structures of the Vikings is no longer coincidental. The discovery of L'Anse aux Meadows proved that social contact existed between the daring Scandinavians and the American aborigines in pre-Columbian times"(100).

Other evidences of Norse/Native North American contact before Columbus were observed by early post-Columbian explorers of the Americas, who noticed that certain northern Native tribes exhibited physical (and even linguistic) characteristics that were reminiscent of the Norse. Thompson explains: "European explorers and settlers [who came to the Americas after Columbus] reported numerous encounters with light-skinned natives....French Governor Sieur DeRoberval described the Iroquois as 'very white, but they paint themselves for fear of heat and sunburning.' In 1604, French explorer Samuel DeChamplain encountered fair-skinned natives living in eastern Canada. Because of their physical features, DeChamplain believed they were of Nordic heritage. In 1698, French Jesuit Pierre Charleviox saw bearded, blond-haired natives on Labrador. In 1779, American explorer George Rogers Clark encountered light-skinned, blond-haired, and blue-eyed warriors. Colonists reported that the Lenni Lenape (or Delaware tribe) were most like Europeans in speech and appearance"(78).

Whoever these light-skinned, bearded, and blond-haired people were, they certainly weren't Native Americans! They absolutely must have been descendants of Europeans who came here long before Columbus set sail.

Norse visitations to North America appear to go back several millennia--to the Bronze Age--and probably had played a major part in an ancient international copper trade. There have been about 5,000 ancient copper mines found around the Great Lakes area (mostly near Lake Superior), which are believed by many to have been involved in this trade. Such mining operations began perhaps as early as the 5th millennium B.C. It's estimated that a total of at least 20 million pounds of copper were removed from the area. Where could it all have gone? Native Americans almost never made use of this metal, so it must have been shipped off elsewhere. And there can be no better assumption made than that it went to Bronze Age Europe, where this commodity was in great demand.

Fortunately, we don't just have to rely on assumption. There is tangible evidence to make a European connection here, as Schoch explains: "In the Mediterranean area, copper was often melted into an ingot shaped like a cured ox hide, called 'reels' by archaeologists because they look like the home-made handles boys use to reel in kite string. Ingots of much the same shape have been found in North American burial mounds. It does seem curious that people separated by thousands of miles of water would have developed the same shape for transporting and storing metal, a shape that has no particular functional value. The reel-shaped ingot is one of those arbitrary similarities that suggests contact"(1). Schoch then went on to describe how copper axe-heads found throughout Europe and the western Great Lakes region are strikingly identical, being of the exact same design.

But how does all of this tie in specifically with the Norse? Schoch proceeds to talk about "the Peterborough stone, a major petroglyph site located about 100 miles northeast of Toronto, near the town of Peterborough in Ontario." He then describes how the petroglyphs on this stone "show a large ship drawn in a style common in Scandinavia, as well as a series of signs that the epigrapher Barry Fell identified as a Norse inscription written in an alphabet from North Africa...." Schoch further explains how this inscription most closely matches a form of writing ancestral to the Libyan and Tifinagh alphabets, referred to as Proto-Tifinagh, and how Fell dated the inscriptions to the second millennium B.C.

"But why," Schoch continues, "...were these early Norsemen writing their own tongue in a North African alphabet? The likely answer has to do with the remarkable ways cultures spread, change, and borrow. North Africa is part of the extensive Middle Eastern-Berber culture that spans land and sea from Mesopotamia to Morocco. Many of these peoples were accomplished navigators and traders, as were the Norse. Trade no doubt brought Berbers and Norse together, and in the course of that contact they learned something of each other's languages. But the Norsemen didn't have a written language. Proto-Tifinagh gave the unlettered Norse the ability not only to record their own language but to produce records intelligible to their Mediterranean trading partners. The Norse took the alphabet to their fjords and then over the Atlantic to the New World, where some unknown artist carved the petroglyphs of Peterborough. In its time and way, Old Norse in Proto-Tifinagh is no stranger than modern Yiddish, a German dialect written in the Hebrew alphabet, or Maltese, the only Arabic tongue written in the Latin alphabet"(Ibid.).

Barry Fell's translation of the Peterborough inscriptions tells the story of a man named Woden-lithi who journeyed to the area in his ship, "Gungnir," and that the purpose of his visit was "for ingot-copper of excellent quality"(82). Was this Norseman a Bronze Age trader? Fell's translation certainly seems to indicate this. Though some have criticized his translation, claiming that the inscriptions are of Algonquin origin, such criticisms are unwarranted. As Patrick Huyghe wrote: "[Barry] Fell [was] well aware that many of the inscriptions at the [Peterborough] site are the work of later Algonquin artists attempting to imitate what the Scandinavians had originally cut into the limestone. But the central Sun god and Moon goddess figures and certain astronomical signs are clearly not Algonquin"(Ibid.).

Though Fell certainly has had his share of detractors, his work was endorsed, for the most part, by a very well-known and respected epigraphic expert, David H. Kelley, formerly with the University of Alberta, who is best known for his 1976 work, *Deciphering the Maya Script*. Kelley gave his endorsement in an article he wrote for the spring 1990 edition of *The Review of Archaeology*. Though he disagreed with Fell on certain aspects, he stated that his translation of the Peterborough inscriptions was "essentially correct"(Ibid.).

It's possible that the Celts had also worked the ancient Michigan copper mines, perhaps at a later date. This assumption is based on the fact that the miners who last operated there left their hammer stone tools behind in these mines, which precisely match, in every detail, hammer stones found in a copper mine in Munster, Ireland(110).

* * * * * * *

As you can see, the similarities between Old and New World cultures are far too numerous, and far too precise, to be written off as mere coincidences. But the information contained herein is only the tip of the iceberg; the intention here was simply to present a few highlights. There is so much proof of pre-Columbian contact between the Old World and the New, in fact, that the notion of Columbus being the "discoverer" of the Americas is frankly quite laughable.

"We are rewriting the textbooks on the First Americans. The peopling of the Americas was never as simple as simple-minded paradigms said. Instead it will tell of an America that beckoned to far-flung people long before the Mayflower or the Santa Maria or the Viking ships [came here]....It is very clear to me that we are looking at multiple migrations through a very long time period--migrations of many different peoples of many different ethnic origins." - Dennis Stanford, president of the anthropology department at the Smithsonian Institution, *Newsweek*, April 26, 1999.

Conclusion

Throughout this study, it has been my aim to avoid providing too much detail on any subject covered. Only general concepts and accompanying minimal amounts of supportive evidence have been presented, in order to hopefully inspire the reader to rethink, and consequently restudy, the early ages of Earth and human history for his or her own self. It also would not have been very practical to delve too heavily into the minute details of every subject covered, since this would have required a volume of much greater size.

No doubt, most will scoff at many of the claims made in this work. But in turn, I also scoff at many of their conclusions. Yet this is not a contest of who is right or wrong. It has more to do with WHAT is right or wrong. If nothing else, I hope that each reader will at least have learned to look more critically at the "official" view of what has been presented to us as "absolute historical/scientific fact." For, if we are ever to find the truth out about any matter, we have to pursue it for ourselves, not being restrained by any preconceived notions, regardless how much endorsement the "authorities" may afford them.

One of the most redeeming aspects of this entire multi-disciplinary study is its ability to explain a whole host of phenomena in one broad sweep--something that mainstream academia has miserably failed at. Let us undertake a brief review of some of these phenomena that our overall thesis provides sound and sensible answers for:

- Existence of the asteroid belt
- Grand-scale cratering and crustal breeching of all solid-bodied planets, moons, asteroids, and comets in the solar system
- Origin and structure of meteorites
- Orbital and axial eccentricities of planetary, satellite, and other celestial bodies throughout our solar family
- Higher percentage of oxygen in the early Earth's atmosphere
- Tropical fossils in arctic regions
- Perfectly-preserved frozen mammoths that show signs of having frozen very quickly in what was then a tropical environment
- Polystrate fossils, such as petrified tree trunks, that stand straight up through dozens of feet of strata that supposedly took millions of years to accumulate
- Existence of hundreds of global-scale flood legends from almost every ancient culture around the world
- Problem of almost a complete lack of silt and sedimentary deposits, as well as impact craters, on the world's ocean floors

- Mass extinctions of dinosaurs and many other species of animals and plants alike
- Similarity of the shapes of the continents and the mid-oceanic ridges
- Mountain range and canyon formation
- Development of the Earth's major igneous, sedimentary, and metamorphic rock formations
- Enigma of "primitive cavemen"
- Sudden emergence of advanced civilizations in the earliest stages of the global-scale archaeological record
- Comprehensive knowledge of world geography in the long-forgotten past
- Megalithic stone calendars and the origin of Sun worship and astrology
- Indications of advanced technological developments during the earliest phases of man's history
- Evidence of man and dinosaurs coexisting
- Similarities between pre-Columbian American cultures and those of other parts of the ancient world

While our thesis accounts for all the above phenomena and many others related thereto, mainstream scholarship has offered us nothing but a confusing, disconnected array of explanations for these same phenomena, which leave far too many questions, and ignore far too much evidence to the contrary. It is therefore high time that we rethink some of the most basic, underlying concepts in the disciplines of geology, astronomy, meteoritics, archaeology, anthropology, ancient world history, paleontology, and a whole host of other disciplines. Doing so might not win us a Nobel Prize, but it will certainly free our minds so that we can develop a better understanding of our history, and that of our planet and its sister worlds.

Appendix A

Flora and fauna that reveal ancient physical connections between trans-oceanic land masses

Certain unique species of flora and fauna are found on land masses that are now separated by vast expanses of ocean water. This point, coupled with the fact that these land masses are shaped like interlocking puzzle pieces, strongly support the position that these same land masses were once connected together.

First let us look at the fauna.

In the Canary Islands off the northwest coast of Africa live a class of earthworms known as oligochete. This type of worm is also found in southern Europe, indicating that these areas must have once been joined.

The monk seal lives today on the east coast of South America, the west coast of Africa, the western Mediterranean, and, until the early twentieth century, the West Indies.

A unique species of freshwater spongefish, *Heteromeyenia ryderi*, is found only in lakes and rivers along the east coast of North America and the west coasts of Ireland and Scotland.

The seal population at Lake Baikal in Siberia is the exact same species of seal that inhabited the Caspian Sea, 2,000 miles west, up until the nineteenth century.

About 3,000 miles southeast of Africa lies Kerguelen Island, which is home to several species of freshwater fish that are also found in New Zealand (over 6,000 miles to the east) and South America (about 6,500 miles to the west).

This same island, along with Heard Island (400 miles south), are both home to a rare wingless fly. Even if these flies did have wings, there's no way that they could have flown the long distance between these two islands they now inhabit. Thus it is clear that these same two water-separated bodies of land must have been part of a single land mass in ancient times.

There are literally thousands of isolated islands spread across the vast expanse of the Pacific Ocean. On just about every one of these islands can be found freshwater mollusks (*Clausiliacea*). Does this mean that these islands, too, were once connected, and later drifted apart to their current locations, as the Earth expanded? Or, better yet, could it be that they are the tips of former mountains that existed on a large land mass that either sank or was inundated by a rise in sea level? [Endnote 67]

And now for the flora.

There are many species of freshwater weeds, such as *Najas flexilis*, that are indigenous to North America and the British Isles.

The Eugenia tree is endemic to Mauritius (off the coast of Madagascar), Marion Island (just north of the limits of the Antarctic icebergs in the southern Indian Ocean), and the Solomon Islands in the distant Pacific. The distances that separate these islands make it impossible for seeds of this tree to have been transported by wind. Nor could they have been carried by ocean currents, since they do not survive for long while immersed in salt water. And, finally, birds could not have been the mechanism for the distribution of these trees either, because their seeds are not used by any bird species for food.

Kerguelen Island, mentioned above, is also home to a unique type of cabbage that is likewise common to the McDonald group of islands (of which Heard Island, also mentioned above, is one). Kerguelen Island also has several types of flora in common with New Zealand (4,000 miles east) and South Georgia Island (nearly 5,000 miles west).

There are many plant species that are endemic to South America and southern Africa, which are also found on the volcanic island Tristan de Cunha, located halfway between these two areas, in the middle of the Atlantic(17).

As would be expected, we find the same type of trans-oceanic biological parallels in the fossil record as well.

For example, Eurypterid (or "sea scorpion") fossils are found in upstate New York and Scotland, right across the Atlantic.

Another example is the Mesosaur, an extinct lizard-like creature whose fossils are found in the bulge of Brazil and the corresponding dent in southwestern Africa, directly across the Atlantic.

Even the rock layers in these very regions of southwestern Africa and Brazil contain parallels--having the exact same sequential arrangements. First there's a layer of sediment that is overlaid by coal beds that contain fossils of the same species of flora--Glossopteris (a seed fern). Next there's a layer of desert deposits that is followed by a final layer of lava. Clearly, these areas must have once been connected.

Appendix B

Possible secondary tilts of the Earth's axis in the post-flood era

Though the major tilting of the Earth took place at the time of the flood-producing asteroid impact, there appears to have been several more tilts that occurred during the ensuing centuries and millennia, obviously less dramatic, but still inflicting significant climatic changes globally. These shifts probably occurred as a result of additional, but less destructive, impacts by asteroids or comets (or else gravitational tugs by them as they passed by, which pulled the Earth out of alignment)[Endnote 68].

This concept of multiple shifts in the Earth's tilt can be supported on a number of grounds. One possible piece of evidence, mentioned earlier, is the variant magnetometer readings obtained in India's Deccan plateau region, which showed that this area's magnetic orientation had once been inclined 64 degrees south, then 60, then 26, and finally, a major shift occurred, flip-flopping this area's inclination 17 degrees north. We saw the same thing with England, which once had a magnetic inclination of 30 degrees, whereas it is now 65 degrees.

Other evidences include indications of dramatic global climatic changes in just the past few thousand years, long after the flood catastrophe. The proceeding quote from Richard Mooney illustrates several examples of such changes: "Rock paintings in the southern regions of the Sahara show a great number of species of animals--antelope, giraffe, and others--which now live much further south. Paintings and artifacts in the ancient urban center of Catal Huyuk in Anatolian Turkey show that the now desolate plains below the Taurus Mountains were once grassy savannahs occupied by huge herds of ungulates. The description in the Old Testament of the Middle East lands as 'flowering with milk and honey' could not possibly refer to recent historical times. The now arid regions of the coastal strip of Peru and Bolivia must once have been very different. Deserts could not have supported the extensive cities with great urban populations whose ruins have been uncovered there. Mayan legends describe the Yucatan as a land of the 'honey and the deer,' yet much of the interior of the Yucatan today is uninhabited and uninhabitable. A little over two thousand years ago, North Africa was the granary of Europe, a well-watered, fertile land bordering on the Mediterranean. Vast wheat fields and dozens of Roman towns and cities lay in this region. The ruins of these cities lie buried under shifting desert sands today. The Gobi Desert also exhibits traces of once flourishing flora and fauna, all of which have now vanished"(5). (Shortly we will be dealing with much more evidence of major and most likely impact-induced climatic changes in recent millennia.)

A further piece of evidence of a post-flood Earth axis tilt comes from ancient Egypt. The Egyptian solar temple of Amen-Ra at Karnak, built around 2000 B.C., was originally constructed so that the Sun would shine directly down its long corridor once a year at sunset, on the day of the summer solstice, and would illuminate the pharaoh who stood at the far end of the temple. Today, however, the Sun no longer shines down this temple's corridor on the summer solstice. Thus the Earth's axis was obviously reoriented sometime after this temple was erected, and in a manner far greater than what can be accounted for by the relatively insignificant precessionary wobble of the Earth's axis(14).

Of course, these evidences, or at least some of them, may simply be indicative of reorientations of the continental land masses as the Earth expanded, which could have occurred either in addition to, or instead of, post-flood tilts of the Earth's axis. However, there definitely were multiple post-flood Earth axis reorientations, as we will soon see demonstrated by a wealth of incontrovertible evidence. In fact, we can name one of them with certainty at this present juncture--the one that brought about a secondary post-flood Ice Age (or perhaps had kicked into higher gear an already-existing Ice Age that began at the time of the flood catastrophe). Remember that Ice Age fossils are not mineralized, but are deposited on top of strata that *are* mineralized, which were clearly laid down during the initial flood disaster. Thus Ice Age fossils (for the most part, anyway) obviously date to the post-flood era. And since Ice Age fossils show undeniable signs of having frozen suddenly, in an instant, we can probably safely say that the Earth axis reorientation that froze these fossils was instigated by an asteroidal impact.

We can also safely deduce that there was a major asteroidal impact that occurred well into the flood catastrophe, perhaps a few weeks or months after the initial impact that spawned the flood in the first place (and possibly resulting in a secondary Earth axis tilt). We know that an impact event occurred at this time because evidence of it can be found sandwiched in between sedimentary layers that were laid down at the time of the flood. This evidence is known as the K-T Boundary clay layer--a layer that has been found in over 80 locations around the world, situated between the Cretaceous and Tertiary strata of sedimentary formations. This layer stands out in stark contrast to the surrounding strata, and it is very high in iridium, which is extremely rare on Earth but relatively common in meteorites. The event that globally deposited this thin layer (it averages about an inch in thickness) must have been significant enough to knock the Earth off its axis, just like the earlier impact that spawned the flood and the later impact that kicked off the post-flood Ice Age.

Another safe conclusion we can draw is that there were several less severe post-flood impacts, provable by the fact that the craters they made are in pristine condition, and thus could not have been made before or during the flood, which would have either severely weathered, destroyed, or buried them. We have already discussed a couple examples of these--one in Antarctica and the other off the coast of the Yucatan. The latter must especially be post-flood, since it is located on the ocean floor, which didn't even exist (or at least wasn't exposed to oncoming debris) until after the flood, when the Earth began to expand.

A further good example of a clearly post-flood undersea crater (given the name "Burckle") was discovered in the middle of the Indian Ocean in 2005, which is 18 miles in diameter and lies 12,500 feet below the ocean's surface. Speaking of this crater, along with several others and the disasters they caused, the November 14, 2006 *New York Times* reported: "At the southern end of Madagascar lie four enormous wedge-shaped sediment deposits, called chevrons, that are composed of material from the ocean floor. Each covers twice the area of Manhattan with sediment as deep as the Chrysler Building is high.

"On close inspection, the chevron deposits contain deep ocean microfossils that are fused with a medley of metals typically formed by cosmic impacts. And all of them point in the same direction--toward the middle of the Indian Ocean where a newly discovered crater, 18 miles in diameter, lies 12,500 feet below the surface.

"The explanation is obvious to some scientists. A large asteroid or comet, the kind that could kill a quarter of the world's population, smashed into the Indian Ocean 4,800 years ago, producing a tsunami at least 600 feet high....The wave carried the huge deposits of sediment to [the] land....

"This year the group [of scientists studying these chevrons] started using Google Earth, a free source of satellite images, to search around the globe for chevrons, which they interpret as evidence of past giant tsunamis. Scores of such sites have turned up in Australia, Africa, Europe and the United States, including the Hudson River Valley and Long Island.

"When the chevrons all point in the same direction to open water, Dallas Abbott, an adjunct research scientist at Lamont-Doherty Earth Observatory in Palisades, N.Y., uses a different satellite technology to look for oceanic craters. With increasing frequency, she finds them, including an especially large one dating back 4,800 years.

"...Surveys show that as many as 185 large asteroids or comets hit the Earth in the far distant past, although most of the craters are on land....[Endnote 69]

"Deposits from mega-tsunamis contain unusual rocks with marine oyster shells, which cannot be explained by wind erosion, storm waves, volcanoes or other natural processes...

"'We're not talking about any tsunami you've ever seen,' Dr. [Ted] Bryant said [who is a geomorphologist at the University of Wollongong in New South Wales, Australia]. '...No tsunami in the modern world could have made these features. End-of-the-world movies do not capture the size of these waves. Submarine landslides can cause major tsunamis, but they are localized. These are deposited along whole coastlines.'

"For example, Dr. Bryant identified two chevrons found over four miles inland near Carpentaria in north central Australia. Both point north. When Dr. Abbott visited a year ago, he asked her to find the craters.

"...Within 24 hours of searching the shallow water north of the two chevrons, Dr. Abbott found two craters [obviously post-flood].

"Not all depressions in the ocean are impact craters, Dr. Abbott said. They can be sink holes, faults or remnant volcanoes. A check is needed. So she obtained samples from deep sea sediment cores taken in the area by the Australian Geological Survey.

"The cores contain melted rocks and magnetic spheres with fractures and textures characteristic of a cosmic impact. 'The rock was pulverized, like it was hit with a hammer,' Dr. Abbott said. 'We found diatoms fused to tektites,' a glassy substance formed by meteors. The molten glass and shattered rocks could not be produced by anything other than an impact, she said.

"'We think these two craters are 1,200 years old,' Dr. Abbott said. The chevrons are well preserved and date to about the same time [indeed, post-flood].

"Dr. Abbott and her colleagues have located chevrons in the Caribbean, Scotland, Vietnam and North Korea, and several in the North Sea.

"Hither Hills State Park on Long Island has a chevron whose front edge points to a crater in Long Island Sound, Dr. Abbott said. There is another, very faint chevron in Connecticut, and it points in a different direction....

"But Madagascar provides the smoking gun for geologically recent impacts. In August, Dr. Abbott, Dr. Bryant and Slava Gusiakov, from the Novosibirsk Tsunami Laboratory in Russia, visited the four huge chevrons [there] to scoop up samples.

"Last month, Dee Breger, director of microscopy at Drexel University in Philadelphia, looked at the samples under a scanning electron microscope and found benthic foraminifera, tiny fossils from the ocean floor, sprinkled throughout. Her close-ups revealed splashes of iron, nickel and chrome fused to the fossils.

"When a chondritic meteor, the most common kind, vaporizes upon impact in the ocean, those three metals are formed in the same relative proportions as seen in the microfossils, Dr. Abbott said.

"Ms. Breger said the microfossils appear to have melded with the condensing metals as both were lofted up out of the sea and carried long distances.

"About 900 miles southeast from the Madagascar chevrons, in deep ocean, is Burckle crater, which Dr. Abbott discovered last year. Although its sediments have not been directly sampled, cores from the area contain high levels of nickel and magnetic components associated with impact ejecta.

"Burckle crater has not been dated, but Dr. Abbott estimates that it is 4,500 to 5,000 years old..."(157).

========================

Fossil evidence of post-flood catastrophic, impact-induced tsunamis

Around the world, in coastal areas and sometimes further inland, there are found enormous fossil bone deposits in caves and fissures. These fossils are obviously of a relatively recent, post-flood age, since they are not in a petrified state. The bones are encased in tightly-compacted sand or silt, and are often broken or smashed as by a violent force. They are also arranged in a chaotic, jumbled fashion, indicating that they, along with their encasing earthy matrix, were carried and dumped in their current locations by exceptionally-powerful tsunamis.

What happened is that titanic waves swept up great masses of soil and living creatures as they rushed over the land. Then, as these waves were drawn back out to sea, many of the bones and much of the silt got caught in rock clefts, and were compacted and crushed by the force of the retreating waters.

When we realize the tremendous Earth thrusts that must have been involved to produce the tsunamis that created the deposits we will now be looking at, the conclusion will be forced upon us that the only mechanism that can account for them is impacts of the very nature that we just read about, in the above-cited article.

- The first example we'll look at is near Plymouth, England, and the surrounding areas where great numbers of silt-encased fossil bones have been discovered wedged within rock fissures near the coast. Immanuel Velikovsky, in his book *Earth In Upheaval*, gives us this description of these deposits: "In the neighborhood of Plymouth on the [English] Channel, clefts of various widths in the limestone formations are filled with rock fragments, angular and sharp, and with bones of animals--mammoth, hippopotamus, rhinoceros, horse, polar bear, bison. The bones are broken into innumerable fragments. No skeleton is found entire. The separate bones, in fact, have been disbursed in the most irregular manner, and without any bearing to their relative position in the skeleton. Neither do they show wear, nor have they been gnawed by beasts of prey, though they occur with the bones of hyaena, wolf, bear and lion. In other places in Devonshire and also in Pembrokeshire in Wales, ossiferous breccia or conglomerates of broken bones and stones in fissures in limestone consist of angular rock fragments and broken and splintered bones with sharp fractured edges in a fresh state, and in splendid condition, showing no traces of gnawing.

"If the crevices were pitfalls into which the animals fell alive, then some of the skeletons would have been preserved entire. But this is never the case. Again, if left for a time exposed in the fissures, the bones would be variously weathered, which they are not. Nor would the mere fall have been sufficient to have caused the extensive breakage the bones have undergone: these, I consider, are fatal objections to this explanation, and none other has since been offered..."(159).

- Looking to France, we find the same type of deposits, presumably from the same period. Velikovsky elaborates: "On the Mediterranean coast of France there are numerous clefts in the rocks crammed to overflowing with animal bones. Marcel de Seres wrote in his survey of the Montagne de Pedemar in the Department of Gard: 'It is within this limited area that the strange phenomenon has happened of the accumulation of a large quantity of bones of diverse animals in hollows or fissures.' De Seres found the bones all broken into fragments, but neither gnawed nor rolled. No coprolites (hardened animal feces) were found, indicating that the dead beasts had not lived in these hollows or fissures"(Ibid.).

- Velikovsky then goes on to talk about similar finds on the Rock of Gibraltar: "The Rock of Gibraltar is intersected by numerous crevices filled with bones. The bones are broken and splintered [again this familiar pattern is repeated]. The remains of panther, lynx, caffir-cat, hyaena, wolf, bear, rhinoceros, horse, wild boar, red deer, fallow deer, ibex, ox, hare...have been found in these ossiferous fissures."

Again we have a situation here of broken and smashed bones that show no signs of having been picked at by predators. This, along with their random arrangements, make it overwhelmingly clear that they were gathered and dumped in these heaps by powerful tsunamic wave actions. What else could account for such conditions?

Human remains and artifacts have also been found among the bone deposits on the rock of Gibraltar. As Velikovsky wrote: "A human molar and some flints worked by Paleolithic man, as well as broken pieces of pottery of Neolithic man, were discovered among the animal bones in some of the crevices of the Rock"(Ibid.).

- Moving on to Sicily, we find yet more of these same kinds of deposits in a cave of Maccagnone. A good description of these finds is recorded in *Paleontological Memoirs and Notes* by Dr. Hugh Falconer: "[T]he cavern had been filled right to the roof, the uppermost layer consisting of a concrete of shells, bone-splinters, with burnt clay, flint-chips, bits of charcoal and hyaena [remains], which was cemented to the roof by stalagmatic infiltration...of contemporaneous origin....A great physical alteration...occurred to the conditions previously existing, emptying out the whole of the loose, incoherent contents, and leaving only...portions agglutinated to the roof"(160).

What we have here is a situation where a cave was filled with bone deposits by a tsunami, and then was cleared out at a later time by yet another tsunami.

- Malta has its own version of caves like this, containing animal and a great many human bones as well. In one of them, according to Graham Hancock, "No complete skeletons came to light, and the bones lay in confusion through the soil...except that occasionally an arm with fingers, and a complete foot, and several vertebrae would be found lying with the parts *in situ*....[U]nrelated bones and also implements were found in the interior of skulls....Animal bones were found mingled with human." It's estimated that this cave network contains the chaotically-scattered remains of 33,000 individual humans(208).

- Arabia is no exception either. The authors of *When the Earth Nearly Died* inform us that "shells and bones of elk, bison, and other 'northern' animals occur profusely in breccias occupying caves and fissures near Beirut, Lebanon, while remains of late Pleistocene elephants, woolly rhinoceros and *Bubalus* (the Indian water-buffalo), mixed with those of 'southern' animals, have been found in the cave of Ksar'Akil, Syria"(17).

These fossils, no surprise, are found in the same jumbled and fractured condition as those that have been uncovered at all the other sites we have been looking at.

- Coming to the United States, we encounter the same thing. Velikovsky wrote:

"In 1912 near Cumberland, Maryland, workmen cutting the way for a railroad with dynamite and steam shovel came upon a cavern or a closed fissure with a peculiar assemblage of animals. Many of the species are comparable to forms now living in the vicinity of the cave; but others are distinctly northern or Boreal in their affinities, and some are related to species peculiar to the southern, or Lower Austral, region"(159).

Would you be shocked to discover that these bones, too, were noted as being positioned haphazardly, and in a condition of fragmentation?

- We will now look at another United States discovery of this sort. This one is most amazing for two reasons--a) it contains human remains, and b) it's located far inland, yet it still exhibits the same jumbled and violently-deposited characteristics that we see in all these other related sites. In *When the Earth Nearly Died*, we read: "Bishop's Cap Cave in the Organ Mountains, New Mexico,...contained scattered human bones chaotically deposited with those of a ground-sloth (*Nothrotherium*), a coyote, a very large wolf (*Aenocyon*), numerous rodents, a horse, bison, an antelope, a camel (*Tanupolama*) and other Pleistocene animals."

The authors of this same book then go on to mention another cave containing similar remains. This cave is also located far inland--Gypsum Cave, near Las Vegas, Nevada(17).

For these tsunami-laid fossil deposits to be situated so far from any shore, there must have been a significant post-flood tilting of the Earth's axis to send ocean waters hurtling such great distances across the North American Continent.

- Hancock informs us that "Ice Age marine features are present along the Gulf Coast east of the Mississippi River, in some places at altitudes that may exceed 200 feet. In bogs covering glacial deposits in Michigan, skeletons of two whales were discovered. In Georgia marine deposits occur at altitudes of 160 feet, and in northern Florida at altitudes of at least 240 feet. In Texas...the remains of Ice Age land mammals are found in marine deposits. Another marine deposit, containing walrus, seals and at least five genera of whales, overlies the seaboard of the north-eastern states and the Arctic coast of Canada. In many areas along the Pacific coast of North America Ice Age marine deposits extend more than 200 miles inland. The bones of a whale have been found north of Lake Ontario, about 440 feet above sea level, a skeleton of another whale in Vermont, more than 500 feet above sea level, and another in the Montreal-Quebec area about 600 feet above sea level"(20).

- There's one final example from the United States that is worthy of our attention here--the La Brea tar pits near Los Angeles, California. Speaking of this fossil-rich formation, Velikovsky wrote: "The animal remains are crowded together in the asphalt pit in an unbelievable agglomeration. In the first excavation carried on by the University of California a bed of bones was uncovered in which the number of saber-tooth and wolf skulls together averaged twenty per cubic yard. No fewer than seven hundred skulls of the saber-toothed tiger have been recovered.

"Among other animals unearthed in this pit were bison, horses, camels, sloths, mammoths, mastodons, and also birds, including peacocks."

As is so typical of finds of this nature, the bones found at La Brea were often "broken, mashed, contorted, and mixed in a most heterogeneous mass, such as could never have resulted from the chance trapping and burial of a few stragglers."

Human remains have also been found in the La Brea pits, mixed in chaotically with all the other bones.

La Brea is not the only such bituminous deposit found in California. Others have been uncovered in Carpinteria and McKittrick(159).

- Moving on to China, Velikovsky stated: "In the village of Choukoutien, near Peking in northern China, in caverns and in fissures in rocks, a great mass of animal bones was found....These rich ossiferous deposits occur in association with human skeletal remains....

"...Mammoths and buffaloes and ostriches and arctic animals left their teeth, horns, claws, and bones in one great melange..."(Ibid.).

- Turning to India, we encounter still more fossil finds of this sort, and far inland at that. Velikovsky explains: "The Siwalik Hills are the foothills of the Himalayas, north of Delhi; they extend for several hundred miles and are 2,000 to 3,000 feet high. In the nineteenth century their unusually rich fossil beds drew the attention of scientists. Animal bones of species and genera, living and extinct, were found there in most amazing profusion. Some of the animals looked as though nature had conducted an abortive experiment with them and had discarded the species as not fit for life." Among the types of creatures found here were "The hippopotamus,...pigs, rhinoceroses, apes [and] oxen"(Ibid.).

Here again we must be dealing with a significant tilting of the Earth's crust to send ocean waters so violently far inland. Perhaps the same tilting that left these deposits is the one that left those in New Mexico and Nevada, which we discussed a few moments ago.

- Australia has its share of such finds as well. The authors of *When the Earth Nearly Died* quote from a New South Wales mining report which states that these very kinds of broken, anarchic deposits can be found "in caverns...at Wellington; at Boree; near the head of the Colo river; at Yesseba, on the Macleay river; at the head of the Coodradigbee; not far from the head of the Bogan, and in other places"(17).

- South America is not left out of the equation either. For instance, authors Allan and Delair tell us that "Numerous limestone caverns scattered across the Brazilian states of Sao Paulo, Minas Geraes, Goyaz and Bahia have yielded rich harvests of organic remains comparable [to others we have looked at]. Of these, the extensive series of caves and deep fissures between the Rio das Velhas and the Rio Paraopeba in Minas Geraes state have yielded evidence of much importance. Staggering quantities of bones filled these apertures, but usually in such broken condition and so promiscuously that complete skeletons were extreme rarities. Near Caxoeira do Campo, for instance, is a cave some 120 feet long, 6-9 feet wide and 30-40 feet high. From this [excavators] procured half a cubic foot of red earth packed with minute bones. Of all the half under-jaws detectable in it, 400 belonged to small opossums and the remainder to no less than 2,000 different kinds of mice, bats, porcupines and small birds"(Ibid.).

With such a wide distribution of fossil arrangements of this type (and we have only looked at but a few examples), the conclusion is inescapable that we are dealing with oceanic turbulence on a monumental scale. Surely these deposits must have been the result of multiple post-flood impacts, perhaps occurring over many centuries, or even millennia.

==========================

There are multitudinous historical records (and additional physical evidences) of specific catastrophic, tsunami-producing events that transpired over the millennia following the flood, which may correspond with some of the impact sites mentioned in the *New York Times* article quoted above. These events were accompanied by either earthquakes or severe weather and climatic anomalies, obviously the result of Earth axis tilts. We will now review some of the fascinating historical records and additional physical evidences of some of these specific events, along with the approximate dates they are believed to have taken place:

- In or around the year 4400 B.C., according to ice-core and tree-ring data, it appears that a calamitous impact had occurred on Earth. Ice-core samples from Greenland show that there was a great build-up of ice around that time, and tree-rings in oak trees from a bog in Ireland reveal very thin and closely-spaced growth rings from the same period, further indicating a severe, temperature-dropping climatic change that stunted growth of the trees.

Further evidence of this crisis comes from ancient Egypt. According to Egyptian texts from this year, or thereabouts, a great comet appeared over the region. Perhaps a piece of this celestial object, or the object itself, impacted the Earth at this time. Also, according to John Anthony West in his book *Serpents in the Sky*, the ancient Egyptian calendar was established around 4240 B.C. Did their calendar need revising at this time, because of the Earth axis tilt that occurred in conjunction with the disaster of this era?

A new civilization arose in conjunction with this catastrophe--the Badari culture, along the Nile River.

Additionally, there was a mass migration of nomadic peoples to Europe from north of the Black Sea at this time, quite clearly in retreat from a sudden onset of incessant frigid-cold weather that was brought on by the cometary impact of this era.

As we saw earlier, the main flood catastrophe served as a catalyst for the development (or redevelopment) of many of the world's great civilizations in very ancient times. But as we are now seeing (and will continue to see as we progress), the same was true with all the lesser post-flood disasters--they spawned the development (or redevelopment) of many civilizations across the globe, all throughout the centuries and millennia following the initial flood calamity.

It also needs to be pointed out here that the destruction of the legendary city of Atlantis (and/or Mu and Lemuria) may have occurred during one of the post-flood disasters we are now discussing, and not during the flood catastrophe itself.

- Around 3150 B.C.--again, according to ice-cores and tree-rings--another temperature-dropping celestial catastrophe seems to have taken place. Some scientists, based on computer calculations and simulations, believe that the asteroid Olijato had passed dangerously close to the Earth at this time. Perhaps this asteroid was once much larger, and that the Earth's powerful tidal forces caused it to break up when it passed too close, resulting in a large piece of it striking our home planet.

As was the case with the 4400 B.C. event being the starting date for the

Egyptian calendar, we find the same thing having occurred with the Mayas, regarding this catastrophic date. The starting point of the Mayan calendar, 3114 B.C., corresponds roughly with the date of this apparent catastrophe.

Meanwhile, in Egypt, things seemed to be going quite well. Pharaoh Narmer had united Upper and Lower Egypt, ending the predynastic era. This perhaps indicates a prosperous and population-surging period for Egypt, since people most likely mass-migrated into the area during this time, to escape more inclement weather to the north.

It was also just after this time, in 3102 B.C., that the Kali yuga (more on yugas shortly) had begun, marking the starting date of the Hindu calendar.

Shortly thereafter, in 3079 B.C., the ancient Vietnamese nation of Van Lang was established by the first Hùng Vuong.

At roughly this same time, the first pottery came into use in Colombia, indicating migration of people to this region to establish a civilization there, clearly in wake of the catastrophe of this period.

- 2345 B.C. appears to be our next major catastrophe date, initially arrived at through the same ice-core and tree-ring data that yielded the above dates, as well as those that follow.

Coinciding with this date were several civilizational declines that occurred over a five-century period between 2500 and 2000 B.C. (the Old Bronze Age). These civilizations included the peoples of the Levant, Anatolia, Greece, Egypt's Old Kingdom, the Akkadian Empire in Mesopotamia, the Helmand civilization in Afghanistan, and the Harappan civilization of the Indus Valley in Pakistan.

These cyclical rises and falls of civilizations, in sync with recurring catastrophes, would help explain why ancient technology, as discussed elsewhere in this work, had also followed this fluctuating pattern. Such periodic catastrophes had often prevented even the memory of technological advancements, as well as the memory of history in general, from being transmitted to succeeding generations. Philo of Alexandria expressed this point vividly, when he wrote: "By reason of the constant and repeated destructions by water and fire, the later generations did not receive from the former the memory of the order and sequence of events"(61). Speaking in reference to ancient Greece specifically, Josephus made this similar statement: "[T]he region about Greece has been invaded by thousands of destructive plagues [catastrophes], which blotted out the memory of past events: and as they were always setting up new modes of life,...each of them supposed that their own was the beginning of all"(110). And finally, the Mayan *Popol Vuh* states that the "First Men [ancient ancestors] who succeeded in seeing, succeeding in knowing all that there is in the world" had lost everything at the hand of the "gods." These gods, apparently in punishment for misdeeds, sent a calamity (or calamities), and thus "all the wisdom and all the knowledge of the First Men, [together with their memory of their] origin and their beginning, were destroyed"(20).

These oscillating periods of advancement and decline, fair weather and severe, are likely the reason why many ancient cultures divided up their history into different ages.

Two cultures that did this, in particular, were the Mayas and Aztecs, which called their successive ages "Suns." Could they have meant by this the changing path of the Sun in the sky that resulted from each Earth tilt that followed every major celestial impact?

Another culture that did this was the ancient Hindus. According to Vedic tradition, the divisions of their history were called Yugas, as already mentioned, and they were delineated by periods of ascent and decline of their civilization. Interestingly enough, these same periods were believed to coincide with celestial events(141).

Let us come back to our discussion of the catastrophe of 2345 B.C.

Not only did this catastrophic event spawn colder temperatures in certain regions, but other regions that had formerly been fertile had suddenly become dry deserts, such as the plains of the Dead Sea, northern India, and the Sahara (which grew dramatically in size at this time--recall the quote cited at the beginning of this appendix, which talked of Saharan rock paintings depicting animals that now live far south of the Sahara). Tremendous flooding in China was also reported during this period, contemporaneous with the reign of Emperor Yao (or Yahou), which began in the year 2357 B.C.

There were other regions affected at this time. The area south of the Sahara, for example, had become subtropical, and North America became wetter and cooler. Mexico, on the other hand, went the opposite direction and turned drier, as did southern Europe. There is also evidence, from during this same era, of changes in ocean currents and disruptions of the Earth's crust, both on a global scale.

The ancient Irish Annals speak of this time thusly: "Nine thousand...died in one week. Ireland was thirty years waste"(174).

There are several craters that have been discovered worldwide, such as a two-mile-wide one in the Al'Amarah region of southern Iraq and several others in Argentina, all dating to this same era, which many scientists believe were collectively the cause of these global climatic and tectonic changes.

An Akkadian stele, dating to the reign of Naram-Sin (around 2300 B.C.--roughly the time of this catastrophe), portrays an outdoor scene with two "suns" hovering overhead. Could this extra "sun" be the cometary or asteroidal object that struck the Earth during this period?

- 1628 B.C. is the next date we arrive at for a major, worldwide catastrophe. Associated with this date are several unusual incidents from the historical record that are perhaps consequential to the disaster of this period.

One of these incidents was the suspension of the "Mandate of Heaven" in the Xia Dynasty of China, which was the equivalent of medieval Europe's "Diving Right of Kings" dogma, which held that kings had divine authority to rule the masses in pretty much any way they pleased. What brought China's version of this to a close at this time may have been spawned by severe displeasure with the then-reigning emperor, because he did not prevent the catastrophe that had then struck. Or perhaps they saw the catastrophe as a sign that "the gods" were displeased with the emperor's performance, and were thus punishing him, and the whole rest of the nation, with "fire from heaven."

Another contemporaneous incident of this nature from the historical record is found inscribed on the Babylonian Venus Tablets. They contain astronomical data from the reign of King Ammizaduga, indicating that the planet Venus had disappeared for nine months and four days. Such a phenomenon would be expected from enshrouding dust hovering in the upper atmosphere, in the wake of a large asteroidal impact. Of course, all stars and planets would have been obscured at this time. But Venus must have been singled out because it was a very important deity in the Babylonian pantheon.

This same period, which was known as the Middle Bronze Age, was marked by a sudden decline and outright collapse of many civilizations in the Eastern Hemisphere, all related to natural calamities that were unquestionably the result of a major impact (or multiple impacts). Commenting on this age of political and geophysical upheaval, Velikovsky wrote that "catastrophes of earthquake and fire were of such encompassing extent that Asia Minor, Mesopotamia, the Caucasus, the Iranian plateau, Syria, Palestine, Cyprus, and Egypt were simultaneously overwhelmed"(159).

Speaking of Egypt, it was during this time that the Second Intermediate Period had come to a close, which separated the Middle Kingdom from the New Kingdom. During this era, Egypt had split up into several small fiefdoms. The Hyksos had invaded and controlled Upper Egypt, local rulers lorded over Middle Egypt, and Nubian invaders took charge of Lower Egypt. We can conclude, then, that the chaos and breakdown resulting from celestial traumas are ripe breeding grounds for political overthrows, amongst other things.

What may also have transpired in conjunction with this tumultuous period in Egyptian history was the escape of the Hebrews from enslavement in this land, as recorded in the biblical book of Exodus. The famous "ten plagues" that occurred at this time, by the way, might very well have been symptoms of a large meteoritic impact.

First we are told of the water turning to blood. This is a phenomenon that is actually quite commonly found in ancient literature, and is frequently associated with large asteroidal impacts. What happened in these instances, one might guess, is that iron from the impacting body vaporized, then oxidized as it mixed with water vapor in the atmosphere, and lastly fell as a reddish-colored rain, looking like blood.

The book of Exodus (chapter 7) relates that "...all the waters that were in the [Nile] river turned to blood. And the fish that [were] in the river died; and the river stank; and the Egyptians could not drink of the water of the river; and there was blood throughout all the land of Egypt."

The only way the Egyptians were able to get drinkable water, the biblical record states, was to dig for it underground. This makes perfect sense, since this water would not have been contaminated by the falling red rains. The Nile, however, along with all other bodies of water in the region, as the biblical record further states, did not clear up for a full seven days.

This event was witnessed by an ancient Egyptian, Ipuwer, who wrote these words on an ancient papyrus: "The river is blood....Plague is throughout the land. Blood is everywhere"(12).

Let's compare this with what a certain "Roger of Wendover" wrote about a parallel phenomenon that played out in the midst of the sixth century: "In the year of grace 541, there appeared a comet in Gaul [France], so fast the whole sky seemed on fire. The same year there dropped real blood from the clouds...and a dreadful mortality ensued."

Turning to the Hindu epic, the *Mahabharata*, we find a tale of fireballs in the sky that were the work of a dreadful bird with one leg, one eye, and one wing, which hovered clumsily overhead as it vomited blood(1).

Mayan records relate a story of a similar nature. They talk about a time when a great cataclysm caused the Earth to quake, interrupting the Sun's motion and turning the waters of the rivers into blood.

Babylonian records talk of a calamity that resulted in red dust and rain falling from the sky. Reference is also made to "bloody rain"(12).

We shall now return to our discussion of the plagues of Egypt.

After the water had turned to blood, we are then informed of a series of infestations of frogs, gnats, and then flies. Later, during the eighth plague, we are told about another infestation, this time of locusts. We could expect such infestations resulting from a major meteoritic impact. With the water supply polluted (or turned to "blood"), the press would have been on for such critters to find fresh water and stored edibles among human populations, frantically swarming them in their search.

Such a calamity would, at the same time, offset the mating season of these creatures, driving them to make desperate attempts at ensuring the survival of their species, multiplying their numbers so completely out of control that they would indeed have attained plague status.

Infestations were connected with a celestial catastrophe in ancient China, during the reign of Emperor Yahou. It is said that an immense tidal wave "that reached the sky" fell upon the land of China. "The water was well up on the high mountains, and the foothills could not be seen at all," says *The Shu King, the Canon of Yao* (Yahou). At this time, the Sun is said to have not set for a span of ten days (a likely exaggeration), the forests were ignited, and a multitude of abominable vermin were brought forth(Ibid.).

Next came the Egyptian plague of the diseased livestock. Again, this is just what would be expected from a crisis of this magnitude, especially with fresh water being a rare commodity. Coupled with the effects of flies and other insects carrying diseases, an epidemic would spread quite swiftly, bringing about an enormous death toll.

We could also expect the next plague to follow--boils, which might have resulted from "fallout" particles emitted by the blast of the impact, or perhaps from the spread of diseases carried by the flies, locusts, etc., which had first affected the livestock.

This plague was followed by a powerful thunder and hail storm, which was later followed by a period of extended pitch-black darkness. These phenomena would unquestionably accompany an asteroidal impact as well.

Regarding the plague of darkness, particularly, archaeologist William F. Albright found an ancient water trough in El Arish bearing hieroglyphs that described a period of prolonged darkness in ancient Egypt, apparently confirming the biblical account(163).

The tenth and final plague, the death of all the firstborn, would also be expected to result from a celestial catastrophe in ancient Egypt. What probably happened here is that foodstuffs became poisoned during the plague of darkness, either by locusts or the black mold *Cladosporium*. After this period of darkness, the firstborn would have been given priority with food rations, according to Egyptian custom at that time. Thus all the firstborn would have been the ones to die from the contamination. Upon seeing the effects of this food poisoning, the others in each household would have refrained from eating the toxic morsels, and thus were themselves saved from death(Ibid.).

It is believed by numerous scholars that the famous Thera (Santorini) eruption (which we discuss elsewhere herein) had taken place at about this same time--1628 B.C. If so, this would explain the "pillar of cloud by day and fire by night" that the Israelites followed, after leaving Egypt--it was the eruptive cloud ejected by Thera. In order for this cloud to have been seen by the fleeing Israelites, it would have needed to be about 24 miles high. Yet most scientists today agree that this plume was likely at least 30 miles in height--6 miles higher than it needed to be for this purpose.

The parting of the Red Sea can also be explained in the context of a celestial disaster. What this may have been was a typical prelude to a tsunami, where the waters retreat far out to sea (or, in this case, the water level of the Red Sea significantly dropped) before the big wave struck. The Red Sea is actually shallow in spots near its head, with sandbar-like formations that stretch across its width. It was probably across one of these that the Israelites passed when the water level suddenly sank. The pursuing Egyptians later drowned, of course, when the waters abruptly returned.

==========================

Ancient lunar impacts and consequent calendar adjustments

We talked earlier in this study about the original length of the pre-flood month being 30 days long, with a total of 360 days in a 12-month year. We went on to discuss how a lunar impact must have been responsible for the change in the month's length, just as an Earth impact must have been responsible for the change in the year's length. But it was not necessarily at the time of the flood that the month changed from its original 30-day length. While there's little room for doubt that the Moon suffered impacts at this time, these impacts don't appear to have been significant enough to alter the length of the month--that seemingly came later, in 1628 B.C., coinciding with the Earth impact of that same date, presumably as the Earth/Moon system passed through a swarm of debris orbiting about the Sun.

Evidence for a change in the month's length at this time, brought on by a major lunar impact, comes to us in the form of a fascinating archaeological find that dates to circa 1600 B.C. Known as the Nebra Sky Disc, it was unearthed in Germany in 1999. The March 2, 2006 London *Times* described this artifact thusly: "The bronze disc is about 30 cm in diameter, has a blue-green patina and is inlaid with a gold sun, moon and 32 stars"(161).

What needs our careful attention here is the function that this object served, as pointed out by the website of an astronomy community known as the New Jersey Night Sky: "A group of German scholars who studied this archaeological gem has discovered evidence which suggests that the disc was used as a complex astronomical clock for the harmonization of solar and lunar calendars...."(162).

Is it not clear that this disc was created in response to a change in the month's length, and thus to help keep the new month in sync with the solar year? Of course, the above article didn't mention anything about a lunar impact. Nor did it make any inference to the Nebra Sky Disc being an attempt to harmonize the new month with a 360-day year (adding 5 days at the end), which was still being observed by most cultures around the world at that time, and for numerous centuries thereafter. But that's to be expected from mainstream scientific thought, which is entirely incapable of "thinking outside the box."

The Nebra-era impact was apparently not the only month-altering lunar strike that history has recorded. Another one seems to have taken place in the eighth century B.C. According to Velikovsky, an Egyptian inscription from this very century makes reference to the Moon having been "disturbed in its movement"(12).

As a result of this lunar impact, which had disturbed the Moon's movement, the month was altered to its current length of just under 30 days. Prior to that time, beginning with the Nebra impact (1628 B.C.), the month's length, according to a variety of ancient sources, was 36 days.

From the Nebra impact to the lunar strike of the eighth century B.C., the week was measured, by many cultures around the world, in 9-day intervals. Among the cultures that did this were the Chinese, the Hindus, the Persians, the Babylonians, the Egyptians, the Greeks, and the Romans(Ibid.). This 9-day week, of course, would have amounted to an even 4 weeks in a 36-day month, and to an even 10 months in a 360-day year.

But once the lunar impact of the eighth century B.C. had occurred, the press was on once more to maintain harmony between yet another new monthly period (of just under 30 days--its current length) and the solar year of 360 days (which was still being observed at that late date, for the most part). Previously this harmonization task was far easier to accomplish, since the 36-day month, as already stated, was compatible with a 360-day year, as was the 30-day month before it. In fact, we can see from this a logical explanation of why so many ancient cultures were desirous of retaining a 360-day year for so long after it had changed to 365 days. However, once the month was altered to under 30 days--a time-frame that is not evenly divided into a 360-day year--great confusion had ensued, and a struggle was underway, as already stated, to iron out this incongruity.

We find a good example of this struggle portrayed in the writings of Plutarch, who declared that in the time of Romulus the people were "irrational and irregular in their fixing of the months," reckoning some months at 35 days and others at 36, "trying to keep to a year of 360 days." Plutarch also said that Numa, Romulus' successor, had corrected the irregularities of the calendar and changed the order of the months(Ibid.).

As you can see, the problem here was not with measuring the length of the year or the month, which all the ancients were quite proficient at. Instead, the real issue was, once again, how to come up with a way to integrate the monthly and yearly cycles so that they could be matched up evenly. But when it became obvious that this could no longer be done, the Julian calendar was eventually universally adopted, with a 365-day year comprised of months whose lengths were arbitrarily assigned, without respect to the lunar cycle.

With the shortening of the month from 36 to just under 30 days, the necessity arose to add an additional two months to the year, which was done under Numa (again, the successor to Romulus). Reference to this adjustment was made in the writings of Ovid, who stated: "When the founder of the city [Rome] was setting the calendar in order, he ordained that there should be twice five months in his year....He gave his laws to regulate the year. The month of Mars was the first, and that of Venus the second....But Numa overlooked not Janus [January] and the ancestral shades [February] and so to the ancient [ten] months he prefixed two"(Ibid.).

The ice-core and tree-ring data do not reveal anything for the eighth century B.C., but neither should we expect them to, since it appears that only the Moon was impacted at that time, and not the Earth. Thus there would have been no enveloping blanket of dust in the atmosphere, nor an Earth axis tilt, and consequently no major climatic changes. However, the effects upon the Earth must have still been rather devastating in regards to major tsunamis, somewhat like those discussed above, except probably not as intense.

So, to recap, here are the historical changes that have taken place in our calendar, at the hand of lunar catastrophes, as revealed in ancient records:

- Before the 1628 B.C. lunar impact:
7-day week
30-day month
12-month/360-day year (with 5 days added at the end of each year)

- After the 1628 B.C. lunar impact:
9-day week (for many cultures)
36-day month
10-month/360-day year (with 5 days added, still, to the end of each year)

- After the eighth century B.C. lunar impact:
7-day week
a month of just under 30-days
12-month/365-day year

========================

- 1159 B.C. is the next date we encounter that marks a major disaster from space. The ice-cores show extremely high levels of acidity for this year, and the tree-rings reveal an abnormal period of growth. In fact, this year, and the several that followed it, show the least amount of growth as recorded in the tree-rings, signifying an unusually cold climate in Ireland. But things appear to have been in the reverse for Turkey and its surrounding regions at this same time. For growth-rings in old oak trees there denote a huge spurt in growth, indicating that the catastrophe under consideration must have caused massive rainfall in this area.

A familiar repeating pattern emerged in the political world at this time. For we see that there was much discord in China (the Mandate of Heaven was again suspended, under the Shang Dynasty), Greece, Turkey, and almost all the Middle East except for Mesopotamia, resulting in the end of the Bronze Age. Egypt had its troubles at this time as well. It was severely weakened by the attacks of armed refugees, enough so that the New Kingdom soon collapsed internally.

Presumably it was to this same disaster that David the Psalmist was referring when he penned these words: "Then the earth shook and trembled; the foundations also of the hills moved and were shaken, because he [God] was wroth. There went up a smoke out of his nostrils, and fire out of his mouth devoured: coals were kindled by it. He bowed the heavens also, and came down: and darkness was under his feet. He made darkness his secret place; his pavilion round about him were dark waters and thick clouds of the skies. At the brightness that was before him his thick clouds passed, hail stones and coals of fire. The Lord also thundered in the heavens, and the Highest gave his voice; hail stones and coals of fire. Yea, he sent out his arrows [meteors?], and scattered them; and he shot out lightnings, and discomfited them. Then the channels of waters were seen [the seabed--is this a reference to a tsunami exposing the ocean floor?], and the foundations of the world were discovered at thy rebuke, O Lord, at the blast of the breath of thy nostrils." Psalm 18:7-9, 11-15. Speaking further on this matter, the Hebrew scriptures state: "...there was a famine in the days of David three years, year after year....So the Lord sent a pestilence upon Israel...and there died of the people...seventy thousand men." II Samuel 21:1; 24:15.

A Chinese text from this period declares: "The earth emitted yellow fog...the sun was dimmed...three suns appeared...frosts in July...the five cereals [grain crops] withered...therefore famine occurred"(174).

Speaking of multiple suns, a Babylonian stone carving shows two suns in the sky, just like the Akkadian stele referenced earlier, except that this carving depicts a crescent moon situated between the two suns. This Babylonian work dates to roughly 1100 B.C.--the very time-frame of the catastrophe currently being discussed. Was this extra sun a representation of the comet or asteroid that struck the Earth around this time?

Coming back to China, we discuss elsewhere herein a major exodus from that land at about this same time, or just thereafter, which resulted from a desire to escape the political turmoil that arose in the wake of the overthrow of the Shang Dynasty. We further mention the possibility of this group of refugees migrating to the Americas and founding the Olmec city of La Venta. Thus, once again we find political disruption and mass migration accompanying a celestial impact disaster.

- We now come to 207 B.C., the next catastrophe date established by ice-cores and tree-rings. And here again we find political turmoil associated with this event. In China, the Mandate of Heaven was suspended once more, this time under the Q'in Dynasty. Great discontent arose as the entire country was plunged into oppressive starvation brought on by the catastrophe. Thus the reigning king was overthrown by a new one, initiating the Han Dynasty. Confirming the celestial cause of all this chaos, one Chinese record from this period noted that "the stars were lost from view for three months"(174).

At this very time, as related by the Irish Annals, there was a devastating die-off of cattle in the land of the Celts. This was particularly problematic for them, since cattle served as their main source of food and income.

According to the Roman historian Livy, "two suns" had appeared in the sky in 206 B.C. This second "sun" was so bright that it could be seen in the daytime. He also reported that "strange lights" were seen in the evening sky two years later, in 204 B.C. Other Roman historians reported showers of stones falling from the sky a year earlier, in 205 B.C. So, clearly there was a lot going on astronomically during this period, which created havoc the world over.

It was also at about this time when the city of Teotihuacan was founded near present-day Mexico City. This metropolis suddenly arose out of nowhere, indicating that it probably came about as the result of a mass migration in wake of the 207 disaster.

Elsewhere in this study we suggest that this city was initially founded by an expedition from China, just shortly before this time. Perhaps this city started off as a small settlement, but soon experienced a major population explosion as many more migrants arrived after the impact occurred.

- 44 B.C. is another date that the tree-ring and ice-core data point to as a time of catastrophe. And, in addition to these physical evidences, as always, we also have historical records to corroborate such data.

Plutarch tells us that, after Caesar's death, "there were earthquakes and the obscuration of the sun's rays: for during all that year its orb rose pale and without radiance."

Virgil stated, also after Caesar's death, that "often we saw [Mount] Etna flooding out from her burst furnaces, boiling over the Cyclopean fields, and whirling forth balls of flame and molten stones..." He also said that "The sun...veiled his shining face in dusky gloom," resulting in what appeared to be an "everlasting night." He further pointed out that a comet had emerged so bright in the sky that it was visible during the daytime.

Chinese records talk of a red daylight comet in May and June of this year. This color, of course, would have been the result of dust in the upper atmosphere. It was also recorded that six consecutive grain harvests had failed in China, beginning in this same year.

By 43 B.C., a year later, Chinese texts stated that it snowed in April, and that frost had killed the mulberries. The Sun was also said to have been pale blue in color, and that it cast no shadows.

Then, by October of that same year, 43 B.C., according to these same Chinese records, the Sun seemed to be shining normally in the sky. However, by the spring of 42 B.C., the Sun, Moon, and stars again appeared "veiled and indistinct"(174).

- 536 A.D. is the next significant date we come to for upsets in the political and meteorological/climatological realms, evidenced, once again, by the historical, ice-core, and tree-ring data.

In Ireland, the grain crop had failed on two occasions, in 536 and 538, bringing about mass starvation in both Ireland and England. In fact, the great famine that occurred during the reign of the legendary King Arthur is believed to correspond with this same catastrophe.

Chinese chronicles of this era tell of a widespread famine in that part of the world as well. It was also around this time that the emperor of China had ordered the abandonment of the imperial capital, Loyang, in the year 534. This was followed by the political collapse of northern China in 535, and the economic and social chaos of southern China during the decade of the 540s.

Looking to Mexico, we find disease and famine hitting hard there during this period as well, leading to the collapse of the city of Teotihuacan. The Mayan civilization also suffered a setback at this point, not picking up again until about 550 A.D.

There was much contention going on in the religious world at this juncture as well. In the year 538, the Goths were driven out of Rome, ending a long period of strife between Arianism and the Church of Rome.

Speaking in reference to the year 540 A.D., Procopius, a Byzantine author of this period, stated that "During this year a most dread portent took place. For the sun gave forth its light without brightness...and it seemed exceedingly like the sun in eclipse, for the beams it shed were not clear..."

Another Byzantine chronicler, John Malalas, wrote about an obvious tsunami that occurred at this time: "[T]he sea advanced on Thrace by four miles and covered it in the territories of Odyssos and Dionysopolis and also Aphrodison. Many were drowned in the waters. By God's command the sea then retreated to its own place"(Ibid.).

John Lydus, also writing about this same disastrous period, recorded these words: "The sun became dim...for nearly the whole year...so that the fruits were killed at an unseasonable time."

Cassiodorus of Italy penned this account: "The sun...seems to have lost its wanted light, and appears of a bluish color. We marvel to see no shadows of our bodies at noon, to feel the mighty vigor of the sun's heat wasted into feebleness, and the phenomena which accompany an eclipse prolonged through almost a whole year. We have had...a summer without heat...the crops have been chilled by north winds...the rain is denied."

Reflecting back, another chronicler, Michael the Syrian, wrote: "[T]he sun became dark and its darkness lasted for eighteen months. Each day it shone for about four hours, and still this light was only a feeble shadow...the fruits did not ripen and the wine tasted like sour grapes"(Ibid.).

As was mentioned earlier, we actually have an eyewitness account on record of the event that caused these worldwide problems at this time--that of Roger of Wendover, who described, as you may recall, the appearance of a comet in France that rained blood and inflicted death upon those in the region.

Similarly, we have this statement made by Zacharias of Mithylene: "[I]n the month of December [of 538-9] a great and terrible comet appeared in the sky at evening-time for one hundred days"(Ibid.).

Another area affected by this calamity was India. For it was about this time (around 540 A.D.) that the Gupta Dynasty, a period marked by superior cultural and artistic brilliance, had come to an abrupt end.

An unusual phenomenon was witnessed at the time of this catastrophe--a brightly-glowing night sky that lasted for months on end, obviously the result of dust from the cometary or meteoritic impact suspended high in the atmosphere, lit perhaps intermittently by the light of the Moon. Commenting on this spectacle, Muirchu's *Life of St. Patrick*, a sixth-century Irish saga, states: "For on that day of his [St. Patrick's] death no night fell...evening did not send the darkness which carries the stars. The people of Ulaid say that to the end of the year in which he departed, the darkness of the nights was never as great as before." Now compare this with a parallel phenomenon that occurred in the wake of the 1908 Tunguska Event (an apparent asteroidal or cometary impact in Siberia). As the 1976 book *The Fire Came By* put it: "[T]he sky throughout Europe was strangely bright. Throughout the United Kingdom...it was possible to play cricket and read newspapers by the glow from the night sky. Photographs were taken at midnight or later, with exposures of about a minute..."(175).

- The final date we will examine is 1178 A.D. It was in this year that a dramatic increase in reported sightings of comets had occurred in Europe.

The same was true in China, where records there stated that a phenomenal ten times the amount of the usual sightings of comets was witnessed around the year 1150.

According to the legends of New Zealand's Maoris, fire had fallen from the heavens around this same time, burning up the forests and killing off many of the large, flightless moa birds. There is actually a set of craters in this region, known as the Tapanui Craters, which date to around 1200 A.D. Additionally, there has also been found black soot in geological samples from the area that support the claim of forest fires having occurred there from the fire that had "fallen from the heavens."

During this same epoch, there was a massive amount of emigration across Polynesia, the pre-Incan civilization of Peru had vanished, and the Aztecs had relocated from the place they called Aztlan (probably an area in or near modern-day Mazatlan on the Pacific coast) to the higher altitude of the Valley of Mexico, possibly to escape tsunamis from the impacts that were occurring.

It was also at about this same time that the Easter Island civilization is believed to have suddenly vanished.

Meanwhile, over in China, things weren't any better. The weather had become unusually severe, resulting in the Yellow River catastrophically flooding in 1194, which further resulted in the complete destruction of Kaifeng, the northern capital of the Song (or Sung) Dynasty. Also as a result of this disaster, the mouth of this river had moved nearly 200 miles south. Because of all of this, the Mandate of Heaven was again suspended, brining this dynasty to a close.

Genghis Khan, at about this same time, said that he saw a sign in the sky (obviously the approaching comet or asteroid) that he understood as a signal that he was to lead the Mongol hordes out of their homeland and across Asia. I need to also mention that the Persian chronicler, Al Juvaini, stated that the weather in Mongolia was so cold during this period (which lasted about two generations) that apples would not grow there.

Apparently the impact of 1178 was the culmination of a series of impacts that may have begun about sixty years earlier. For we find that the *National Geographic* website reported on May 24, 2007, that "In a study published today in the journal Geophysical Research Letters, a team from Arizona and Colorado found that the Southwest suffered a six-decade megadrought from 1118 to 1179....

"The new findings came from a study of growth rings in trees from the upper Colorado River..."(187).

Collisions were not only occurring on Earth during this time. We will now scrutinize a fascinating record of a powerful impact that was seen to happen on the Moon in this same year. This lunar strike was witnessed by five individuals who were sitting outside, just after dark, in Canterbury, England, on June 25, 1178. Their experience was later related to a Canterbury monk named Gervase, who wrote: "The upper horn [of the crescent Moon] split in two. From the midpoint of the division a flaming torch sprang up, spewing out, over a considerable distance, fire, hot coals, and sparks. Meanwhile the body of the moon which was below writhed, as if it were in anxiety, and to put it in the words of those who reported it to me and saw it with their own eyes, the moon throbbed like a wounded snake. Afterwards it resumed its proper state. This phenomenon was repeated a dozen times or more, the flame assuming various twisting shapes at random and then returning to normal. Then, after these transformations, the moon from horn to horn, that is, along its whole length, took on a blackish appearance. The present writer was given this report by men who saw it with their own eyes and are prepared to stake their honor on an oath that they have made no addition or falsification in the above narrative"(45).

Judging from its fresh appearance and location near the Moon's edge (as seen from Earth), most scientists are in agreement that the lunar scar known as

Bruno Crater is the impact site of this twelfth-century witnessed event.

Another eyewitness account of a lunar impact, this one from the lower Congo in Africa, which dates to unknown antiquity, states that "long ago the sun [a bright impacting body?] met the moon and threw mud at it [ejecta blanket?], which made it less bright. When this meeting happened there was a great flood [tsunami?]..."(20).

This account bears a striking resemblance to the 1178 record, and may in fact be a reckoning of the selfsame event.

As a final footnote to this entire 1178 disaster (the one on the Earth and the Moon), attention should be drawn to the fact that there was a major supernova explosion in 1054, 124 years earlier, the remnant of which we now call the Crab Nebula, in the constellation Taurus (which we discuss elsewhere herein). It is located about 6.5 light yeas away. Is it possible that material from this explosion could have been responsible for the 1178 catastrophe? Scientists estimate that material still escaping from this exploded star is moving away from the center point of the explosion at a rate of half the speed of light. Thus, 124 years is far more than enough time for debris from there to have reached Earth by then.

This explosion was a major one. Chinese Annals tell us that this "guest star" was so bright that it could be easily seen during the day. The Sung-Shih (of the Sung Dynasty) gives this account: "On the first year of the Chi-ho reign period, fifth month, chi-chou [July 4, 1054], a guest star appeared approximately several [inches] to the south-east of Tian-kuan [Aldebaran, or Zeta Tauri]. After a year and more it gradually vanished"(194, 205).

"Each of the planets, according to them [the Chaldeans], has its own particular course, and [their] velocities and periods of time are subject to change and variation [because of impacts and gravitational tugs, obviously]." - Diodorus of Sicily.

"The worlds [asteroidal debris] are unequally distributed in space; here there are more; there fewer; some are waxing, some are in their prime, some waning: coming into being in one part of the universe, ceasing in another part. The cause of their perishing is collision with one another." - Democritus.

* * * * * * *

As should be most apparent by now, there clearly were multiple disasters that occurred after the flood. Though quite obviously not as severe as the flood itself, they nevertheless must have inflicted extensive devastation, sometimes on a global scale. No wonder ancient Egyptian priests told Solon, in Plato's *Timaeus*: "You remember but one deluge, though many catastrophes had occurred previously....There have been, and will be again, many destructions of mankind arising out of many causes; the greatest have been brought about by the agencies of fire and water, and other lesser ones by innumerable other causes....after long intervals." And it is not to be doubted that some, if not all, of these periodic post-flood catastrophes resulted in additional Earth tilts.

One last catastrophe worthy of our attention here, which apparently involved a tilting of the Earth, is the famous story of the extended day (or "the day the Sun [and Moon] stood still"), as recorded in the biblical book of Joshua. The account in question is found in chapter 10, verse 13, and reads as follows: "And the sun stood still, and the moon stayed [still],... So the sun stood still in the midst of heaven, and hasted not to go down about a whole day."

Though many question the veracity of this story, an astonishing confirmation of it comes from across the Atlantic, on the other side of the globe, tucked away in an ancient legend from the Andes region of South America. What's so astonishing about it is the fact that it mentions a prolonged night, which is just what we would expect, since it would have been nighttime in the Americas when it was daytime in the Middle East. This legend states that the Sun had stayed away for about the equivalent of a modern 20-hour period, and that this event took place during the reign of Titu Yupanqui Pachacuti II, around 1400 B.C.--a time contemporaneous with the biblical Joshua(25).

There exists yet another record from the Americas of this long night--the Mexican Annals of Cuauhtitlan--the history of the empire of Culhuacan in Mexico, written in Nahua-Indian in the sixteenth century. Therein it is related that during a cosmic catastrophe that occurred in the remote past, the night did not end for a long time(12).

There are also several records from back on the other side of the globe, which, like the biblical record, talk of an extended day.

Take, for instance, the Chinese Annals, which declare that, during the reign of Emperor Yao (or Yahou), the Sun stood above the horizon for such a long time that it was feared the world would be set on fire.

Another example of this--the last we will look at--comes from the Latin poet Ovid, who said that a day was once lost, and that the Earth was in terrible danger from the intense heat of the Sun at this time.

So what type of mechanism could have facilitated such a celestial spectacle? What could have made the Sun (and the Moon) appear to stand still in the sky? Obviously there must have been some sort of disruption in the Earth's rotation, perhaps caused by a large passing asteroid that exerted a powerful gravitational tug on the Earth. In this case, the tilting would not have been a sudden one, like what was experienced from the impact that spawned the flood catastrophe. Instead, the tilting that produced Joshua's long day was more gradual, occurring over a day's time. Thus the tectonic and tsunamic effects of this tilting would have been comparatively minimal.

I want to point out that the account given by Joshua of this event might be misunderstood. Joshua did not necessarily state that the Sun stood still in the same spot for the duration of a whole day, but only that it "stood still in the midst of heaven, and hastened not to go down about a whole day." This suggests that the Sun seemed to stand still only for a time, with the real emphasis being on the fact that it remained *somewhere* in the sky for almost a full 24 hours.

What appears to have happened here is that the Earth underwent a complete polar reversal, so that the Sun rose at one end of the horizon, then, after trekking a good distance across the sky, it seemed to hover in roughly the same spot for a time, and finally reversed direction and set back in roughly the same area that it rose from. There are actually many ancient legends, some of which we cited earlier, that imply this very thing--that the Sun changed its course, although we applied these references to the time of the flood-producing catastrophe. Yet these very legends, or at least some of them, could just as easily have been referring to a much later time--to "the day the Sun stood still."

At any rate, there are numerous ancient references to an Earth axis tilt which we can be absolutely sure had occurred after the asteroid impact that spawned the first Earth tilting and the consequential flood. How can we know this? Because these particular references mention a sudden change that took place in the seasons (remember that there were no seasons prior to the flood-producing asteroid strike). And a sudden change in the seasons is just what we would expect to happen if the Earth tilted sometime in the ancient post-flood world. For it would have happened that whatever hemisphere (northern or southern) was tilted toward the Sun at that time, and thus experiencing summer, would have been suddenly tilted away from the Sun, thus experiencing winter, and vice versa. Let us now have a look at some examples of ancient references to this phenomenon.

In the *Texts of Taoism* we find this statement: "The breath of heaven is out of harmony....The four seasons do not observe their proper times."

The Taoist text of Wen-Tze contains this narrative: "When the sky, hostile to living beings, wishes to destroy them, it burns them; the sun and the moon lose their form and are eclipsed; the five planets leave their paths; the four seasons encroach one upon another; daylight is obscured; glowing mountains collapse; rivers are dried up; it thunders then in winter, hoarfrost falls in summer; the atmosphere is thick and human beings are choked; the state perishes; the aspect and the order of the sky are altered..."

An Egyptian document, known as Papyrus Anastasi IV, states: "The winter is come as [instead of] summer; the months are reversed and the hours are disordered."

In *The Birth of the War-God*, the ancient Hindu poet Kalidasa related how "The seasons have forgotten how to follow one another now; they simultaneously bring flowers of autumn, summer, spring"(12).

A fascinating statement along this line is found in the biblical book of Daniel, where we read: "And he [God] changeth the times [years] and the seasons..." Daniel 2:21.

The final ancient account of this nature that we will look at is from a Mesoamerican legend that tells of a time when "the order of the seasons was altered"(12)[Endnote 70].

Some have argued, and rightly so, that the gravitational force of an asteroid--even a large one--seems inadequate to tilt the entire Earth off its axis, especially resulting in a complete polar reversal. But the truth is, in order to effect a polar reversal, flipping the entire planet over may not necessarily have been required. As was originally suggested by Charles Hapgood, all that may have been needed is for the Earth's rigid outer crust to have been tugged at, and then to slide over its underlying liquid rock mantle (the Crustal Displacement theory). If, during post-flood times, a sizable asteroid passed over a large protrusion of the Earth's lower crust (sea floor), such as a mid-oceanic ridge, the entire crust of the Earth, on a global scale, could very well have been moved without affecting the rest of the underlying rotating planet (or at least not affecting it to a serious degree).

This theory, it should be pointed out, was endorsed by Albert Einstein, a contemporary of Hapgood. In the Foreword to Hapgood's 1958 book, *Earth's Shifting Crust*, Einstein summarized Hapgood's hypothesis by stating that the only thing that would be needed to cause the Earth's crust to shift over the rest of the planet would be large accumulations of polar ice (causing an imbalance in the crust's center of gravity), which wouldn't have to necessarily involve the gravitational influence of another celestial body at all (although I personally believe that such an influence was involved). Here's how Einstein explained it: "In a polar region there is continual deposition of ice, which is not symmetrically distributed about the pole. The earth's rotation acts on these unsymmetrically deposited masses, and produces centrifugal momentum that is transmitted to the rigid crust of the earth. The constantly increasing centrifugal momentum produced in this way will, when it has reached a certain point, produce a movement of the earth's crust over the rest of the earth's body"(152)[Endnote

71].

Proof that the cause of Joshua's prolonged day was due to a near-Earth passage of a celestial object is found right within the book of Joshua itself. Turning back a couple verses before the passage cited above, we find a reference to an event that occurred just prior to the lengthened day. This verse states: "...the Lord cast down great stones from heaven upon them unto Azekah, and they died: they were more which died with hailstones than they whom the children of Israel slew with the sword." Joshua 10:11.

Here we see that the long day was preceded by a rain of stones from the heavens, i.e. a meteor shower. {The word "hailstones" in this verse can also be translated simply as "stones.") What must have happened here is that the Earth had passed through an orbiting swarm of meteoritic debris from the extinct trans-Martian planet, a large piece of which tugged on the Earth as it passed by, while at the same time smaller pieces from this swarm had actually fallen to Earth, producing a meteor shower, or "stones from heaven."

The tree-ring and ice-core data do not record significant changes for this period of about 1400 B.C. But this may be due either to a miscalculation of the time that Joshua lived, or else to the fact that no large piece (or pieces) of this swarm of debris had struck the Earth around 1400 B.C., and thus there wasn't enough of a major climatic effect to show up in the ice-cores or tree-rings. Even though the seasons at this time would have changed from summer to winter, or vice versa, the effects would not have been anywhere near as noticeable as if there was a major impact on Earth, which would have darkened the skies and dropped the temperature drastically over a large part of the Earth, lasting for many years.

"Behold, the Lord maketh the earth empty, and maketh it waste, and turneth it upside down [a reference to the reversal of its polar axis?]..." - Isaiah 24:1.

"The sun and moon stood still in their habitation..." - Habakkuk 3:11.

"And it shall come to pass in that day, saith the Lord God, that I will cause the sun to go down at noon, and I will darken the earth in the clear day." - Amos 8:9.

"[God] shaketh the earth out of her place, and the pillars thereof tremble. [He] commandeth the sun, and it riseth not; and sealeth up the stars." - Job 9:6, 7, 9, 10.

* * * * * * *

As well as the ancients informing us of celestial catastrophes in the past, many of them also warned of a major future calamity of this very sort. Apparently they understood the cyclical nature of these events, realizing that they reoccurred whenever orbiting debris crossed the Earth's path. They were, after all, expert astronomers.

One ancient civilization that made a prediction of this kind was the Mayas. In fact, the Mayas gave an exact date for a major cosmic catastrophe that would amount to the "end of the world"--December 21, 2012. What is significant here is not so much the date, but the fact that they saw complete destruction as imminent. The Aztecs, too, made a prediction of this same caliber. They referred to it as the end of the "Fifth Sun," when the Earth will be moved (knocked off its axis?) and all of mankind will perish. Did these two Mesoamerican civilizations know of a specific swarm of cometary or asteroidal material that the Earth would pass through in the future?

Apparently so. For in the Mayan *Book of Chilam Balam*, in Katun 13, the following prophetic prediction is made: "This is a time of total collapse where everything is lost. It is the time of the judgment of God. There will be epidemics and plagues and then famine. Governments will be lost to foreigners and wise men and prophets will be lost."

Please notice how this prediction contains the same climatological and even governmental calamitous elements that we discussed earlier, which accompanied many celestial disasters of the past.

Another prediction like this was made by the Roman poet Seneca, who wrote these verses around the time of Christ:

Dark grows the sun,
Brothers shall fight and kill...
Axe-time, sword-time,
Shields are sundered,
Wind-time, wolf-time,
Till the world falls dead....

Mountains dash together...
And heaven is split in two,
The sun grows dead--
The earth sinks into the sea,
The bright stars vanish
Fires rage and raise their flames
As high as heaven(18).

The seventeenth century French seer, Nostradamus, also foresaw an approaching catastrophe of this nature. Here are several samples of what he wrote along this line:

Century 1, Quatrain 17:

> For forty years the rainbow will not appear;
> For forty years it will be seen every day:
> The arid earth will grow more dry,
> And great floods when it will be seen.

Century 1, Quatrain 56:

> You will see, sooner and later, great changes made,
> Extreme horrors and vengeances:
> For as the moon is thus led by its angel,
> The heavens draw near to the reckoning.

Century 4, Quatrain 67:

> In the year when Saturn and Mars are equally fiery,
> The air is very dry, a long comet:
> From hidden fires a great place burns with heat,
> Little rain, hot wind, wars and raids.

Century 2, Quatrain 62:

> Mabus will soon die, then will come
> A horrible undoing of people and animals,
> At once one will see vengeance,
> One hundred powers, thirst, famine, when the comet will pass.

Century 1, Quatrain 46:

> Very near to Auch, Lectoure and Mirande [southwestern France],
> Great fire in the sky for three nights will fall:
> A very stupendous and marvelous event will happen,
> Very soon after the earth will tremble.

Century 6, Quatrain 5:

> Very great famine (caused) by a plague-ridden wave [tsunami?],
> Will extend through long rain the length of the Arctic pole:
> "Samarobryn" one hundred leagues from the hemisphere,
> They will live without law, exempt from politics.

The Buddhist book, *Visiddih-Magga*, makes this forecast:

> This world will be destroyed;
> And the mighty ocean will dry up;
> And this broad earth will be burnt up(12).

A similar foreboding is given in the biblical book of Revelation, where we are warned of a series of plagues that will befall the Earth, which closely parallel those that fell upon the land of Egypt in the days of Moses, as we discussed earlier. In Revelation chapters 15 and 16, the apostle John wrote:

> And I saw another sign in heaven, great and marvelous, seven angels having the seven last plagues; for in them is filled up the wrath of God....And I heard a great voice out of the temple saying to the seven angels, Go your ways, and pour out the vials of the wrath of God upon the earth. And the first went, and poured out his vial upon the earth; and there fell a noisome and grievous sore...And the second angel poured out his vial upon the sea; and it became as the blood of a dead man: and every living soul died in the sea. And the third angel poured out his vial upon the rivers and fountains of waters; and they became blood....And the fourth angel poured out his vial upon the sun; and power was given unto him to scorch men with fire. And men were scorched with great heat....And the fifth angel poured out his vial...[and there were] pains and sores....And the sixth angel poured out his vial upon the great river Euphrates; and the water thereof was dried up....And the seventh angel poured out his vial into the air....And there were...lightnings; and there was a great earthquake, such as was not since men were upon the earth, so mighty an earthquake, and so great. And...the cities of the nations fell....And every island fled away, and the mountains were not found. And there fell upon men a great hail out of heaven, every stone about the weight of a talent: and men blasphemed God because of the plague of the hail; for the plague thereof was exceeding great. Revelation 15:1; 16:1-4,

8-10, 12, 17-21.

We are actually overdue for such a catastrophe. According to the tree-ring and ice-core data, the Earth hasn't been met with an impact from an asteroidal or cometary body since 1178 A.D. Perhaps we should take the warnings of Seneca and John the Revelator, as well as others like them, to heart.

Appendix C

Races of giant humans in the post-flood world

We have already seen that the people who lived before the flood were of a much greater height than those living on Earth today. But we find that a wealth of written and archaeological evidence exists to support the assertion that some people (probably certain extinct races) continued to grow to large sizes for a considerable time after the flood disaster.

One example of written evidence of this is found in the biblical book of Genesis (chapter 6, verse 4). We actually cited part of this verse before, where it says, referring to the time just before the flood catastrophe, that "there were giants in the earth in those days." But then this passage goes on to say, "and also after that," meaning that these giants continued to live on after those days of the pre-flood world. And indeed, as we look in later passages in the Bible, we find that there truly were giant people on Earth after the flood, living in scattered communities. Here are a few examples of such passages:

- "And there we saw the giants, the sons of Anak, which come of the giants: and we were in our own sight as grasshoppers, and so we were in their sight." - Numbers 13:33.

- "That also was accounted a land of giants: giants dwelt therein in old time; and the Ammonites call them Zamzummims." - Deuteronomy 2:20.

- "And the border went up by the valley of the son of Hinnom unto the south side of the Jebusite; the same is Jerusalem: and the border went up to the top of the mountain that lieth before the valley of Hinnom westward, which is at the end of the valley of the giants northward." - Joshua 15:8.

- "And Ishbibenob, which was of the sons of the giant, the weight of whose spear weighed three hundred shekels of brass in weight..." - 2 Samuel 21:16.

- "And yet again there was war at Gath, where was a man of great stature...and he also was the son of the giant." - I Chronicles 20:6.

Goliath, the giant that King David is said to have felled with his sling, is mentioned in 1 Samuel 17:4 as having been "six cubits and a span" in height, which, in modern measurements, would mean that he was somewhere between seven and nine feet tall.

The Hebrews were not the only ones who talked of giants after the flood. Many other cultures the world over had tales of extraordinarily tall people who lived in post-flood times.

The Norse talked about a giant race known as the Jotunn.

The Irish tell of a giant people called the Fomorians.

The Greeks spoke of the Titans and the Cyclops.

In Chad, the Kotoko tribe speaks of a race of black giants that lived long ago, who were "so tall that they could look over the trees"(132).

The Chaldeans talked about the Izdubar race of giants.

Ancient records from India speak of three long-gone races of giants: the Danavas, the Daityas, and the Rakshasas.

Legends of Tibet mention giants who inhabited that land in ancient times. The men were said to have averaged 15 feet in height, whereas the women averaged about 12 feet.

The Toltecs of Mexico had a legend of the Quinametzin race of giants who inhabited their land after the great flood.

An ancient Peruvian tradition tells of the Chavin--giant people who inhabited a large strip of land from the Pacific to the sources of the Amazon.

We could go on almost endlessly with this list of post-diluvian legendary references to giants. But let us now move on to the archaeological evidence.

There are a number of ancient Sumerian artworks that depict incredibly tall people standing aside much shorter ones. In these representations, both size-groups are made up of full-grown, normal-looking (non-midget) adults.

Several reliefs from ancient Egypt portray pharaohs that were of a most unusually large stature. Most scholars claim, without evidence, that the reason for this was purely symbolic--to get the idea across that the pharaohs were superior to ordinary men. But it could also be argued that their size difference was literal, and that the Egyptians chose their kings from the existing races of giants, because of their imposing appearance and superior strength.

Back in 1925, 8 giant human skeletons, 8 to 9 feet tall, were found in Walkerton, Indiana. All of the skeletons were dressed in copper armor.

Near Lake Delavan, Wisconsin, in 1912, multiple skeletons of giant humans were uncovered in a burial mound.

In 1911, several red-haired mummies, ranging from 6 ½ to 8 feet in height, were discovered in a cave in Lovelock, Nevada.

Twenty years later, in 1931, also in Lovelock, Nevada, human skeletons were found in the Humboldt Lake bed that were between 8 ½ and 10 feet in length.

During the year 1974, remains of a 7-foot-tall woman were found inside of a cave in Chalk Mountain, Texas.

While digging a pit for a powder magazine in Lompock Rancho, California, in 1933, soldiers came upon a layer of cemented gravel, underneath which they found a 12-foot human skeleton, carved shells, huge axes, and stone blocks with strange symbols carved on them.

Ivan T. Sanderson, a well-known zoologist and frequent guest on the late Johnny Carson's "Tonight Show" in the 1960s (usually appearing with an exotic animal), once told a curious story about a letter he received regarding an engineer who was stationed on the Aleutian island of Shemya during World War II. While building an airstrip there, his crew bulldozed a group of hills and discovered some unusual human remains. The Alaskan hills were in fact burial mounds containing gigantic human remains. The crania measured from 22 to 24 inches from base to crown.

These remains wound up in the hands of the Smithsonian, after which time they managed to "disappear." Commenting on this strange disappearance, Sanderson asked: "[I]s it that these people cannot face rewriting all the text books?"(Ibid.)

In 1872 in Seneca Township, Noble County, Ohio, in what is now called "Bates' Mound," three giant skeletons were found that were at least eight feet long.

A cemetery of skeletons averaging 8 feet in height was discovered in Sayopa, Sonora, Mexico, in 1930.

Skeletons of 8-foot giants were found in a cave in Manta, Ecuador, in 1928.

In the ancient city of Tiahuanaco in Bolivia, skulls were excavated that belonged to people who were at least 10 feet tall. One skull measured 14 X 14 inches.

On the island of Java, bones of men that had a stature of 9 feet were found in 1944.

The December 1895 issue of the British magazine *Strand* carried an article about the remains of a giant recovered during a mining operation in County Antrim, in Ireland. A photograph of this find was included with the article, which showed it inside of a coffin-shaped crate, leaning up against a railroad car at the London and North-Western Railway Company's Broad Street Goods Depot. The height of this gargantuan man was 12 feet 2 inches. After being placed on exhibit for a time in Liverpool and Manchester, the corpse unfortunately disappeared, like so many other finds of its kind.

In the late 1950s, during a road construction project in the Euphrates

Valley of southeast Turkey, many tombs were uncovered that contained the skeletal remnants of giants that ranged between 14 and 16 feet tall(Ibid.).

Of course, even in modern times people have grown, on occasion, for whatever reason, to unusual heights. In the Museum of the Guinness Book of World Records there stands a life-size model of a man (Robert Wadlow) that stood 8 feet 11 inches tall--almost a full 9 feet. So the tales and other indications of giant races that lived in ancient times should not come as such a shock to us.

Appendix D

Ancient records of plasmic discharges in the heavens

As mentioned elsewhere in this study, the tails of comets appear to be electrified plasmas instead of streaming trails of melted and vaporized ice blown behind the comets by the solar wind. Most of the planets and their moons have such plasma tails (resulting from their plasmaspheres, or ionospheres, being blown behind them by the solar wind), but they usually aren't visible to the naked eye. Venus is a good example. Though its plasmasphere and tail are not visible today, they apparently were quite visible in the distant past (and thus must have been much more highly electrically charged). For it happens that there are numerous ancient records, found all over the world, which tell of how Venus once had a visible tail.

The traditions of ancient Mexico, for example, mention that Venus "smoked." Said Alexander Humboldt in his work *Researches*: "The star that smoked, *la estrella que humeava*, was Sitlae choloha, which the Spaniards call Venus." In this regard, Sahagun, the sixteenth century missionary-priest who was an authority on Mexico, stated that the Mexicans referred to a comet as "a star that smoked." So, in comparing these statements, does it not become clear that the ancient Mexicans likened Venus to a comet because it "smoked," or once had a visible tail?

For comparison, in the Hindu Vedas Venus is said to have looked like "fire with smoke."

Likewise, the Talmud (in the Tractate Shabbat) contains this statement: "Fire is hanging down from the planet Venus." The Talmud further says of this planet: "The brilliant light of Venus blazes from one end of the cosmos to the other end." Is this not precisely what we would expect to hear if Venus' long plasma tail was indeed visible in ancient times--that it would have stretched across the entire sky?

The Chaldeans talked of Venus having a "beard," which was also what these same people said about comets.

The Arabs called Venus "Zebbaj," which means "one with hair."

Similarly, "Chaska," the ancient Peruvian name for Venus, which is still in use today, means "wavy-haired."

So powerful was Venus' plasma sheath in ancient times, that it was visible in broad daylight. An archaic Chinese astronomical text from Soochow states that "Venus was visible in full daylight and, while moving across the sky, rivaled the sun in brightness."

Along this same line, as recently as the seventh century B.C. an Akkadian king, Assurbanipal, said that Venus was "clothed with fire and bears aloft a crown of awful splendor."

Under Seti, the Egyptians similarly referred to Venus as a "circling star which scatters its flame in fire...a flame of fire in her tempest."

Coming back to Venus' tail, there were other planets, as we might expect, that were mentioned as having once had visible tails as well. Pliny, for instance, said that "sometimes there are hairs attached to the planets." We can conclude from Pliny's use of the word "sometimes" that, by his time, the plasmaspheres of planets (and therefore their tails as well) had already begun to lose a lot of their energy, and thus were only occasionally visible--perhaps whenever there was an increase in solar activity.

So if Venus, and indeed the other planets, had more prominent and highly-charged plasma tails in the distant past, it is conceivable that these tails may have served as conduits to transmit powerful electrical discharges from one planet to another [Endnote 72]. We do, after all, have ancient chronicles that give fancified eyewitness accounts of such discharges. One obvious example is delivered to us from the ancient Greeks. In fact, in the account of Apollodorus that follows, we might find here a record of what really happened to the destroyed trans-Martian planet (here called Typhon, but also known as Typhoeus to the Greeks)--that it was split asunder by a powerful plasmic electrical discharge from Zeus (Jupiter): "[Typhon] out-topped all the mountains, and his head often brushed the stars. One of his hands reached out to the west and the other to the east [plasmic discharges?], and from them projected a hundred dragons' heads [finger-like plasmic projections?]. From the thighs downward he had huge coils of vipers [plasma filaments often coil, looking like snakes] which...emitted a long hissing....His body was all winged...and fire flashed from his eyes. Such and so great was Typhon when, hurling kindled rocks, he made for the very heaven with hissing and shouts, spouting a great jet of fire from his mouth....

"[Zeus pursued Typhon], rushing at heaven. Zeus pelted Typhon at a distance with thunderbolts [plasmic discharges], and at close quarters struck him down....[Again] Zeus pelted Typhon with thunderbolts...a stream of blood gushed out..." And so marked the end of Typhon(12).

Let us also recall, as we discussed early on in this work, Plato's reference in *Timaeus* to how Phaeton (which, for all practical purposes, is the same as Typhon, or Typhoeus) was "destroyed by a thunderbolt" in a celestial battle.

Hesiod, in his *Theogony*, thusly described this same battle (between Zeus, or Jupiter, and Typhoeus, or the trans-Martian planet), which also involved a devastating "thunderbolt" strike upon the Earth, at the hand of Zeus (again,

Jupiter):

And the heat and blaze from both of them
were on the dark-faced sea,
from the thunder and lightning of Zeus
and from the flame of the monster (Typhoeus),
from his blazing bolts and from the scorch and breath of the stormwinds....

The boundless sea rang terribly around,
and the earth crashed loudly;
Wide heaven was shaken and groaned,
and high Olympus reeled from its foundations
under the charge of the undying gods.
From the deep sound of the gods' feet,
and the fearful onset of their hard missiles,
the heavy quaking reached even far Tartarus....

From the skies, opposite Mount Olympus,
he [Zeus] came forthwith, hurling his lightning.
The bolts flew thick and fast from his strong hand,
thunder and lightning together,
whirling as an awesome flame.
The fertile earth crashed around in burning,
and the vast wood crackled aloud with fire all about.
All the land seethed,
as did the sweetwater streams and the salty sea....
The hot vapor lapped around the Titans,
of Gaea born;
Flame unspeakable rose bright to the upper air.
The flashing glare of the Thunder-Stone,
its lightning, blinded their eyes--
so strong it was.
Astounding heat seized chaos....
It seemed as if earth and wide heaven above
had come together;
A mighty crash, as though earth was hurled to ruin....

Also the winds brought rumbling,
earthquake and duststorm,

thunder and lightning....

Zeus thundered hard and mightily,
and the earth around resounded terribly,
as did the wide heaven above and the sea and the watery streams,
even the nether parts of the earth....

Through the two of them [Zeus and Typhoeus],
through the thunder and lightning,
heat engulfed the dark-blue seas;
And through the fire...
And the scorching winds and blazing thunderbolt,
the whole earth seethed, and sky and sea.
Great waves raged along the beaches....
And there arose an endless shaking....

...A great part of huge earth was scorched
by the terrible vapor,
melting as tin melts when heated by man's art....
In the glow of a blazing fire
did the earth melt down(108).

Here we can see that the Earth was absolutely devastated by this Jovian lightning strike. Was this a recollection of what had originally spawned (or helped to spawn) the flood catastrophe, shortly after the destruction of the trans-Martian planet? Did a massive thunderbolt from Jupiter accompany the flood-producing impact from a large piece of Typhoeus?

As it turns out, many other ancient cultures described this and other celestial battles that involved thunderbolt strikes on the Earth from space [Endnote 73].

Strabo tells us that the Arimi (Aramaeans or Syrians) had witnessed the battle between Zeus and Typhon, and that Typhon fled in fear, burrowing underground (i.e. a piece of Typhon struck the Earth, penetrating below the surface of the ground). This narrative also hints at the fact that some pieces of Typhon may have sent plasmic discharges to Earth, for Strabo describes how Typhon, or a piece thereof, cut burrows into the Earth and formed the beds of the rivers (soon we will be discussing how dendritic river beds look as though they may very well have been carved out by powerful celestial lightning blasts).

Even some ancient place names hint at extraterrestrial discharges in ancient times. The Egyptian shore of the Red Sea, for example, was known as Typhonia, and was believed to be one of the sites where such a strike took place.

Coming back to ancient mythological accounts of cosmic lightning strikes upon the Earth, the Egyptian god Ra, we are told, was the victor who vanquished Apep in a battle that involved, once again, thunderbolts on Earth from the "realm of the gods" (outer space).

The victor in the Hindu version of this same battle was Indra, who vanquished Vritra (or Ahi) with bolts of lightning. The Vedic record states: "Indra, whose hand wielded thunder, rent piecemeal Ahi who barred up the waters....Loud roared the mighty hero's bolt of thunder, when he, the friend of man, burnt up the monster." This event is even connected with the great flood catastrophe. For the text goes on to say: "...when thou first wast born, oh Indra, thou struckest terror into all the people. Thou, Maghavan, rentest with thy bolt the dragon who lay against the water floods of heaven"(190). The thunderbolts of Indra were also said to have carved out clefts in the Earth, just as Strabo implied about the thunderbolts that shot out at Earth from the battle between Zeus and Typhon. The Vedic texts state of Indra that he "slew the Dragon...and cleft the channels of the mountain torrents...his heavenly bolt of thunder...fashioned [these things]"(208).

Hebrew scriptures speak on this matter as well: "The voice of thy [God's] thunder was in the heaven: the lightnings lightened the world: the earth trembled and shook." Psalm 77:18. The book of Job makes further reference to this: "The pillars of heaven shook and were astounded at his [God's] roar: By his power he stilled the sea, and by his understanding he smote Rahab [the dragon]....By his wind the heavens were made fair, his hand pierced the twisting serpent [Remember Apollodorus's mention of "coils of vipers," cited above?]....Lo, these are but the outskirts of his ways; and how small a whisper do we hear of him....But the thunder of his power who can understand?" Job 26:11-14.

Babylon and Sumer had Marduk, of course, the hero who saved the world from the evil Tiamat, by zapping him with--you guessed it--thunderbolts.

Akkadian mythology talked about Ninurta as the hero who slaughtered the dreadful dragon named Anzu, again, with lightning of a not-so-ordinary variety.

Ancient Germanic mythology had its hero Thor, who also battled a dragon, slaying it with thunderbolts.

Iroquois Indians have an ancient legend that tells of a hero named Heno, who vanquished a great serpent with a bolt of lightning. Other North American tribes, such as the Zuni and the Pawnee, have similar legends(190).

In Mexico, the god Tlaloc shot thunderbolts and carried a stick in the form of lightning. An equivalent deity, the Mayan Chac, likewise held a thunderbolt in his hand and rode on the back of a serpent (dragon), perhaps signifying victory in a battle against this serpentine beast.

The Chinese version of this lightning-bearing divine being was known as Lei Kung, who was also portrayed riding on a dragon(90).

The ancients were unquestionably keenly aware of the ability of planets to emit electrical discharges, and sometimes wrote about this phenomenon from a more scientific perspective. One person who did this in a most eloquent fashion was Pliny. In his *Historia Naturalis*, he wrote: "Most men are not acquainted with a truth known to the students of science from their arduous study of the heavens, that thunderbolts are the fires of the three upper planets [Mars, Jupiter, and Saturn]....Heavenly fire is spit forth by the planet as crackling charcoal flies from a burning log. If such a discharge falls on the earth, it is accompanied by a very great disturbance in the air [shockwave?]."

The mistake should not be made that Pliny was here implying that ordinary storm lightning originated on other planets. In fact, he made a clear distinction, in this same work of his, between these two types of lightning bolts--between ordinary storm lightning and the much more powerful brand that originated from planets.

Another Classical Roman author who made this same clear distinction was Seneca. He said that there was a difference between the "lesser bolts which struck houses and the lightning bolts of Jupiter"(12).

An additional point I wish to emphasize from the last Pliny quote cited is that planetary thunderbolts were not very common in Pliny's time. These spectacles in the realm of the planets were seemingly more common the further one looks back in history, toward the time of the flood.

Seneca (cited above) made some other interesting remarks, from a scientific perspective, about planetary electrical discharges in the distant past. He stated that if Saturn "has Mars in conjunction, there are lightning bolts." He also wrote that when planets come into conjunction, or simply approach each other, "the space between the two planets lights up and is set aflame by both planets and produces a train of fire"(190).

Yet another ancient account like this is found in *The Birth of the War-God*, written by the Hindu poet Kalidasa. Although no planet is mentioned as the culprit behind the thunderbolt described in the following verses, it is clear from the context that it must have indeed originated from out in space:

> There fell, with darting flame and blinding flash
> Lightning the farthest heavens, from on high
> A thunderbolt whose agonizing crash

Brought fear and shuddering from a cloudless sky.

There came a pelting rain of blazing coals
With blood and bones of dead men mingled in;
Smoke and weird flashes horrified their souls;
The sky was dusty grey like asses' skin.

The elephants stumbled and the horses fell,
The footmen jostled, leaving each his post,
The ground beneath them trembled at the swell
Of ocean, when an earthquake shook the host(12).

The mention here of how this thunderbolt came from the "farthest heavens" and how it occurred under a "cloudless sky" together point to an extraterrestrial origin for this phenomenon. Also, the references to the "pelting rain of blazing coals," "smoke and weird flames," the ensuing "dusty grey" skies, and the earthquake and swelling ocean clinch the case that this was not, by any means, an ordinary jolt of lightning.

Such celestial lightning zaps, it seems, left tell-tale signs of their strikes on several planets and moons throughout the solar system, looking just like formations made in laboratories when electrical charges are sent through solid objects, like acrylic blocks. These lab-produced patterns have a branching, or dendritic shape to them. Grand-scale patterns like these are seen primarily all over Earth, such as the Grand Canyon formation. These features are thought to have resulted from running water carving them out. Even this current study proposes that residual flood waters, trapped for a time on the continental land masses, may have been responsible for formations like the Grand Canyon. However, this may not have been the case. After all, just because water flows in these crevices today does not mean that they originated from water-flow [Endnote 74]. Celestial lightning blasts may indeed have been responsible for their existence. Let's not forget that Strabo described how Typhon, or a piece of Typhon, cut burrows into the Earth and formed the beds of the rivers. Mars, incidentally, has similar dendritic patterns, especially around the Valles Marineris region, which may also be the result of electrical discharges from space. Even Saturn's moon Titan has such patterns on its surface, which shouldn't surprise us, since it has a notable plasmasphere. Though liquid methane may flow in these dendritic gullies today, they may very well have originally formed as a result of electrical blasts, perhaps in the remote past.

Some think Valles Marineris itself, as a whole, might be the result of a major electrical blast that occurred long ago, not only because of its physical appearance, but because of statements made in ancient records. A good example of this is found in Greek mythology, where we are told that Ares (Mars) suffered a major battle wound from a lightning weapon (thunderbolt). In addition, Hercules, also associated with Mars by the Greeks, is said to have received a traumatic wound to his hip-joint, again, by a lightning weapon.

In Hindu mythology, both Indra and Ravana, we are told, suffered major wounds from bolts of lightning. Perhaps both of the stories that relate these wound inflictions on these gods referred to one and the same event. Such was often the case with mythology--the stories mutated as time went on. New stories would break out based on old themes, new names were assigned to old gods, or the name of one god assigned to one particular planet would be reassigned to a different planet. Yet, through all the changes throughout the ages, the basic themes and stories remained essentially the same--at least enough so as to recognize the original events they represented.

Other examples of scars left from powerful interplanetary electrical discharges may be a great many of the craters we see on the planets and moons of our solar system. While it's not to be doubted that a large number of them resulted from impacts of comets and asteroids, there are a good many that appear to have been caused by what are called "electric discharge machining," or "electric arc machining." This is a process whereby a discharge hovers over the same area for a time, carving out a circular depression that looks like an impact crater, but has a flat bottom. Such flat crater floors are usually interpreted as old craters that got filled in by dust, lava, or some other material. But they could just as easily be explained by electrical forces, especially when they are accompanied by secondary craters in close proximity that are of comparable size, since such discharges, as demonstrated in the lab, will often jump from one spot to another, carving out additional circular depressions nearby. These flat-bottomed craters are also often found in association with "dark spots" on the nearby terrain--features that have also been produced in lab experiments--which indicate a charring of the surface by high-voltage charges.

Sinuous rilles, particularly common on the Moon, Mars, and somewhat common on Venus, seem to be other good examples of electric arc machining. They are usually connected with a crater, or a series of craters, where the streaming column of plasma that created them had begun at one spot, remained there for a time as it created a crater, and then began to wander, tracing out a twisted linear depression on the surface while it snaked along. Sometimes the column would pause periodically, etching with every pause an additional crater within the rille. These serial craters are usually elongated and often overlap one another, forming the exact same patterns that have been reproduced by electric arc machining experiments in the lab. Mars has quite a few such serial-cratered rilles, as does its moon Phobos, our own Moon, and Earth. (Even lightning strikes on Earth today have been known to produce small-scale breaches in the ground that look much like sinuous rilles, minus the crater-like features within)[Endnote 75].

The column-type discharges that formed these rille features still occur on Earth and Mars to this day, but they are much weaker than they were in the past. Today they create mere streaks across the surfaces of both planets, rather than long and deep linear ditches. We call these modern-day columns "dust devils." The electrical nature of these mysterious phenomena were only recently discovered. As MSNBC reported on April 20, 2004: "Whirling dust devils on Mars probably generate high-voltage electric fields and associated magnetic fields....

"The conclusion is based on studies in Arizona and Nevada, where researchers raced across the deserts to catch dust devils and drive right through them. They found unexpectedly large electric fields exceeding 4,000 volts per meter....

"On Mars, dust devils can be five times wider and soar 5 miles, much higher than even full-blown tornadoes on Earth"(193). It's not known why Mars, being half the size of Earth, would produce dust devils so much larger than those produced here.

Sometimes it happens that the trails left by dust devils contain circular formations within them, just like what we see in the sinuous rilles, except that the circles in the rilles, of course, are plasma-carved craters. Isn't it obvious, then, that if today's dust devils were more highly electrically charged, as they clearly were in ancient times, they would carve out rilles with circular depressions within them, looking just like the serial-cratered sinuous rilles described above? [Endnote 76]

It should be pointed out, too, that pictures of the Martian surface that display multiple streaks from dust devils bear a striking resemblance to the linear markings on the surface of Jupiter's moon Europa. It is proposed elsewhere in this study that these Europa streaks are cracks caused by crustal spreading, and that may still be true for a good many of them. Nevertheless, some may indeed have been created by electrical discharges from Jupiter, especially when considering how profoundly similar they look to the smaller-scale dust devil streaks on Mars.

Could it have been electrified dust devils, although of much greater size and strength, appearing in the sky as glowing columns of plasma and dust, that the Old Testament referred to with such labels as "a pillar of a cloud" or "a pillar of fire," etc.? Note the following references:

- "pillar of a cloud" by day, "pillar of fire" by night (Exodus 13:21, 22)[Endnote 77]
- "cloudy pillar" (Nehemiah 9:12)
- "pillar of the cloud" (Nehemia 9:19)
- "pillars of heaven" (Job 26:11)

And could the following Old Testament references also be alluding to plasmic discharges, where mention is made of fire coming down from heaven, along with similar phrases?

- "fire from the Lord out of heaven" (Genesis 19:24)
- "the mountain burned with fire unto the midst of heaven" (Deuteronomy 4:11)
- "there came down fire from heaven" (2 Kings 1:10)
- "the fire of God is fallen from heaven" (Job 1:16)

Similar type phenomena have also been observed in more relatively recent times. In the Russian Far East in the year 1111 A.D., a fiery pillar appeared from the ground and rose several miles into the sky. Lightning shot out all around the pillar, and it was accompanied by a thunderous noise.

During the course of an evening in January of 1319, over most areas of Russia, numerous witnesses observed "fiery pillars" similar to those sighted in 1111, which extended from the ground toward the sky. Some people also sighted a "heavenly arc."

We talked about Jupiter (or Zeus) a short time ago, as being a major dispenser of electric discharges in ancient times, and how some, or at least one, of these outbursts apparently struck the Earth. Well, Jupiter is still at it, although not on such a grand scale. A major flux tube currently exists between Jupiter and its innermost large satellite, Io. This tube, in the shape of a ring, runs perpendicular to the plane of Io's orbit, conducting a powerful current that runs between the poles of Io and Jupiter. In addition to this, there is also a powerful electrified plasma torus that exists all along the orbital path of Io. It has actually been photographed in visible light by spacecraft, achieved by leaving the camera shutter open long enough to capture its faint glow. The flux tube is roughly 400,000 volts, 5 million amps, and 2 quadrillion watts(192). So strong are the electrical discharges of Io's flux tube, in fact, that Io, when photographed in Jupiter's shadow by NASA's Galileo spacecraft, was seen to literally glow in the dark in a powerful electrified plasmic cloud that entirely enveloped it [Endnotes

78 and 79].

The aurorae that we see here on Earth today, in the extreme northern and southern latitudes, are caused by the same force as the glow around Io--a plasmasphere [Endnote 80], except that Io's is much stronger, existing visibly on a global scale. But it may have been that the Earth was once fully enveloped in a visibly-glowing plasmasphere as well, seeing that both its plasmasphere and magnetosphere were much stronger long ago. This having been the case, the whole night sky (and perhaps the day sky as well) must have once been aglow with plasmic activity, all over the planet [Endnote 81].

In the ancient skies, the plasmas that appeared must have assumed many unusual and awe-inspiring shapes. We could imagine the ancients having portrayed these shapes in their rock art, as "messages from the gods." And indeed, we find that many archaic motifs depicted on stone, found at locations around the world, match with uncanny accuracy the patterns that have been produced with plasmic discharges in the laboratory.

For example, a common universal theme in ancient rock art is what is called the "squatting man," found in such diverse places as Arizona, New Mexico, Armenia, Guiana, Spain, the Alps, United Arab Emirates, Italy, and Venezuela. It's a stick figure with circles or dots on either side of its torso. This exact same design, in plasmic form, has been produced in the lab. It appears whenever donut-like toroids (toruses), stacked one upon another, are bent by induced magnetic fields. From the viewing perspective of the ancient onlookers who beheld these marvels in the sky, the edges of the upper toroid of each figure may have appeared to point upward (forming the "arms" of the "squatter"), whereas the lower toroid may have appeared to point downward (forming the "legs").

Some rock art glyphs, found in places like Arizona and Australia, depict squatter-type designs that have multiple "arms" and "legs." This same pattern has also been plasmically reproduced in the lab, as a system of stacked toroids. The rock art representations of these features sometimes have two "feet" at their bases in the form of circles within circles, looking like owl eyes.

This "owl eyes" configuration is often drawn separately in rock art, such as on Easter Island, and is also found on the Sumerian goddess Inanna (which was associated with Venus) and the eye goddess of the Columbia River region of the American Northwest. Computer simulations of plasmic toroids have produced this exact same owl-eye pattern, which must have been seen by the ancients in the sky, and served as their inspiration for such artwork as just described. There is even a pair of galaxies (NGC 2207 and IC 2163) that take on the exact appearance of "owl eyes." The ancients may very well have seen a plasmic configuration just like this in the skies of the distant past, although on a vastly smaller scale and at a far more closer range, perhaps created by two asteroids in close proximity to each other, instead of two far-away galaxies.

In many ancient works of art that depict the epic thunderbolt battle between the hero deity and the conquered dreadful dragon, weapons are usually pictured in the hand of the vanquishing hero, through which the deadly lightning blows were delivered. For instance, in an Akkadian work, Ninerta is pictured, as he is about to kill Anzu the dragon, with a weapon in each hand that has three prongs on either end. These miniature double pitchfork weapons, associated with the dissemination of lightning bolts, look strikingly like laboratory-produced bipolar plasmic discharges that have "pinches" along their axes, and pitchfork-like filaments extending out of both symmetrical ends. Could such discharges have been seen in the sky by the ancients, who viewed them as the weapons of their heroic deities, and thus incorporated them into their art and mythology?

Interestingly, there are several deep-sky nebulae that take on this very type of appearance, such as the Butterfly Nebula and the Ant Nebula. Perhaps these are large-scale versions of what appeared to viewers of the Earth's ancient skies, except that the nebulae they saw were probably around nearby planets or asteroids, instead of distant stars.

* * * * * * *

With all the evidence presented in this appendix, there can be no doubt that electrical discharges played a major role in the flood and its aftermath. Only when we take this factor into consideration can we make sense of so many ancient legends, myths, and artworks from around the world, as well as geologic formations from around the solar system, which would otherwise remain unexplained.

Endnotes

1. Another unusual feature of the early Earth to make note of, which is also described in the biblical book of Genesis, is the fact that it was apparently never darkened by gloomy, cloudy skies. In chapter two, verse five, we are told that originally there was no rainfall on Earth. But if this were the case, how, the obvious question arises, were plants watered? The Genesis record goes on to explain that "...there went up a mist from the earth, and watered the whole face of the ground." Genesis 2:6. In other words, there appears to have been a global subterranean water table that sent a mist up from underground to water the Earth's flora (perhaps there was a global network of artesian springs--more on this later).

We are also told in chapter 9 of the book of Genesis that rainbows were unknown at this early time. Additionally, we find in this same chapter further confirmation that clouds, or at least major cloud systems, were also non-existent: "I [God] do set my bow in the cloud, and it shall be for a token of a covenant between me and the earth. And it shall come to pass, when I bring a cloud over the earth, that the bow shall be seen in the cloud." Genesis 9:13, 14.

For comparison, the Babylonian tale of Gilgamesh states as well that rainbows were unknown to those who lived in the remote past. Beyond this, many ancient legends from all over the Americas tell this same story(151). One Native American document that records this point, in particular, is the Mayan *Book of Chilam Balam of Chumayel.*

2. Compelling scientific evidence exists to support the ancient water canopy hypothesis. One important piece of such evidence comes to us from the observations of NASA satellites, which have confirmed that there is far more hydroxyl in the Earth's hydrosphere than what prevailing models had predicted. How does this provide evidence for the water canopy theory? The parent molecule of hydroxyl (OH) is water (H2O). Since ultraviolet radiation from the Sun breaks down water in the Earth's upper atmosphere into hydroxyl and hydrogen, the current high level of hydroxyl at this high atmospheric layer tells us that a much larger amount of water must have previously existed there. This former abundance of water in the hydrosphere is now generally acknowledged by most scientists(115).

3. Roland T. Bird, a paleontologist from the American Museum of Natural History, speaking of these human footprints in Glen Rose, wrote: "Yes, they apparently were real...the strangest things of their kind I had ever seen....[They were] perfect in every detail. But each imprint was 15 inches long!"(10)

4. Regarding ancient India's 360-day calendar year, author Richard Mooney wrote: "In all the ancient classic writings of the Hindu Aryans, there is a year of 360 days. The *Aryabhatiya*, the ancient Indian mathematical and astronomical work, says: 'A year consists of 12 months. A month consists of 30 days. A day consists of 30 nadis. A nadi consists of 60 vinadikas'"(5).

In reference to ancient Rome's reckoning of the calendar year as being 360 days in length\, Mooney wrote: "The ancient Romans...had a year of 360 days. Plutarch, in his life of Numa, wrote that in the time of Romulus the year was made up of twelve 30-day months"(Ibid.).

The biblical book of Genesis provides some interesting insight on this matter. We are told in Genesis 7:11 that the great flood of Noah had begun on the 17th day of the 2nd month. Then we are told that the Ark rested upon dry land on the 17th day of the 7th month--exactly 5 months later (Genesis 8:4). We are then told, in Genesis 7:24; 8:3, that this duration of 5 months was actually 150 days. This would give us an even 30 days per month before the flood, which would of course equal out to 360 days in a 12-month year.

We find the same thing occurring in the New Testament. For instance, in Revelation 12:6 we are told of a woman being in the wilderness for 1260 days. Then we are told, in verse 14, that this same woman was there in the wilderness for 3 ½ years. This would make each year exactly 360 days long. So again we can see the old 360-day year being utilized, here in the form of prophetic symbolism, obviously harkening back to an earlier time when the year was originally this shorter length.

5. The Earth's speed of rotation (and thus the length of its day) was also likely affected at this time, but we will discuss this later.

6. Some might argue the point that, if the Earth was situated slightly closer to the Sun in ancient times, this may not necessarily have significantly contributed to its overall warmer climate at that time. For, as the argument goes, we find that currently the Earth, because of the eccentric shape of its orbit, varies in its distance from the Sun throughout a given year by a factor of 3 million miles--a variation which has little if any effect on its global temperature. In fact, during the winter months, the Earth is actually closer to the Sun (about 91.5 million miles) than it is during the summer months (about 94.5 million miles). Thus it is true, by today's standards, that the Earth's overall temperature is not altered by a varying distance from the Sun of several million miles. However, such may not have been the case in ancient times. With the effects of the aquatic canopy, which trapped probably 100% of all the solar heat that the Earth received, a closer position to the Sun would undoubtedly have resulted in noticeable temperature increases on a global scale.

7. Egyptian archaeology confirms this as well. An old star chart was found on the ceiling of Senmut's tomb (Senmut was the architect of Queen Hatshepsut). This chart shows the sky in the complete reverse of what it is today, with the north in the south and the east in the west. The constellations are also shown to move from west to east, rather than east to west.

8. Most people scoff at the idea of a global flood. However, the evidence of such a disaster is undeniable and overwhelming. We will be looking at much of this evidence as we proceed. But for now, consider the fact that roughly 70% of the Earth's rock formations are sedimentary in nature, which were laid down by water--lots of water. Also, the Himalayas are the tallest mountain range in the world. Mount Everest, the tallest of these mountains, is over five miles high. And yet, all up and down its slopes are found fossils of saltwater fish and shells, testifying that this colossal structure was once below sea level. In fact, the same holds true for pretty much all other mountains, everywhere.

Whether the sea was once higher or the land was once lower doesn't matter at this point. What does matter is that the evidence clearly points in the direction of a complete inundation of all points of dry land in the distant past, regardless their current elevations.

9. Actually, the Earth's core might be made up of a completely different material, as we shall see later.

10. It should be pointed out here that evidence does exist for multiple shifts in the Earth's

rotational axis in the remote past, which may indeed have been responsible for these multiple reorientations in the Earth's core and its magnetic pole (see Appendix B for more details).

11. Actually, there are quite a few more known moons than this, but they are of no substantial size, and probably represent simple fragments of larger moons, or captured interplanetary debris.

12. Jupiter's moon Io is also known for having a lot of volcanoes. However, this particular feature of this satellite does not appear to be the result of a past catastrophe, since Io is currently volcanically active. Thus its volcanic nature is most likely the result of current tidal forces acting upon it, being pulled by Jupiter on one side, and by Jupiter's other three large moons on the other side--Callisto, Ganymede, and Europa.

13. In regards to this proposed exploded planet that once existed between Mars and Jupiter, it is significant to point out at this time the existence of a 4,500-year-old Akkadian cylindrical seal, now in the Vorderasiatisches Museum in Berlin, Germany, which depicts all the planets of our solar system (except Pluto, which, as already stated, is not really a planet), in correct relative position and size. Not only does this seal show planets that cannot be seen with the naked eye (Uranus and Neptune), but, most strikingly, it also shows the existence of a planet between the orbits of--you guessed it--Mars and Jupiter. Actually, there were several ancient cultures that were aware of the existence of this long-lost planet. According to Greek mythology, for instance, it was known as Phaeton, and to the Sumerians it was known as Tiamat(25).

14. It could also be that the chunk of debris that hit this now-extinct planet came from a dust lane in a spiral arm of our Milky Way galaxy, which our Sun passed through in its orbit around the galactic center.

15. We will proceed in this study under the assumption that this extinguished planet orbited the Sun by itself, without being accompanied by any moons. However, let the reader be aware that if this planet did possess one or more satellite companions, the amount of debris being ejected out into the solar system would obviously have been far more substantial, thus inflicting even more damage.

Actually, the ancient Sumerians did assign this extinct planet with a satellite, and the name they gave it was Kingu.

16. Today the asteroid belt collectively contains only about 1/500 the mass of the Earth, indicating that the bulk of the material from the destroyed trans-Martian planet must have been violently thrown into distant regions of the solar system, creating havoc everywhere it struck.

17. This crater in the Gulf of Mexico (known as the Chicxulub Crater) is roughly 112 miles across and is believed to have been made by an object roughly 6 miles in diameter. But the discovery of an even larger crater in Antarctica was announced in early June of 2006, known as the Wikes Land Crater. Buried under a half-mile-thick sheet of ice, it measures 300 miles across and is believed to have been caused by the impact of an object that was around 30 miles in diameter(128). Yet the impact that formed this crater probably isn't the one that spawned the flood catastrophe either (for reasons already stated).

18. Between the 16th and the 20th of July, 1994, two dozen fragments of comet Shoemaker-Levy 9, which broke up while approaching the planet Jupiter (due to the tremendous tidal forces of its gravity), struck this planet in succession, creating large holes in its upper atmosphere as each piece hit. The fragments ranged between 0.62 and 5.5 miles across and impacted Jupiter at an average speed of 37 miles per second. The energy of the impacts averaged 1.7 million megatons for each 0.62 mile's worth of material (miles in diameter). These tremendous bursts of energy caused plumes of debris to be thrown hundreds of miles above Jupiter's cloud tops. Needless to say, if one of the larger pieces of this comet had struck the Earth, the results would have been devastating, but probably not as much as the impact that spawned our flood catastrophe.

19. This darkening of the sky by dust from the asteroidal impact, which blocked out the Sun, would have made a significant contribution to the onset of the Ice Age.

20. The intense heat from grand-scale fires described here, resulting from an impact by a celestial body, has been confirmed by modern science. As the May 20, 2007 *Guardian* reported: "Scientists will outline dramatic evidence this week that suggests a comet exploded over the Earth nearly 13,000 years ago, creating a hail of fireballs that set fire to most of the northern hemisphere.

"Primitive Stone Age cultures were destroyed and populations of mammoths and other large land animals, such as the mastodon, were wiped out. The blast also caused a major bout of climatic cooling that lasted 1,000 years and seriously disrupted the development of the early human civilisations that were emerging in Europe and Asia.

"'This comet set off a shock wave that changed Earth profoundly,' said Arizona geophysicist Allen West. 'It was about 2 km-3 km in diameter and broke up just before impact, setting off a series of explosions, each the equivalent of an atomic bomb blast. The result would have been hell on Earth. Most of the northern hemisphere would have been left on fire.'

"The theory is to be outlined at the American Geophysical Union meeting in Acapulco, Mexico. A group of US scientists that include West will report that they have found a layer of microscopic diamonds at 26 different sites in Europe, Canada and America. These are the remains of a giant carbon-rich comet that crashed in pieces on our planet 12,900 years ago, they say. The huge pressures and heat triggered by the fragments crashing to Earth turned the comet's carbon into diamond dust. 'The shock waves and the heat would have been tremendous,' said West. 'It would have set fire to animals' fur and to the clothing worn by men and women. The searing heat would have also set fire to the grasslands of the northern hemisphere. Great grazing animals like the mammoth that had survived the original blast would later have died in their thousands from starvation. Only animals, including humans, that had a wide range of food would have survived the aftermath'"(185).

21. A plasma, incidentally, is a gaseous cloud of electrically-charged atoms (ions), which may or may not be visible to the naked eye.

22. Regarding the Kentucky human footprint fossil find, Professor W.G. Burroughs, head of the department of geology at Berea College in Berea, Kentucky, reported in 1938: "During the beginning of the Upper Carboniferous Period [the end of the Devonian to the beginning of the Permian], creatures that...had human-like feet left tracks on a sand beach in Rockcastle County, Kentucky."

Geologist Albert G. Ingalls, in a 1940 edition of *Scientific American*, gave the following response to the discovery of these tracks that is so typical of mainstream science regarding such finds: "If man...existed as far back as in the Carboniferous Period..., then the whole science of geology is so completely wrong that all the geologists will resign their jobs and take up truck driving. Hence,...science rejects the attractive explanation that man made these mysterious prints in the mud of the Carboniferous with his feet"(39).

This is a most telling admission. It's the old "sweep it under the rug" tactic. Whenever finds don't conform to preconceived ideas, they are either suppressed, reinterpreted, or scoffed at as fakes. As author Michael Cremo put it: "[S]cience does not always operate according to its high ideals. The way science works, we are normally told, is on the basis of free and open discussion of evidence and ideas. But...we see elements of the scientific community restricting access to evidence and preventing open discussion of it. Yes, there is in fact a knowledge filter"(40).

23. Coal seams, especially those that are dozens of feet thick (which are fairly common), were quite clearly formed under catastrophic conditions. As Harold Coffin put it: "The thickness of peat needed to produce one foot of coal depends on a number of factors, such as the type of peat, the amount of water in the vegetable matter, and the type of coal. The scientific literature on coal gives figures ranging from a few feet to as many as twenty. Let us assume that ten feet would be near the average figure. On this basis, a coal seam thirty feet thick would require the compression of 300 feet of peat. A 400-foot seam of coal would be the result of a fantastic 4,000 feet of peat.

"There are few peat bogs, marshes, or swamps anywhere in the world today that reach 100 feet. Most of them are less than 50 feet. A much more reasonable alternative theory is that the vegetable matter has been concentrated and collected into an area by some force, undoubtedly water....

"The concept of a global deluge that eroded out the forests and plant cover of the pre-flood world, collected it in great mats of drifting debris, and eventually dropped it on the emerging land or on the sea bottom is the most reasonable answer to this problem of the great extent and uniform thickness of coal beds"(34).

24. Another unusual feature observed in fish fossils from the Green River Formation (and in other fish fossils from similar formations throughout the world) is the manner in which some specimens are found with smaller fish sticking out of the mouths of larger fish. In other words, these larger fish were buried alive while in the process of eating. Their burial was so sudden that they didn't even have time to finish their last meal. Can it be denied that fossils like these were formed under catastrophic conditions?

A further fossil discovery of a similar nature involves is an ichthyosaur from England, which was preserved in the process of giving birth(149).

25. A similar account to this is found in the apocryphal Book of Jasher. It states that "the Lord caused the whole earth to shake, and the sun darkened, and the foundations of the world raged, and the whole earth was moved violently, and the lightning flashed, and the thunder roared, and all the fountains in the earth were broken up, such as was not known to the inhabitants before."

Still another Hebrew record of this sequence of events is found in ancient Talmudic material, which was derived from even older Babylonian traditions. Commenting on this issue, Louis Ginzberg wrote in his *Legends of the Jews*: "The flood was produced by the union of the male waters, which are [were] above the firmament [the ancient aquatic canopy], and the female waters issuing from the earth [spewing up from the underground global water chamber]. The upper waters rushed through the space [the puncture hole] left when God removed two stars [chunks of debris from the exploded planet?] out of [the direction of] the constellation Pleiades....There were other changes among the celestial spheres during the year of the flood"(158).

26. Other apparent references in ancient legends to the breakup of the aquatic canopy and the release of its waters might be the frequent allusions to the sky falling, which are common the world over.

The Cashinaua Indians of Brazil, for instance, recall a time of deafening thunder, powerful lightning strikes, and a "collapsed sky"(153).

Similarly, the African Ovaherero tribe has an ancient legend stating that the sky "fell down."

In the Mayan *Chilam Balaam*, we find this account: "During the Eleventh Ahau Catoun it occurred...when the earth began to waken. And a fiery rain fell, and ashes fell, and rocks and trees fell down....And then, in one watery blow, came the waters...the sky fell down and the dry land sank. And in a moment the great annihilation was finished"(18).

According to ancient Aztec mythology, "The sea was thought to extend outward and upward until--like the walls of a cosmic house--it merged with the sky [a reference to the ancient aquatic canopy?]....The sky, therefore, was known to contain waters which might in perilous times descend in deluges, annihilating man"(44).

The people of Easter Island in the remote Pacific preserve this legend: "In the days of [King] Rokoroko He Tau, the sky fell...from above on to the earth"(209).

Other cultures whose ancient traditions tell this same story include the Celts, the Eskimos of Greenland, the Lapps of Finland, and the ancient peoples of Tibet, China, and Samoa.

27. Evidence of the continents having once been joined together comes in many other forms, aside from the fact that their contours, and those of the mid-oceanic ridges, line up perfectly. One of the most striking bodies of such evidence is the close relationship that exists between the flora and fauna that are found at or near coastal regions of land masses that are now separated by vast bodies of ocean water, but were at one time connected before the Pangaea breakup. For a brief treatment of this subject, see Appendix A.

28. The lower crust layer--the ocean floors--averages about 4 to 5 miles thick, as compared to the upper crust layer--the continental land masses--which averages about 19 miles thick.

29. The Bible is another ancient document that appears to make reference to the Earth having expanded in ancient times. In Isaiah 42:5, for example, we read: "Thus saith God the Lord, he that created the heavens, and stretched them out; he that spread forth the earth..." A similar statement is found in Isaiah 44:24, where God is recorded as saying: "I am the Lord...that stretcheth forth the heavens...[and] spreadeth abroad the earth..." In both of these texts we see that the "spreading forth" of the Earth is equated with a "stretching out" of the heavens. Are these passages alluding to the fact that the ancient atmosphere was also stretched out as the Earth expanded? Other texts that mention this stretching out of the heavens are as follows: Isaiah 45:12; 51:13; Jer. 10:12; 51:15.

30. It so happens that the continental land masses are actually set down into the ocean floors (their bases are sunk beneath the surface of the ocean floors). So how could they possibly have glided across the ocean floors if they're stuck in place?

31. As far as how the mountain formations on the continents rose to their current high elevations, we will wait until later to deal with this issue.

32. A passage similar to this one, which seems to refer to the expanding Earth as having been the mechanism for the waters receding after the flood, is Psalm 136:6, which says: "To him [meaning God] that stretched out the earth above the waters..."

33. If the size and speed of this impacting asteroid was significant enough, it could have either slowed down or speeded up the rotation rate of Earth, depending on which side of the rotating planet it struck. If it hit the side that was rotating toward it, it would have slowed down the Earth's rotation. But if it hit on the side that was rotating away from it, the Earth's rotation would have been speeded up.

It would appear that what actually wound up happening is that the impacting body that hit the Earth had struck the side that was rotating toward it, slowing down the Earth's rotation rate. For we find that calculations made of the Earth's mass and angular momentum, as compared to those of other objects in the solar system, indicate that its day must once have been (and still should technically be) about 30 hours long.

I should mention, however, that such calculations are based upon the Earth having always been its current size. But if the Earth was about 40% smaller before the flood catastrophe, this would render these calculations null and void. With a smaller Earth, we would expect its rate of rotation to have been more rapid, because it would have had a different angular momentum. Thus the original length of an Earth day may have been shorter--perhaps around 14 to 16 hours.

34. Some might question how the Earth could have been pushed further from the Sun by a piece of debris flown from the direction of an exploded planet outside the orbit of Mars. Common sense would dictate that a large asteroid from this direction, far beyond the Earth's orbit, would have knocked the Earth closer to the Sun. However, if, at the time of impact, the Earth was located on the opposite side of the Sun from the direction of this approaching piece of debris, hitting the Earth from inside its orbit, the result may very well have been to knock it slightly further away from the Sun.

35. Or it may be that a discharge (or discharges) of this nature, except to a much more powerful degree, struck the Earth from Venus. This is no mere idle speculation. Today, Venus has an invisible plasma tail that reaches out almost as far as the Earth's orbit. It appears, however, that in ancient times it was much stronger, and thus did reach as far as Earth. (For more information on this, see Appendix D.)

36. The Moon also has a polar wobble, oscillating roughly 15 meters about its axis in a 3-year interval (by contrast, the Earth's wobble, or precession, varies across a range of about 45 degrees in a 26,000-year cycle).

But polar wobbles, of course, are not the only type of precessionary aberrations in the movements of celestial bodies in the solar system. There is also what is called orbital precession, which happens in two ways: One is the slight fluctuation in the inclination of the orbital plane that occurs over a long period of time. The other, also occurring over a long time period, involves the major axis of an elliptical orbit precessing within the orbital plane, in response to gravitational tugs from other celestial bodies (which results in the orbiting body tracing out a spirographic pattern over many hundreds or thousands of years). Pretty much all planets and moons exhibit both of these types of precession, which are probably all consequences of the exploded trans-Martian planet disaster.

37. Earlier we reviewed some of the more well-known pieces of evidence that dinosaurs continued to live beyond the flood catastrophe. This being the case, these dinosaurs may have grown to a smaller size than they did before this calamity, having adapted themselves to the new post-flood environment of less gravity, air pressure, and atmospheric oxygen content.

38. This appears to have happened with the planet Mercury. The surface of this planet exhibits enormous escarpments (or elongated ridges), some up to hundreds of kilometers in length and as much as three kilometers high. Certain ones cut through the rings of craters, and look as though they were formed by horizontal compression when Mercury, like the Earth, slightly shrank as its interior cooled in the prehistoric past. It is estimated that the surface area of Mercury shrank by about 0.1% (or a decrease of about 1 km of the planet's radius).

39. Here are just a few notable examples of such ancient submarine ruins:

1. The BBC reported on December 7, 2001: "A team of explorers working off the western coast of Cuba [in the Guanahacabibes Peninsula] say they have discovered what they think are the ruins of a submerged city built thousands of years ago.

"Researchers from a Canadian company used sophisticated sonar equipment to find and film stone structures more than 2,000 feet (650 metres) below the sea's surface....

"The explorers first spotted the underwater city last year, when scanning equipment started to produce images of symmetrically organized stone structures reminiscent of an urban development.

"In July, the researchers returned to the site with an explorative robot device capable of highly advanced underwater filming work.

"The images the robot brought back confirmed the presence of huge, smooth blocks with the appearance of cut granite.

"Some of the blocks were built in pyramid shapes, others were circular, researchers said.

"They believe these formations could have been built more than 6,000 years ago, a date which precedes the great pyramids of Egypt by 1,500 years"(102).

2. One of the most amazing underwater mysteries in the world is the Yonaguni megalithic structure, found off the coast of Japan in 1985 by Kahachiro Aratake, a professor at the University of the Ryukyus, while looking for schooling hammerhead sharks in the area. The 300-foot-long structure is situated 90 feet below the ocean's surface, and is comprised of massive blocks of stone that were obviously cut and set in place by an ancient civilization that existed long before this area was engulfed by seawater.

3. On January 19, 2002, the BBC reported: "The remains of what has been described as a huge lost city may force historians and archaeologists to radically reconsider their view of ancient human history.

"Marine scientists say archaeological remains discovered 36 metres (120 feet) underwater in the Gulf of Cambay off the western coast of India could be over 9,000 years old.

"The vast city--which is five miles long and two miles wide--is believed to predate the oldest known remains in the subcontinent by more than 5,000 years"(103).

40. Two other areas where large inland seas are believed to have once existed are the Amazon Basin and the Sahara Desert. The trapped flood waters that once resided in these locations must have, for the most part, simply dried up, since their locations were so near the equator. But some of these waters, at least initially, probably drained out to sea, just like the waters in the other, more northerly extinct lakes that resulted in the formation of the Grand Canyon and the Scablands, as described above.

41. Mars and Earth are not the only bodies in the solar system that have (or have had) underground global water chambers. Another such world is Jupiter's moon Europa. The website of NASA's Jet Propulsion Laboratory says of this world that "Evidence points strongly to a global subsurface ocean beneath an ice shell"(203).

Saturn's moon Titan may also have a subterranean water system. Says the website of the same establishment just cited: "NASA's Cassini spacecraft has discovered evidence that points to the existence of an underground ocean of water and ammonia on Saturn's moon Titan"(204).

Other possible candidates for underground oceans are Neptune's Triton and Saturn's Enceladus, both of which shoot up geysers from deep in their interiors.

Still other suspects are Jupiter's Ganymede and Callisto.

Even our own Moon may have had a major underground water supply. On July 9, 2008, the BBC reported: "US scientists have found evidence that water was held in the Moon's interior....

"...a new study in *Nature* magazine shows water was delivered to the lunar surface from the interior in volcanic eruptions [long ago]....

"The discovery came from lunar volcanic glasses, pebble-like beads collected [on the lunar surface] and returned to Earth..."(212).

Perhaps we can conclude from all of this that global underground oceans are (or were) features of ALL solid-surfaced celestial bodies.

42. Another factor that makes liquid water on Mars almost impossible today is its weak gravity, which is not able to retain water vapor. However, if Mars was smaller in the past, like Earth appears to have been, it would have had an increased gravitational force and a denser atmosphere, which would have made conditions on Mars much more conducive to liquid water existing on its surface.

43. Scientists have always known that the north polar cap of Mars was comprised of water ice, and it was believed that its southern cap was mostly carbon dioxide ice. However, it was discovered in mid-March of 2007 that this cap, too, is predominantly composed of water ice. On March 15 of that year, the Yahoo News website reported: "A spacecraft orbiting Mars has scanned huge deposits of water ice at its south pole so plentiful they would blanket the planet in 36 feet of water

if they were liquid, scientists said....

"The scientists used a joint NASA-Italian Space Agency radar instrument on the European Space Agency Mars Express spacecraft to gauge the thickness and volume of ice deposits at the Martian south pole covering an area larger than Texas.

"The deposits, up to 2.3 miles thick, are under a polar cap of white frozen carbon dioxide and water, and appear to be composed of at least 90 percent frozen water, with dust mixed in, according to findings published in the journal *Science*"(178).

44. It was discovered in 2007 that Mars may still have a great deal of its original flood waters from its catastrophe frozen underground, all over the planet. Some of it appears to be very deep underground, while other pockets lie much closer to the surface. The BBC reported on May 2, 2007: "Scientists in the US say that initial data from a new way of scanning Mars has shown [that] up to half of the Red Planet's surface may contain ice.

"The new method of scanning for water offers vastly more accurate readings than before, they say....

"The new data shows wide variation as to how deep below the surface ice exists.

"The deposits--far beyond the ice that is known to exist in the planet's North Pole [and South Pole]--could be so large that were they to melt, they would deluge the planet in water forming an ocean"(184).

45. For comparison, the impact that created the Chicxulub Crater in the Gulf of Mexico seems to have reeked havoc on the other side of the planet, just like Hellas and its two related impacts did on Mars (although the Martian version occurred on a much larger scale). For it happens that just opposite the Gulf of Mexico is the Deccan Plateau in India, where, as stated earlier, lava deposits 10,000 feet thick are found.

A similar arrangement exists on the planet Mercury. Just opposite the giant Caloris impact basin, exactly on the other side of the planet, lies a patch of bumpy terrain that may have resulted from the Caloris blast.

46. Apparently, in the post-flood world, not only was it a new innovation to eat raw flesh, but to eat flesh period. The book of Genesis reveals that the diet of man in the paradisal pre-flood world was vegetarian. In chapter 1, verse 29, we read: "And God said [to the inhabitants of Eden], Behold, I have given you every herb bearing seed, which is upon the face of all the earth, and every tree, in the which is the fruit of a tree yielding seed; to you it shall be for meat [food]." In the very next verse, we discover that there were no carnivores in the animal kingdom either: "And to every beast of the earth, and to every fowl of the air, and to every thing that creepeth upon the earth, wherein there is life, I have given every green herb for meat: and it was so."

In the immediate aftermath of the flood, because plant life was scarce, survivors were forced to resort to eating flesh out of desperation. Thus we have God making this statement to the flood survivors: "And the fear of you and the dread of you shall be upon every beast of the earth, and upon every fowl of the air, upon all that moveth upon the earth, and upon all the fishes of the sea; into your hand are they delivered. Every moving thing that liveth shall be meat [food] for you; even as the green herb have I given you all things." Genesis 9:2, 3. And then we find this related account in the apocryphal Book of Adam and Eve: "They sought for food everywhere. For nine days they roamed about, but found none such as they used to eat in Paradise [the pre-flood world]. 'We used to have angels' food then,' they complained"(17).

47. In early September of 2006, a team of British archaeologists came to this very conclusion--that civilization arose in the wake of a catastrophic change in the environment. The online British news journal that covered this story, Eurekalert.org, stated: "Severe climate change was the primary driver in the development of civilisation, according to new research by the University of East Anglia [in eastern England]....

"In a presentation to the BA Festival of Science on September 7, Dr. Nick Brooks will challenge existing views of how and why civilisation arose. He will argue that the earliest civilisations developed largely as a by-product of adaptation to climate change and were the products of hostile environments.

"'Civilisation did not arise as the result of a benign environment which allowed humanity to indulge a preference for living in complex, urban, "civilized" societies,' said Dr. Brooks.

"'On the contrary, what we tend to think of today as "civilisation" was in large part an accidental by-product of unplanned adaptation to catastrophic climate change. Civilisation was a last resort--a means of organising society and food production and distribution, in the face of deteriorating environmental conditions'"(131).

No doubt, the earliest post-flood civilizations arose in the Middle East--an area that was quite plush at that time, perhaps due to the melting of glaciers in more northern latitudes, which provided a huge supply of rain to the Nile region. Bearing this in mind, we would do well to take into account the writings of Diodorus, who said that the "first genesis of living things [after the flood] fittingly attach[ed] to this country [Egypt]." He further went on to explain that "the moisture from the abundant rains [caused by melting glaciers]...was mingled with the intense heat which prevails in Egypt itself...[and] the air became very well tempered for the first generation of living things [after the flood]"(20).

48. Perhaps autistic savants can provide us some insights into the superior brain functions that pre-flood man appears to have had. Savants are capable of performing incredible mental feats. For example, they can solve long, complicated mathematical equations faster than one can type them into a calculator, and with perfect accuracy at that. They can also memorize astonishing amounts of information, and then recite it all back without making a single mistake. Are these geniuses in possession of mental powers that the rest of us are incapable of exercising? Or can it be that we all have the potential for such abilities, but are not able to tap into them so easily as savants are?

Perhaps, before birth, these gifted individuals experienced a surge of blood flow to the brain, maximizing the amount of oxygen reaching that organ, resulting in certain areas of the brain being more developed than others. Had more oxygen reached our brains at key points of their developmental stages, perhaps we too would be able to perform such amazing feats. And maybe this is what happened with pre-flood man. Because there was more oxygen in the atmosphere, and thus in the bloodstream, antediluvian man could very well have had the genius of savants, and even more so.

49. Don't you find it rather strange how the ancients recognized their early ancestors as having been more advanced than themselves, while modern scientists, for the most part, insist that the earliest progenitors of the human race were primitive brutes?

50. While we're on the subject of the origin of the earliest civilizations, it seems an appropriate time to mention the origin of languages. Most modern languages are believed to have been

derived from one common protolanguage. As the November 27, 2003 edition of the *Washington Post* reported: "Indo-European--the mother tongue of such modern and extinct languages as American English, Haitian Creole, Gaelic, Punjabi and ancient Hittite--originated 8,000 to 10,000 years ago, perhaps among ancient farmers in what is now Turkey, according to new research released today"(107). With this thought in mind, it is of profound interest to draw attention to the fact that the biblical book of Genesis reveals this same thing--that in the early post-flood era, man had spoken one language: "And the whole earth was of one language, and of one speech." Genesis 11:1. It would appear that this single language had originated before the flood, and that multiple languages did not come into use until after this catastrophe.

I also want to draw attention to the fact that the above-cited *Washington Post* quote reveals that this original protolanguage has been traced to ancient Turkey, the very place where we are told that Noah, the famous flood survivor, had settled after the flood waters dissipated--right in the mountains of Ararat in eastern Turkey. As the book of Genesis puts it: "And the ark rested in the seventh month, on the seventeenth day of the month, upon the mountains of Ararat." Genesis 8:4.

The Bible is not the only ancient record that speaks of a single, original language that all men once spoke. The Mayan *Popol Vuh* states: "Those who gazed at the rising of the sun [the ancients]...had but one language....This occurred after they had arrived at Tulan, before going West. Here the language of the tribes was changed. Their speech became different. All that they had heard and understood when departing from Tulan had become incomprehensible to them....For the tongue[s]...had already become different....Alas, alas, we have abandoned our speech! Why did we do this?...Our language was one when we departed from Tulan, one in the country where we were born"(18).

The Navajo tribe says that in long ages past, all men "spoke one tongue," but that soon after the great flood there "came many languages"(17).

Likewise, the Babylonian historian/priest Berossus, in the third century B.C., and the Greek historian Polyhistor, in the first century B.C., had both talked about this ancient single language that was once universally spoken(111).

51. Then again, it may be that this hidden power in mercury HAS been rediscovered, but that the knowledge of it is being suppressed. Wikipedia has this to say about a mercury compound known as "red mercury": "Samuel Cohen, the 'father of the neutron bomb,' has been claiming for some time that red mercury is a powerful explosive-like chemical known as a ballotechnic. The energy released during its reaction is enough to directly compress the secondary without the need for a fission primary. He claims that he has learned that...Soviet scientists perfected the use of red mercury and used it to produce a number of softball-sized 'pure fusion' bombs, which he claims were made in large numbers.

"He goes on to claim that the reason this is not more widely known is that elements within the US power structure are deliberately keeping it 'under wraps' due to the scary implications such a weapon would have on nuclear proliferation. Since a red mercury bomb would require no fissile material, it would seemingly be impossible to protect against its widespread proliferation given current arms control methodologies. Instead of trying to do so, they simply claim it doesn't exist, while acknowledging its existence privately"(186).

52. Cloaking devices are certainly not out of the realm of possibility. In fact, such a technology has been under development for a long time in our modern world, with much progress having been made in the early twenty-first century. And, as you might guess, the primary motivating factor for

the development of this technology, as appears to have been the case in ancient times, is its military applications. Speaking on this matter, the March 1, 2005 edition of *The Scotsman* stated: "A cloaking device that makes objects invisible is being developed by researchers....

"[E]lectronic engineers at the University of Pennsylvania are working on a[n]...invisibility shield called a 'plasmonic cover.'

"The development, which works by preventing objects from reflecting and scattering light, could have widespread use in the military as it would be more effective than current stealth technology"(166).

There has also been a lot of research conducted to develop "invisibility cloaks" (or coats) that enable people to become "invisible." The progress made in this area is actually quite astonishing. As the February 7, 2003 *USA Today* revealed: "A University of Tokyo professor claims he and his research team have developed a system that can make you 'invisible.' Engineering Professor Susumu Tachi is in the early stages of technology that he says will eventually enable camouflaged objects to be virtually transparent by wearing an optical device. Professor Tachi demonstrated the technology....In a photo of graduate student Kazutoshi Obana, it appears as if three men walking in the background can be seen 'through' Obana's green overcoat"(165).

This "invisibility coat" sounds so reminiscent of something that the ancient Greeks are said to have had--a "magical" helmet that rendered the wearer invisible. Likewise, ancient Tibetan documents speak of the secret of "antima,"described as the "cap of invisibility"(63). It would appear, then, that the ancients were a considerable degree ahead of us in this area.

53. Other advanced weapons mentioned in ancient texts from India, which weren't necessarily employed from the air, include:

1. Chemical and biological agents: One text talks of how, when one particular weapon was unleashed, "poison was produced, covering the earth with deadly fumes" which "stupefied....[It was] the poison that would have destroyed the world."

Another text describes "Sumhara" (a missile that caused crippling) and "Mohamastra" (a weapon that completely paralyzed its victims).

An "Avidiastra" was a weapon that supposedly attacked the nervous system of its human targets.

A "Prasvapana" was said to induce sleep.

And, finally, a "Sikharastra" seems to have produced napalm-like effects. A similar weapon is described in the *Mahavira*, where we are told how Rama used "a fire weapon capable of reducing the great army of Kumbhakarana to ashes."

2. Beam weapons: A weapon called "Kapilla's Lance" is described as having been capable of turning 50,000 men to ashes in just a few moments of time(63). Was this some type of laser canon?

54. On a similar note, the *Ramayana* states that when the god Rama was threatened by an "army of monkeys," he put a "magic arrow" into action. This produced a flash of lightning "stronger than the heat from a hundred thousand suns," turning everything to dust. The document then went on to say that the hair of survivors fell out, and that their nails disintegrated(109).

55. There's no question that the ancients had ventured to the polar regions. Many archaic documents from around the world describe distant "wonderlands" where conditions existed that prevail only in these parts of the world.

One document in this class is the *Mahabharata*. It speaks of a land where "the night can hardly be distinguished from the day....The day and night are together equal to a year to the residents of the place."

Another text from ancient India, the *Surya Siddhanta*, says that "The gods beheld the sun, after it has once arisen, for half a year."

The seventh Mandala of the *Rigveda* refers to a place where there were many days "between the first beams of the dawn and actual sunrise."

And lastly, the Iranian *Zend Avesta* mentions a land (Airyana Vaejo) in which "the stars, the moon and the sun are only once a year seen to rise and set, and a year seems only as a day."

56. We talked earlier about how the ice sheet that covers Antarctica today is the result of flood waters that froze and were thus unable to drain back out to sea. This is true, for the most part. But as we can see, there were parts of Antarctica--some of its coastal regions--that were not covered with ice at the time of the post-flood world surveys (as the Piri Reis map, and other maps like it, reveal). So where did the rest of this ice come from, which now covers the coastal areas of Antarctica? It came from post-flood asteroidal impacts that resulted in Earth tiltings and tsunamis that sent massive amounts of water over the coastal regions that quickly froze from the Sun being blocked out, thereby expanding the size of Antarctica's ice cap. This coastal Antarctic ice has never melted, as it did in other areas of the world, because Antarctica, of course, is perpetually too cold. (See Appendix B for more information on post-flood Earth impacts and tiltings.)

57. The fact that this map shows the Azores to be larger, as well as showing a huge nearby island that is no longer there, is not so unusual, as this area is highly geologically active. Even in recent times islands have been seen to arise and subside in a short time. For example, let us take a second look at a Charles Berlitz quote cited earlier: "In 1808 a volcano on Sao Jorge in the Azores crested several thousand additional feet and in 1811 a large volcanic island appeared in the Azores which, after being given a name--Sambrina--and charted on maps, suddenly returned to the sea"(29).

58. In a similar vein of thought, Mooney wrote that "Ancient Nordic maps...show Greenland as three separate islands. The French polar expedition (1947-1949) led by Paul-Emile Victor undertook a seismic survey of Greenland, which showed that under the ice cap Greenland actually was [at one point] composed of three separate islands. In shape and area, the ancient maps were extremely accurate"(5).

These maps must also obviously date to the early post-flood world survey period, before the Ice Age covered Greenland in a thick blanket of ice.

59. It was probably the Polynesians who introduced early South Americans to the panpipe. Specimens found in Peru, Bolivia, Colombia, and Panama are exact duplicates of those unearthed in the Solomon Islands. They are even tuned to the same pitch(110).

60. Regarding this Phoenician expedition, Schoch wrote: "[This] voyage, recounted in one terse paragraph by Herodotus, began in the Red Sea and took three years. The Phoenicians sailed all

summer, then put in as the weather turned, sowed a crop, and resumed traveling after the harvest. In this seasonal way they traveled down the east coast of Africa, rounded the Cape of Good Hope, and came up the long western reach of the continent, heading back into the Mediterranean through the Pillars of Hercules, or the Strait of Gibraltar"(1).

The total distance traversed in this voyage was over 15,000 miles. So the question must be asked, If this ancient culture was able to successfully undertake such a lengthy journey, why is it so hard for modern scholars to comprehend journeying across the Atlantic to the Americas by this same ancient group of people--a journey which is only a couple thousand miles or so? (Actually, at their closest points, the west coast of Africa [at Freetown, Sierra Leon] is only 1,800 miles from the eastern-most coast of South America [at Recife, Brazil].)

61. There is even evidence from the other side of the Atlantic that the Phoenicians reached Brazil. In a Phoenician tomb, an artifact was found that was made of Brazilian "Pau" (or axe-breaker) wood(209).

62. So obvious is the African influence on these giant stone Olmec head carvings, that even some mainstream academic works have admitted that these figures do indeed seem to indicate pre-Columbian trans-oceanic contact between the Old and New Worlds. The following quote from *World Civilizations*, a standard collegiate world history textbook, is one such example: "The origins of the Olmecs remain shrouded in mystery, but some of the enormous stone sculptures seem to have definite African features that indicate trans-Atlantic contact"(92).

63. The hoard uncovered in Venezuela was housed in an ancient ceramic jar on a beach. The jar contained hundreds of coins which ranged from the reign of Augustus (31 B.C. to 14 A.D.) right up to 350 A.D. This find, incidentally, is now in the Smithsonian. Was this the collection of an ancient coin enthusiast?(86)

64. One of Arabia's most famous historians and geographers, Al-Mas'Udi, who lived from 895 to 956 A.D., wrote about this advanced seafaring knowledge of Solomon. He stated in his work *Histories* that Solomon was in possession of maps that "showed the constellations, the stars, the earth with her continents and oceans, the inhabited landmasses, her plants and animals and many other wondrous things"(209).

65. Ogham scripts have actually been found all over the United States. In southeastern Colorado, for example, in a narrow cave known as Crack Cave, Barry Fell translated an Ogham inscription to read: "Strikes here on the day of Bel." Interestingly, at dawn, on the equinox, the Sun's first light strikes this inscription and the two marks etched above it(88).

66. It would appear that the Vikings made it to Peru, the land of the Incas. Or perhaps I should say that the Incas made it to Norway, either on their own or by tagging along aboard a Viking ship on a return voyage home. The basis for this proclamation comes from the July 2, 2007 edition of *The Norway Post*: "The remains of two elderly men and a baby were discovered during work in a garden, and one of the skulls indicates that the man was an Inca Indian.

"There is a genetic flaw in the neck, which is believed to be limited to the Incas in Peru, says archaeologist Mona Beate Buckholm.

"The Norway Post suggests that maybe the Vikings traveled even more widely than hitherto believed? Why could not the Viking settlers in Newfoundland have strayed further down the coast on one of their fishing trips?"(189)

67. It is in this very area--the Pacific basin--that many believe the lost continent of Mu (or Atlantis, or Lemuria) was once located. Speaking of the destruction of this extinct world, a pre-Columbian Mayan manuscript, the *Troana Codex*, says: "[T]here occurred terrible earthquakes, which continued without interruption until the thirteenth Chuen. The country of the hills of mud, the land of Mu, was sacrificed: being twice upheaved it suddenly disappeared during the night, the basin being continually shaken by the volcanic forces. Being confined, these caused the land to sink and to rise several times in various places. At last the surface gave way and ten countries were torn asunder and scattered. Unable to stand the force of the convulsion, they sank with their 64,000,000 of inhabitants"(17).

The Native people of Easter Island have a different name for this lost continent, Hiva. They say of it: "Hiva is a land that is gone. Now it is below the Pacific Ocean"(154).

68. If the Earth was indeed struck specifically by a comet (or by more than one comet) in the post-flood world, this would explain why so many cultures in the ancient world associated the scene of a comet in the night sky with a bad omen, striking the onlookers with terror.

69. A good example of such a land-based crater is the famous Barringer Meteor Crater in Canyon Diablo, Arizona. This perfectly-preserved impact hole is about a mile wide and 570 feet deep, and must have created a tremendous amount of havoc when it was formed, especially in the western hemisphere. This crater has got to be post-flood as well, since it remains in such a perfect state of preservation, which the violent flood conditions would not have allowed. It is, in fact, the world's best-preserved impact crater.

70. We find another interesting reference to sudden seasonal changes, this time in ancient China, from the historical memoirs of Se-Ma Ts'ien, who stated that Emperor Yahou sent astronomers to the "Valley of Obscurity" and to the "Sombre Residence" to observe the new movements of the Sun and the Moon, and "to investigate and to inform the people of the order of the seasons." It needs to be pointed out, however, that Yahou's reign was some 900 years before Joshua's time, and thus this text must be referring to an earlier post-flood axis tilt that changed the seasons.

71. Mars may also have undergone crustal displacement in its remote history. In the December 1985 issue of *Scientific American*, Peter H. Schultz talked about unique deposits on Mars that "show many of the processes and characteristics of today's poles, but they lie near the present-day equator." The reason for this, he said, appeared to be "the movement of the entire lithosphere, the solid outer portion of the planet as one plate....[which took place] in rapid spurts followed by long pauses."

The same seems to have been the case with Jupiter's moon Europa. According to a May 14, 2008 *New Scientist* article posted on this journal's website: "The icy outer shell of Jupiter's moon Europa may have slipped about 80° [in the distant past], carrying the moon's polar regions towards its equator..."(210).

72. Ordinarily, the protective plasmaspheres around the planets would have shielded them, in ancient times, from electrical discharges from any other planet. But if something wandered close

enough to a planet, a powerful discharge would have resulted, perhaps destroying the wanderer and setting that planet up, by lessening its charge, for a future deadly blast from a nearby planet, when it passed through its plasma tail. This, of course, would have resulted in the creation of a major scar on the surface of the target planet.

In the case of the trans-Martian planet's destruction, a piece of debris (again, probably from a supernova explosion) may have wandered close enough to this planet that it created an imbalance in its charge as compared to other nearby planets, which left it vulnerable to a deadly blast from a sister world (namely Jupiter). And once this planet was destroyed, its debris, flung in every direction, set off a chain reaction all throughout the solar system, enabling all the other planets, along with their moons, to strike with, or be struck by, such deadly bolts. These discharges were much like the static electric shocks that we are all accustomed to, which balance out inequities in charges, except on a much smaller scale, of course.

73. Pliny stated that Mars, in the distant past, had discharged a powerful electrical jolt at the Earth, adding that it specifically hit Bolsena, a town in Tuscany. And there just so happens to be a large lake in that town today, bearing the same name as the town, the basin of which may very well have been created by this event. Pliny stated that, according to the Babylonians, Mars had "moved the earth off its hinges." He also wrote that "The theory of the Babylonians deems that even earthquakes and fissures in the ground are caused by the force of the stars [meaning planets] that is the cause of all other phenomena, but only by that of those three stars [planets] to which they assign thunderbolts"(12). The three planets referenced here were Mars, Jupiter, and Saturn.

Speaking of these three planets being sources of ancient powerful electrical discharges, an April 15, 1951 *New York Times* article mentioned these same three planets as being responsible for modern electrical disturbances in the Earth's upper atmosphere (although clearly to a much lesser extent than was the case long ago). This article mentioned "a strange and unexpected correlation between the positions of Jupiter, Saturn, and Mars, in their orbits around the sun, and violent electrical disturbances in the earth's upper atmosphere. This would seem to indicate that the planets and the sun share in a cosmic/electrical balance mechanism that extends a billion miles from the center of our solar system. Such an electrical balance is not accounted for in current astrophysical theories." It is also noteworthy that shortwave radio frequencies get disturbed whenever these same three planets are lined up.

74. Of course, it was also proposed that Earth expansion may have been what created the Grand Canyon and similar formations around the world. It could be that several mechanisms were collectively responsible for such features.

75. Electric discharges can produce a large variety of unusual shapes on planetary surfaces, and have apparently done just that. Venus has several unusual geological features known as "pancake domes," most of which overlap one another. It is difficult to imagine these formations having come about in any other way than plasmic discharges.

76. Another phenomenon on Mars that appears to be the result of plasmic discharges in the past is the hoard of what are called "blueberries," which were found by NASA's Opportunity rover. They are small spheroids of hematite that have most scientists baffled as to how they were formed. But plasma cosmologists have found, through laboratory experiments, that plasma discharges can produce small nodules that look just like their Martian counterparts.

77. Recall that this particular phenomenon, as mentioned in Appendix B, may have instead been the volcanic plume from the eruption of Thera (Santorini).

78. According to some, the "volcanoes" of Io may actually be caused by powerful, yet invisible, electrical "lightning bolts" from Jupiter striking Io, ejecting material from its surface out into space in the process. One of the reasons for drawing this conclusion is that the eruption plumes are too symmetrical in shape--perfectly-formed domes, and their ejecta blankets are also perfectly-formed rings. How could a volcano erupt in flawless dome-shaped plumes and selectively deposit ejecta in symmetrical ring shapes? Also, Io's "volcanoes" often appear to be mere holes in the ground, rather than being cone-shaped. Another problem is that one of Io's "volcanoes," Prometheus, relocated its position 46.5 miles west during the interval between the Voyager 2 and Galileo spacecraft visits. How could a volcano move? All of these anomalies force doubt upon us as regarding the volcanic model for Io. Yet every one of these anomalies are exactly what we would expect from electric discharge machining.

79. Speaking of Jupiter's powerful electrical discharges, it may have been such a discharge that was responsible for the breakup of Comet Shoemaker-Levy 9 into numerous pieces, before it actually hit the planet back in 1994.

80. The reason we only see aurorae at the poles on Earth today is because all the lines of force of our planet's magnetic field intersect at these points, and thus the energy is concentrated significantly enough in these regions to produce visible plasma.

81. The general rule of thumb seems to be that the larger a celestial object is, the stronger its plasmasphere, and thus the brighter it glows. Plasma cosmologists believe that stars glow because of their powerful plasmaspheres that envelop their exteriors, and not because of any internal nuclear chain reactions, as is commonly thought. This may very well be true, since sunspots, which allow us to see below the surface of the Sun, have been measured to be much cooler in temperature than the Sun's surface. Thus stars may have relatively cold bodies underneath their hot plasma sheaths.

Plasmas are the most common, and apparently the most powerful, phenomenon in the universe. They are, it turns out, the cause of the ghostly glows of galaxies and nebulae.

References

1. Robert Schoch, *Voyages of the Pyramid Builders: The True Origins of the Pyramids from Lost Egypt to Ancient America.* New York: Putnam, 2003.

2. William F. Warren, *Paradise Found, or the Cradle of the Human Race.* Fredonia Books, 2002.

3. John Whitcomb and Henry Morris, *The Genesis Flood].* Phillipsburg, NJ: Presbyterian and Reformed Publishing Co., 19961.

4. Raymond and Rose Flem-Ath's *When the Sky Fell: In Search of Atlantis.* St. Martin's Paperbacks, 1995.

5. Richard E. Mooney, *Colony: Earth.* Greenwich, CT: Fawcett Publications, Inc., 1974.

6. http://minerals.cr.usgs.gov/gips/0amber.htm

7. Dennis Petersen, *Unlocking the Mysteries of Creation.* El Dorado, CA: Creation Resource Publications, 2002.

8. Carl Baugh, *Dinosaur.* Orange, CA: Promise Publishing Co., 1991.

9. Rene Noorbergen, *Secrets of the Lost Races.* New York: Bobbs Merrill Company, 1977.

10. "Thunder In His Footsteps," *Natural History*, May 1939.

11. S.A. Cranfill, *They Came From Babel.* Reno, Nevada: The Write House, Ltd., 1994

12. Immanuel Velikovsky, *Worlds in Collision.* Garden City, New York: Doubleday and Company, Inc., 1950.

13. W.B. Wright, *The Quarternary Ice Age.* London, 1937.

14. Paul D. Ackerman, *It's a Small World Afterall.* Grand Rapids, Michigan: Bakker Book House, 1986.

15. David Hatcher Childress, *Technology of the Gods*. Kempton, Ill: Adventures Unlimited Press, 2000.

16. Roland Dixon, "Achomowial Astugewi Tales," *Journal of American Folk-Lore*, 1908.

17. D.S. Allan & J. B. Delair, *When the Earth Nearly Died*. Bath, England: Gateway Books, 1995.

18. Charles Berlitz, *Mysteries from Forgotten Worlds*. Golden City, New York: Doubleday, 1972.

19. Giorgio De Santillana and Hertha Von Dechend, *Hamlet's Mill*. Boston: David R. Godine Publishing, Inc., 1977.

20. Graham Hancock, *Fingerprints of the Gods*. New York: Crown Publishers, Inc., 1995.

21. Harry Robert Turney-High, *Ethnology of the Kutenai*. Mildwood, NY: Draus Reprint Co., 1974.

22. Hiram Bingham, *Lost City of the Incas*. New York: Duell, Sloan and Pearce, 1948.

23. J.P. Kennett and N.D. Watkins, "Geomagnetic Polarity Change, Volcanic Maxima and Faunal Extinction in the South Pacific," *Nature*, Vol. 227, 1970.

24. Donald Wesley Patten, *The Biblical Flood and the Ice Epoch*. Seattle, WA: Pacific Meridian Publishing Co., 1966.

25. Hans Zillmer, *Darwin's Mistake*. Netherlands: Pioneer Publishing, 2002.

26. J.D. Mulholland, "Movements of Celestial Bodies--Velikovsky's Fatal Flaw," *Scientists Confront Velikovsky*, D. Goldsmith, ed., Ithaca and London, 1977.

27. Raymond Van Over, ed., *Sun Songs: Creation Myhos from Around the World*. New York: A Mentor Book of the New American Library, 1980.

28. James Mooney, *Myths of the Cherokee*, American Bureau of Ethnology Annual Report Part I. Washington, D.C.: U.S. Government Printing Office, 1900.

29. Charles Berlitz, *Atlantis: The Eighth Continent*. New York: G.P. Putnam's Sons, 1984.

30. Barry Fell, *Bronze Age America*. Boston: Little, Brown and Company, 1982.

31. William D. Stansfield, *The Science of Evolution*. New York: Macmillan, 1977.

32. Curt Teichert, *Bulletin of the Geological Society of America*, Vol. 69, Jan. 1958.

33. Robert Lee, "Radiocarbon, Ages In Error," *Anthropological Journal of Canada*, Vol. 19, No. 3, 1981.

34. Harold Coffin, *Origin by Design*. Hagerstown, MD: R & H Publishing, 1983.

35. *Encyclopedia Britannica*, under the heading "Geology," Vol. 10, 1956.

36. O. D. von Engeln and K. E. Caster, *Geology*. New York: McGraw-Hill, 1952.

37. Edmund M. Spieker, "Mountain-Building Chronology and Nature of Geologic Time-Scale," *Bulletin American Association of Petroleum Geologists*, Vol. 40, August 1956.

38. Walt Brown, *In the Beginning*. Phoenix, AZ: Center for Scientific Creation, 2001.

39. Michael Cremo\and Richard Thompson, *The Hidden History of the Human Race: The Condensed Edition of Forbidden Archaeology*. Los Angeles, CA: Bhaktivedanta Book Publishing, 1999.

40. Michael Cremo, "Forbidden Archaeology," a chapter in *You Are Being Lied To*, Russ Kick, ed. New York: The Disinformation Company, Ltd., 2001.

41. Wilfrid Francis, *Coal, its Formation and Composition.* London: Edward Arnold Ltd., 1961.

42. Hugh Miller, *The Old Red Sandstone.* Boston: Gould and Lincoln, 1857.

43. "Workers Find Whale in Diatomaceous Earth Quarry," *Chemical and Engineering News*, Oct. 11, 1976, p. 40.

44. Burr Cartwright Brundage, *The Fifth Sun: Aztec Gods, Aztec World.* Austin, TX: University of Texas Press, 1979.

45. Graham Hancock, *The Mars Mystery: The Secret Connection Between Earth and the Red Planet.* New York: Three Rivers Press, 1998.

46. Hugh Miller, *The Testimony of the Rocks.* New York: John B. Alden, 1992.

47. Ignatius Donnelly, *Atlantis: The Antediluvian World.* New York: Gramercy Publishing Company, 1949.

48. *The Mythology of All Races*, Vol. 10. New York: Cooper Square Publishers, Inc., 1964.

49. http://myweb.cableone.net/subru/Zoroastrianism.html

50. Eduard Suess, *The Face of the Earth*, Vol. I, London, 1904.

51. *Earth Story.* Hagerstown, MD: R & H Publishing, 1977.

52. Charles Hapgood, *The Path of the Pole.* Philadelphia, Chilton Company, 1970.

53. Lindsey Williams, *The Energy Non-Crisis.* Kasilof, Alaska: Worth Publishing Co., 1980.

54. Bal Gangadhar Tilak, *The Arctic Home in the Vedas.* Poona City, India: Kesari, 1903.

55.
http://www.classiqs.com/Latin%20Pages/Roman%20Culture/Eruption%20of%20

Vesuvius/eruption_of_vesuvius.htm

56. http://www.experiencewashington.com/Page_pid-105300.html

57. J. Eric S. Thompson, *The Rise and Fall of Maya Civilization*. Norman, OK: University of Oklahoma Press, 1954.

58. William K. Hartmann, *A Traveler's Guide to Mars: The Mysterious Landscapes of the Red Planet*. New York: Workman Publishing, 2003.

59. Alan and Sally Landsburg, *In Search of Ancient Mysteries*. New York: Bantam Books, 1974.

60. John Holmes, "Neanderthals Linked to West Europeans," *Insight*, September 11, 1989.

61. Andrew Tomas, *We Are Not the First*. New York: Bantam Books, 1971.

62. *Feats and Wisdom of the Ancients*, Library of Curious and Unusual Facts. New York: Time-Life Books, 1990.

63. Jonathan Gray, *Dead Men's Secrets*. Thorsby, Australia, 1998.

64. Henry Hodges, *Technology In the Ancient World*. New York: Barnes and Noble Books, 1970.

65. Peter James and Nick Thorpe, *Ancient Inventions*. New York: Ballantine Books, 1994.

66. Carl Baugh, *Panorama of Creation*. Hearthstone Publishing, 1994.

67. *Josephus' Writings*, Book 1, Chapter 3.

68. Lao Tzu, *Tao Te Ching*, Chapter 15.

69. http://www.pureinsight.org/pi/articles/2003/10/27/1899.html

70. http://www.pureinsight.org/pi/articles/2003/10/27/1899.html

71. Dick Teresi, *Lost Discoveries: The Ancient Roots of Modern Science--from the Babylonians to the Maya*. New York: Simon and Schuster, 2002.

72. John Anthony West, *The Serpent in the Sky*. New York: Harper and Row, 1984.

73. P. J. Wiseman, *New Discoveries in Babylonia*. As quoted by Richard E. Mooney in *Colony: Earth*. Greenwich, CT: Fawcett Publications, Inc., 1974.

74. William F. Allman with Joannie M. Schrof, "Lost Empires of the Americas," *U.S. News & World Report*, April 2, 1990, p. 46.

75. *Scientific American*, April 1968.

76. *Scientific American*, July 1935.

77. Orville Hope, *6000 Years of Seafaring*. Lakemont, GA: Copple House, 1988.

78. Gunnar Thompson, *American Discovery*. Seattle, Washington: Misty Isles Press, 1992.

79. Charles Hapgood, *Maps of the Ancient Sea Kings*. Philadelphia: Chilton Company, 1966.

80. Geoffrey Ashe, Thor Heyerdahl, et al, *The Quest for America*. New York: Praeger Publishers, 1971.

81. Ivan Van Sertima, *They Came Before Columbus*. New York: Random House, 1976.

82. Patrick Huyghe, *Columbus Was Last*. New York: Hyperion Press, 1992.

83. Gunnar Thompson, *Nu Sun*. Fresno, CA: Pioneer Publishing Co., 1989.

84. Barbara Walker, *Women's Encyclopedia of Myths and Secrets*, Harper San Francisco, 1983.

85. *Mysteries of the Ancient Americas: The New World Before Columbus*. Pleasantville, NY: Reader's Digest, 1986.

86. David Hatcher Childress, *Lost Cities of North and Central America.* Stelle, IL: Adventures Unlimited Press, 1992.

87. *The World's Last Mysteries.* Pleasantville, NY: Reader's Digest, 1976.

88. Barry Fell, *America B.C.* (New York, NY: The New York Times Book Company, 1989.

89. R. A. Jairazbhoy, *Rameses III--Father of Ancient America.* London, England: Karmak House, 1994.

90. R. A. Jairazbhoy, *Asians in Pre-Columbian Mexico.* Enfield, Middx., England: Pika Print Ltd., 1976.

91. R.A. Jairazbhoy, *Ancient Egyptians and Chinese in America.* London: George Prior Associated Publishers Ltd., 1974.

92. Peter N. Stearns, Michael Adas, and Stuart B. Schwartz, *World Civilizations*, Vol. 1. New York: HarperCollins College Publications, 1996.

93. Urana Clarke, "How Our Music Began," *The Book of Knowledge*, Vol. 18, New York, 1957.

94. George F. Carter, et al, *The Diffusion Issue.* Santa Barbara, California: Stonehenge Viewpoint Press, 1991.

95. Alexander Adams, *Millennia of Discoveries.* New York: Vantage Press, 1994.

96. http://physicsweb.org/articles/news/4/12/2

97. Jeffrey Kargel, *Mars: A Warmer Wetter Planet.* Chichester, UK: Praxis Publishing, 2004.

98. Michael Carr, *Water On Mars.* New York: Oxford University Press, 1996.

99. Frederick Pohl, *Atlantic Crossings Before Columbus.* New York: W.W. Norton & Co., 1961.

100. William McNeil, *Visitors to Ancient America*. Jefferson, North Carolina: McFarland & Company, Inc., Publishers, 2004.

101. J.E. Weckler, *Polynesians: Explorers of the Pacific*. Washington, D.C.: Smithsonian Institution, 1943.

102. http://news.bbc.co.uk/1/hi/world/americas/1697038.stm

103. http://news.bbc.co.uk/1/hi/world/south_asia/1768109.stm

104. W.C. Bryant, S.H. Gay, N. Brooks, *Scribner's Popular History of the United States*, Vol. 1. Charles Scribner's Sons, 1897.

105. Salvatore Michael Trento, *The Search for Lost America: The Mysteries of the Stone Ruins*. Chicago: Contemporary Books, Inc., 1978.

106. http://www.nwcreation.net/anomalies.html

107. http://www.washingtonpost.com/ac2/wp-dyn/A16616-2003Nov26?language=printer

108. Zecharia Sitchin, *The Wars of Gods and Men*. New York: Avon Books, 1985.

109. http://www.world-mysteries.com/aa_4.htm#Nuclear

110. Jim Bailey, *Sailing To Paradise: The Discovery of the Americas by 7000 B.C.* New York: Simon and Schuster, 1994.

111. Zecharia Sitchin, *The Twelfth Planet*. Avon, 1999.

112. William Corliss, *Ancient Man: A Handbook of Puzzling Artifacts*. Glen Arm, MD: The Sourcebook Project, 1978.

113. Michael Cremo and Richard Thompson, *Forbidden Archaeology: The Hidden History of the Human Race*. San Diego, CA: Bhaktivedanta Institute, 1993.

114. http://en.wikipedia.org/wiki/Bombardier_beetle

115. Robert Matthews, *New Scientist*, July 1997, pp. 26, 27.

116. Charles Berlitz, *Doomsday 1999 A.D.* Garden City, NJ: Doubleday & Co., 1981.

117. http://faculty.ucr.edu/~legneref/archeol/fuentema.htm

118. http://www.geocities.com/Tokyo/Bay/7051/poko2.htm

119. http://www.colostate.edu/Depts/Entomology/courses/en570/papers_2000/wells.html

120. http://www.metmuseum.org/toah/hd/teot/hd_teot.htm

121. http://www.colorado.edu/Conferences/chaco/tour/bonito.htm; http://www.ssc.msu.edu/~laej/historypapers/Burley3/Burley3text.html

122. http://www.world-mysteries.com/mpl_6.htm

123. http://en.wikipedia.org/wiki/Great_Pyramid_of_Giza

124. http://www.mystae.com/streams/science/russcrew.html
http://www.mystae.com/streams/science/russcrew2.html

125. http://www.phoenicia.org/america.html

126. http://www2.privatei.com/~bartjean/chap11.htm

127. http://www.hope-of-israel.org/hebinusa.htm

128. http://www.msnbc.msn.com/id/13089686

129. Richard Firestone, Allen West, Simon Warwick-Smith, *The Cycle of Cosmic Catastrophes*. Rochester, VT: Bear & Company, 2006.

130. Michael Bradley, *Guide to the World's Greatest Treasures*. New York: Barnes and Noble, 2005.

131. http://www.eurekalert..org/pub_releases/2006-09/uoea-ccr090406.php

132. Johathan Gray, *Lost World of Giants*. Brushton, NY: Teach Services, 2006.

133. Brian Sykes, *Nature*, Vol. 352, August 1,1991, p. 381.

134. *New Scientist*, October 17, 1992, p. 15.

135. http://www.newscientist.com/article.ns?id=dn7195
http://www.cbsnews.com/stories/2005/03/24/tech/main683019.shtml
http://www.usatoday.com/tech/science/2005-03-24-t-rex_x.htm?csp=34

136. Buddy Davis, Mike Liston, and John Witmore, *The Great Alaskan Dinosaur Adventure*. Green Forest, AZ: Master Books, 1998.

137. Samuel N. Kramer, *History Begins At Sumer*. Garden City, NY: Doubleday, 1959, pp. 170-81.

138. Helen L. Hoke, *Dragons, Dragons, Dragons*. New York: Watts, 1972, p. 179.

139. James Legge, *The Chinese Classics*, Vol. V, Book X, Year XXIX, par. 4, p. 729.

140. http://science.nasa.gov/newhome/headlines/ast29apr99_1.htm

141. Walter Cruttenden, *Lost Star of Myth and Time*. Pitsburgh, PA: St. Lynn's Press, 2006.

142. http://mars.jpl.nasa.gov/MPF/science/rotational.html

143. http://www.space.com/scienceastronomy/mars_moons_origin_030729.html

144. http://www.science-frontiers.com/sf060/sf060p02.htm

145. http://www.aoi.com.au/matrix/NuteeriatAll.pdf

146. Joseph Christy-Vitale, *Watermark: The Disaster that Change the World and Humanity 12,5-- Years Ago*. New York: Paraview Pocket Books, 2004.

147. J. Darmesteter (translator), *Zend-Avesta: The Vendidad*, 3 volumes. Oxford, 1883.

148. http://learning.indiatimes.com/gobble/scifi/sciencefront.htm

149. http://palaeo.gly.bris.ac.uk/Palaeofiles/Fossilgroups/euryaps/fossils.html

150. J.R. Jochmans, *Strange Relics From the Depths of the Earth*. Lincoln, NE: Forgotten Research Society, 1979.

151. Harold T. Wilkins, *Mysteries of Ancient South America*. Secaucus, NJ: Citadel Press, 1974.

152. Charles Hapgood, *Earth's Shifting Crust: A Key to Some Basic Problems of Earth Science*. New York: Pantheon Books, 1958.

153. R.W. Williamson, *Religious and Cosmic Beliefs of Central Polynesia*, Vol. 1, 1933, p.41.

154. David Hatcher Childress, *Lost Cities of Ancient Lemuria and the Pacific*. Stelle, Ill: Adventures Unlimited Press, 1988.

155. Henry Fountain, "When Giants Had Wings and 6 Legs," *New York Times*, February 3, 2004.

156. Brilliant computer animations of the expanding Earth, as well as other planets and moons throughout the solar system, are available at http://www.nealadams.com/nmu.html

157. http://www.nytimes.com/2006/11/14/science/14WAVE.html?pagewanted=1&ei=5070&en=c81c4b5f24ea5297&ex=1164258000&emc=eta1

158. Louis Ginzberg, *Legends of the Jews*. Philadelphia, PA: Jewish Publication Society of America, 1920, Vol. 1.

159. Immanuel Velikovsky, *Earth In Upheaval*. New York: Dell Publishing Co., Inc., 1955.

160. Hugh Falconer, *Palaeontological Memoirs and Notes*. London, 1868.

161. http://www.timesonline.co.uk/article/0,,13509-2064962,00.html

162.
http://www.njnightsky.com/nuke/html/modules.php?name=News&file=article&sid=411

163. http://en.wikipedia.org/wiki/Ten_plagues

164. http://www.nature.com/news/2006/061127/full/444534a.html

165.
http://www.usatoday.com/tech/news/techinnovations/2003-02-07-see-through_x.htm

166. http://thescotsman.scotsman.com/international.cfm?id=226392005

167. W. Raymond Drake, *Gods and Spacemen in Greece and Rome*. Signet (New American Library), 1974. See also Frank Edwards, *Stranger than Science*. Carol Publishing Corporation, 1992 (ISBN: 0821625136).

168. Brinsley Le Poer Trench, *The Flying Saucer Story*. Tandem Books, Ltd., 1966. See also
http://en.wikipedia.org/wiki/Ufo

169. David Hatcher Childress, The Anti-Gravity Handbook. Stelle, IL: Adventures Unlimited Press, 2003.

170. Harold T. Wilkins, *Flying Saucers On the Attack*. Citadel Press, 1954.

171. Zecharia Sitchin, *The Twelfth Planet*. New York: Avon Books, 1976.

172. Alexander von Wuthenau, *Unexpected Faces in Ancient America*. New York: Crown Publishers, Inc., 1975.

173. http://www.hope-of-israel.org/hebinusa.htm

174. Mike Baillie, *Exodus to Arthur: Catastrophic Encounters with Comets*. London: B.T. Batsford, Ltd., 1999.

175. J. Baxter and T. Adams, *The Fire Came By*. London: Macdonald and Jane's, 1976.

176. Nigel Davies, *Voyagers to the New World*. Albuquerque: University of New Mexico Press, 1979.

177. Ellen Lloyd, *Voices from Legendary Times*. New York: iUniverse, Inc., 2005.

178. http://news.yahoo.com/s/nm/20070315/ts_nm/mars_water_dc

179. Gunnar Thompson, *Secret Voyages to the New World*. Seattle, WA: Misty Isles Press, 2006.

180. http://www.hfml.sci.kun.nl/froglev.html

181.
http://www.space.com/businesstechnology/060215_technovel_antigravity.html

182. http://www.americanantigravity.com/hutchison.html

183. Hugh G. Owen, *Atlas of Continental Displacement: 200 Million Years to the Present*, Cambridge University Press, 1983.

184. http://news.bbc.co.uk/1/hi/sci/tech/6617851.stm

185. http://observer.guardian.co.uk/world/story/0,,2083758,00.html

186. http://en.wikipedia.org/wiki/Red_mercury

187. http://news.nationalgeographic.com/news/2007/05/070524-drought.html

188. Zecharia Sitchin, *Divine Encounters*. Bear and Company, 2002.

189. http://www.norwaypost.no/cgi-bin/norwaypost/imaker?id=87357

190. David Talbott and Wallace Thornhill, *Thunderbolts of the Gods*. Portland, OR: Mikamar Publishing, 2005.

191. Donald Scott, *The Electric Sky*. Portland, OR: Mikamar Publishing, 2006.

192. Donald W. Patten, *Catastrophism and the Old Testament*. Seattle, WA: Pacific Meridian Publishing Co., 1988.

193. http://www.msnbc.msn.com/id/4789219/

194. http://super.colorado.edu/~astr1020/sung.html

195. Paul A. Violette, *Earth Under Fire: Humanity's Survival of the Ice Age*. Rochester, VT: Bear & Company, 2005.

196. Graeme R. Kearsley, Mayan Genesis: South Asian Myths, Migrations and Iconography in Mesoamerica. London: Yelsraek Publishing, 2001.

197. R.A. Jairazbhoy, *Ancient Egyptians in Middle and South America*. London: Ra Publications, 1981.

198. http://news.yahoo.com/s/afp/20080220/sc_afp/spaceastronomymarswater_080220183433;_ylt=Aus2eMddFzvSgJxYO2SGuLXPOrgF

199. R.A. Jairazbhoy, *Ancient Egyptian Survivals in the Pacific*. London, England: Karnak House, 1990.

200. Paul R. Cheesman and Millie F. Cheesman, *Ancient American Indians: Their Origins, Civilizations & Old World Connections*. Bountiful, Utah: Horizon Publishers, 1991.

201. Donald W. Hemingway, *Christianity in America Before Columbus?* Salt Lake City, Utah: Hawkes Publishing Inc., 1988.

202. Diane E. Wirth, *Parallels: Mesoamerican and Ancient Middle Eastern*

Traditions. St. George, UT: Stonecliff Publishing, 2003.

203. http://www.jpl.nasa.gov/events/lectures/nov07.cfm

204. http://saturn.jpl.nasa.gov/news/press-release-details.cfm?newsID=826

205. John Spencer Carroll, *A Search for Qetzacoatl*. Santa Barbara, CA: Stonehenge Viewpoint Press, 1987.

206. http://www.pbs.org/wgbh/nova/newton/alchemy.html

207. *Underground!*, ed. Preston Peet. New York: The Disinformation Company, 2005.

208. Graham Hancock, *Underworld: The Mysterious Origins of Civilization*. New York: Three Rivers Press, 2002.

209. Graham Hancock, *Heaven's Mirror: Quest for the Lost Civilization*. Doubleday Canada, 2002.

210. http://space.newscientist.com/article/dn13903-jupiter-moons-poles-wandered-far-and-wide.html?feedId=online-news_rss20

211. http://en.wikipedia.org/wiki/Acoustic_levitation

212. http://news.bbc.co.uk/2/hi/science/nature/7497715.stm

213. Maurice Cotterell, *The Supergods*. London: Harper Collins, 1997.

17105488R00212

Printed in Great Britain
by Amazon